MEDICAL AND HEALTHCARE ROBOTICS

Medical Robots and Devices:
New Developments and Advances

MEDICAL AND HEALTHCARE ROBOTICS

New Paradigms and Recent Advances

Edited by

OLFA BOUBAKER

University of Carthage, National Institute of Applied Sciences and
Technology, Tunis, Tunisia

Series Editor

OLFA BOUBAKER

ELSEVIER

ACADEMIC PRESS
An imprint of Elsevier

Academic Press is an imprint of Elsevier
125 London Wall, London EC2Y 5AS, United Kingdom
525 B Street, Suite 1650, San Diego, CA 92101, United States
50 Hampshire Street, 5th Floor, Cambridge, MA 02139, United States
The Boulevard, Langford Lane, Kidlington, Oxford OX5 1GB, United Kingdom

Notices
Knowledge and best practice in this field are constantly changing. As new research and experience broaden our understanding, changes in research methods, professional practices, or medical treatment may become necessary.

Practitioners and researchers must always rely on their own experience and knowledge in evaluating and using any information, methods, compounds, or experiments described herein. In using such information or methods they should be mindful of their own safety and the safety of others, including parties for whom they have a professional responsibility.

To the fullest extent of the law, neither the Publisher nor the authors, contributors, or editors, assume any liability for any injury and/or damage to persons or property as a matter of products liability, negligence or otherwise, or from any use or operation of any methods, products, instructions, or ideas contained in the material herein.

ISBN 978-0-443-18460-4

For information on all Academic Press publications
visit our website at https://www.elsevier.com/books-and-journals

Publisher: Mara E. Conner
Acquisitions Editor: Sonnini R. Yura
Editorial Project Manager: John Leonard
Production Project Manager: Prem Kumar Kaliamoorthi
Cover Designer: Matthew Limbert

Typeset by STRAIVE, India

Working together
to grow libraries in
developing countries

www.elsevier.com • www.bookaid.org

Contents

Contributors   ix
Preface   xiii
Introduction   xvii

1. Assistive robotic technologies: An overview of recent advances in medical applications   1

Md Samiul Haque Sunny, Md Mahbubur Rahman, Md Enamul Haque,
Nayan Banik, Helal Uddin Ahmed, and Mohammad Habibur Rahman

1.1 Introduction   1
1.2 Current state of healthcare robotics   2
1.3 Robots in medical applications   2
1.4 Future trends   18
1.5 Conclusions   19
References   20

2. Soft robotics in medical applications: State of the art, challenges, and recent advances   25

Mostafa Kaviri, Ali Jafari Fesharaki, and Soroush Sadeghnejad

2.1 What is soft robotics?   25
2.2 Why soft robotics?   27
2.3 Rehabilitation   28
2.4 Soft wearable robots   30
2.5 Cardiac assist devices   35
2.6 Surgical robots   37
2.7 Challenges and state-of-the-art methods   41
2.8 Conclusions   53
Conflict of interest statement   55
Acknowledgments   55
References   55
Further reading   61

3. Rehabilitation robotics: History, applications, and recent advances   **63**

Soroush Sadeghnejad, Vida Shams Esfand Abadi, and Bahram Jafari

3.1   Introduction   63
3.2   The history of rehabilitation robotics   66
3.3   Applications and recent advances   70
3.4   Conclusions   81
Acknowledgments   82
Conflict of interest statement   82
References   82

4. Gait devices for stroke rehabilitation: State-of-the-art, challenges, and open issues   **87**

Thiago Sá de Paiva, Rogério Sales Gonçalves, Giuseppe Carbone, and Marco Ceccarelli

4.1   Introduction   87
4.2   Gait devices for stroke rehabilitation   88
4.3   Robot-assisted gait training   90
4.4   Safety aspects of gait devices   111
4.5   Control techniques used on gait rehabilitation devices   112
4.6   Challenges and open issues   113
4.7   Conclusions   116
References   117

5. Robotic devices for upper limb rehabilitation: A review   **123**

Kishor Lakshmi Narayanan, Tanvir Ahmed, Md Mahafuzur Rahaman Khan, Tunajjina Kawser, Raouf Fareh, Inga Wang, Brahim Brahmi, and Mohammad Habibur Rahman

5.1   Introduction   123
5.2   End effector-type upper limb rehabilitation robots   126
5.3   Exoskeleton-type upper limb rehabilitation robots   135
5.4   Design specs of some notable rehabilitation robots   148
5.5   Research challenges and future directions   151
5.6   Conclusions   151
References   152
Further reading   156

6. Failure of total hip arthroplasty (THA): State of the art   **157**

Atef Boulila, Lanouar Bouzid, and Mahfoudh Ayadi

6.1   Introduction   157
6.2   Failures related to total hip arthroplasty   165

6.3 Future directions 175
6.4 Conclusions 177
References 177

7. Artificial intelligence in robot-assisted surgery: Applications to surgical skills assessment and transfer **183**
Abed Soleymani, Xingyu Li, and Mahdi Tavakoli

7.1 Introduction 183
7.2 Surgical skills assessment 184
7.3 Surgical skills transfer 192
7.4 Future directions 197
7.5 Conclusions 197
References 198

8. A virtual-based haptic endoscopic sinus surgery (ESS) training system: From development to validation **201**
Soroush Sadeghnejad, Mojtaba Esfandiari, and Farshad Khadivar

8.1 Introduction 201
8.2 History and state of the art 202
8.3 Development of an endoscopic sinus training system 205
8.4 Experimental procedures and evaluation 210
8.5 Results 212
8.6 Conclusions and final remarks 215
Conflict of interest statement 217
Acknowledgments 217
References 217

9. Medical and healthcare robots in India **221**
Kshetrimayum Lochan, Ashutosh Suklyabaidya, and Binoy Krishna Roy

9.1 Introduction 221
9.2 Types of healthcare robots 223
9.3 Commercial and research areas of surgical robots 225
9.4 Disadvantage/downside of robotics in health 233
9.5 The future of robotic surgery in India 234
9.6 Conclusions 235
References 235

10. Trend of implementing service robots in medical institutions during the COVID-19 pandemic: A review   **237**

Isak Karabegović, Lejla Banjanović-Mehmedović, Ermin Husak, and Mirza Omerčić

10.1 Introduction   237
10.2 Trend of implementing service robots in medical institutions   237
10.3 Implementation of service robots in medical institutions during the
     Covid-19 pandemic   241
10.4 AI-based mobile disinfection robots against COVID-19   255
10.5 Challenges and open issues for service robotics   257
10.6 Conclusions   259
References   260
Further reading   262

11. Conclusions   **263**

Olfa Boubaker

Index   *265*

Contributors

Vida Shams Esfand Abadi
Bio-Inspired System Design Laboratory, Department of Biomedical Engineering, Amirkabir University of Technology (Tehran Polytechnic), Tehran, Iran

Helal Uddin Ahmed
Biorobotics Laboratory, University of Wisconsin-Milwaukee, Milwaukee, WI, United States

Tanvir Ahmed
Biorobotics Laboratory, University of Wisconsin-Milwaukee, Milwaukee, WI, United States

Mahfoudh Ayadi
University of Carthage, National Engineering School of Bizerte, Menzel Abderrahman, Tunisia

Nayan Banik
Department of Computer Science, University of Wisconsin-Milwaukee, Milwaukee, WI, United States

Lejla Banjanović-Mehmedović
Faculty of Electrical Engineering, University of Tuzla, Tuzla, Bosnia and Herzegovina

Olfa Boubaker
University of Carthage, National Institute of Applied Sciences and Technology, Tunis, Tunisia

Atef Boulila
University of Carthage, National Institute of Applied Sciences and Technology, Tunis, Tunisia

Lanouar Bouzid
University of Tunis El Manar, Tunis Faculty of Medicine, Tunis, Tunisia

Brahim Brahmi
Biorobotics Laboratory, University of Wisconsin-Milwaukee, Milwaukee, WI, United States; Electrical Engineering Department at the College of Ahuntsic, Montréal, Canada

Giuseppe Carbone
Department of Mechanical, Energy, and Management Engineering, University of Calabria, Rende, Italy

Marco Ceccarelli
Department of Industrial Engineering, University of Rome Tor Vergata, Rome, Italy

Mojtaba Esfandiari
Bio-Inspired System Design Laboratory, Department of Biomedical Engineering, Amirkabir University of Technology (Tehran Polytechnic), Tehran, Iran

Raouf Fareh
Electrical Engineering Department, University of Sharjah, Sharjah, United Arab Emirates

Ali Jafari Fesharaki
Bio-Inspired System Design Laboratory, Department of Biomedical Engineering, Amirkabir University of Technology (Tehran Polytechnic), Tehran, Iran

Rogério Sales Gonçalves
School of Mechanical Engineering, Federal University of Uberlândia, Uberlândia, Brazil

Md Enamul Haque
Department of Mechanical Engineering, Milwaukee, WI, United States

Md Samiul Haque Sunny
Department of Computer Science, University of Wisconsin-Milwaukee, Milwaukee, WI, United States

Ermin Husak
Technical Faculty of Bihać, University of Bihać, Bihać, Bosnia and Herzegovina

Bahram Jafari
Bio-Inspired System Design Laboratory, Department of Biomedical Engineering, Amirkabir University of Technology (Tehran Polytechnic), Tehran, Iran

Isak Karabegović
Academy of Sciences and Arts of Bosnia and Herzegovina, Sarajevo, Bosnia and Herzegovina

Mostafa Kaviri
Bio-Inspired System Design Laboratory, Department of Biomedical Engineering, Amirkabir University of Technology (Tehran Polytechnic), Tehran, Iran

Tunajjina Kawser
Department of Anatomy, Shaheed Tajuddin Ahmad Medical College, Gazipur, Bangladesh

Farshad Khadivar
Bio-Inspired System Design Laboratory, Department of Biomedical Engineering, Amirkabir University of Technology (Tehran Polytechnic), Tehran, Iran

Md Mahafuzur Rahaman Khan
Biorobotics Laboratory, University of Wisconsin-Milwaukee, Milwaukee, WI, United States

Xingyu Li
Electrical and Computer Engineering Department, University of Alberta, Edmonton, AB, Canada

Kshetrimayum Lochan
IIT Palakkad IHub Foundation, Indian Institute of Technology Palakkad, Palakkad, Kerala, India

Kishor Lakshmi Narayanan
Neuro-Rehabilitation Lab, School of Electronics Engineering, Vellore Institute of Technology, Vellore, India

Mirza Omerčić
iLogs Gmbh, Klagenfurt, Austria

Thiago Sá de Paiva
School of Mechanical Engineering, Federal University of Uberlândia, Uberlândia, Brazil

Md Mahbubur Rahman
Department of Mechanical Engineering, Milwaukee, WI, United States

Mohammad Habibur Rahman
Department of Computer Science; Biorobotics Laboratory, University of Wisconsin-Milwaukee; Department of Mechanical Engineering, Milwaukee, WI, United States

Binoy Krishna Roy
Department of Electrical Engineering, National Institute of Technology Silchar, Silchar, Assam, India

Soroush Sadeghnejad
Bio-Inspired System Design Laboratory, Department of Biomedical Engineering, Amirkabir University of Technology (Tehran Polytechnic), Tehran, Iran

Abed Soleymani
Electrical and Computer Engineering Department, University of Alberta, Edmonton, AB, Canada

Ashutosh Suklyabaidya
Department of General Surgery, Silchar Medical College and Hospital, Silchar, Assam, India

Mahdi Tavakoli
Electrical and Computer Engineering Department, University of Alberta, Edmonton, AB, Canada

Inga Wang
Biorobotics Laboratory; Rehabilitation Outcomes Research Lab, University of Wisconsin-Milwaukee, Milwaukee, WI, United States

Preface

This book, *Medical and Healthcare Robotics*, is the inaugural volume in the new book series *Medical Robots and Devices: New Developments and Advances*. It starts this unique series on robotics and artificial intelligence dedicated to the field of medicine and biomedical applications.

Robotics for medical applications is one of the fastest-growing sectors in engineering and technology. The key challenge remains building reliable robotic systems with high levels of autonomy and safety. Medical and healthcare robots include all robotic devices to assist doctors and/or guarantee the care of a patient with health difficulties while establishing a diagnosis for medical, surgical, or functional and social rehabilitation treatment. These systems are also essential tools for student medical training. In addition, health robotics profoundly influences the lives of a large part of the population such as the elderly and disabled, using assistance and rehabilitation robots to improve the quality of their lives and boost their independence. The objective of this inaugural volume is to offer a comprehensive state of the art of important achievements and open issues of this promising domain of engineering and technology.

As a founder and editor of the emerging book series *Medical Robots and Devices: New Developments and Advances* in which this volume is published, I was delighted and honored to receive a kind invitation from Elsevier to edit this inaugural volume. The basic idea behind this editorial project is to give a taste of the works that will follow in the same series. On this point, I am immensely grateful to the ever-keen Elsevier Acquisitions Editor, Sonnini Yura, who was behind the launch of the initial project of the book series and who continues to expend much effort to make this project a success. Special thanks go to John Leonard (Editorial Project Manager) and Prem Kumar Kaliamoorthi (Senior Project Manager) for their patience and reactivity and also to everyone on the Elsevier editorial team who believed in this series.

The book presents the most recent innovations, trends, and concerns in medical robotics. It provides exclusive insights into recent advances and exposes practical challenges encountered and solutions devoted to this growing field. It covers major areas of medical robotics including robotic devices for surgery, exploration, diagnosis, therapy, and training, and healthcare robotics including rehabilitation devices, artificial organs, assistive technologies, service robotics, and companion robots.

The book is a timely and comprehensive reference guide for undergraduate and graduate students, and researchers in the fields of electrical engineering, mechanical engineering, mechatronics, control systems engineering, and biomedical engineering. This volume will be useful for master's programs, leading consultants, and industrial companies. It will also be of high interest for physicians and physiotherapists and all technical people in medical and biomedical fields.

The chapters are written by well-known authors in medical robotics and related engineering fields coming from 11 countries (Australia, Bangladesh, Bosnia and Herzegovina, Brazil, Canada, India, Iran, Italy, Tunisia, the United Arab Emirates, and the USA). Invited leading contributors of this volume include active scientists in various fields of engineering and medical robotics disciplines with a focus on promoting inclusive collaboration across all dimensions. Most of the contributions are presented in the form of investigative chapters well-suited for use in teaching as well as in research. The survey chapters of this book are organized in the table of contents from a general overview of the field to the most insightful details and innovative applications. Each chapter is closed by a section describing open problems and new trends in the subfield.

Overall, 11 chapters written by dynamic researchers in the field of medical and healthcare robotics are collected in this book to provide a complete picture and a state of the art of recent advances in this growing field. Each chapter is structured in several sections including an introduction, a state of the art, and open and future directions in the subfield. The book begins with an introduction that presents basic definitions and foundations for medical and healthcare robotics, highlighting the richness of this research field. It then covers the most innovations in the field in the form of survey chapters. The book ends with a conclusion presenting a summary of achievements as well as an overall picture of typical problems and prospects related to this growing field.

The following researchers are particularly acknowledged here for their considerable efforts to make this book a success:

- Professor Mohammad Habibur Rahman, Director of BioRobotics Lab at University of Wisconsin-Milwaukee, USA, and his international collaborative team, for their valuable contributions in Chapter 1, "Assistive Robotic Technologies: An Overview of Recent Advances in Medical Applications," and Chapter 5, "Robotic Devices for Upper Limb Rehabilitation."

- Dr. Soroush Sadeghnejad, Head of Biomedical Engineering Innovation Center and Director of Bio-Inspired System Design Laboratory at the Biomedical Engineering Department of the AmirKabir University of Technology, Iran, and Vice-President of the Federation of International Robot Sports Association, for a fruitful collaboration. I particularly thank him for leading the valuable contributions of his team in Chapter 2, "Soft Robotics in Medical Applications: State-of-the-art, Challenges and Recent Advances," and Chapter 3, "Rehabilitation Robotics: History, Applications, and Recent Advances." I also thank him for the experimental results presented in Chapter 8, "A Virtual-based Haptic Endoscopic Sinus Surgery Training System: From Development to Validation."

- Professor Giuseppe Carbone, Chair of IFToMM TC on Robotics and Mechatronics, and Editor-in-Chief of the journal *Robotica*, from the University of Calabria, Italy, for leading the valuable contribution of his team in Chapter 4, "Gait Devices for Stroke Rehabilitation: State-of-the art, Challenges, and Open Issues."

- Professor Mahdi Tavakoli, Director of Telerobotic and Biorobotic Systems Group, University of Alberta, Canada, for leading the contribution of his team provided in Chapter 7, "Artificial Intelligence in Robot-Assisted Surgery: Applications to Surgical Skills Assessment and Transfer." He is particularly acknowledged for comments and reviews.
- Professor Atef Boulila, Head with department of Physical Engineering and Instrumentation at the National Institute of Applied Sciences and Technology of the University of Carthage, Tunis, for writing Chapter 6, "Failure of Total Hip Arthroplasty: State of the Art," in collaboration with the Faculty of Medicine of Tunis.
- Professor Binoy Krishna Roy, Head with department of Electrical Engineering at the National Institute of Technology, India, for his great efforts to contribute Chapter 9, "Medical and Healthcare Robots in India," the first survey work dealing with the state of play of this booming field in India, in collaboration with Silchar Medical College and Hospital.
- Professor Isak Karabegović and his team for their essential work in Chapter 10, "Trend of Implementation of Service Robots in Medical Institutions During the Covid-19 Pandemic: A Review," which deals with the significance of service robotics during the challenging situations caused by the Covid-19 pandemic.

Finally, I would like to express my deep thanks to all authors and reviewers of this book for their valuable contributions and useful criticism to make the inaugural volume of this new book series a reality.

Olfa Boubaker
University of Carthage, National Institute of Applied Sciences
and Technology, Tunis, Tunisia

Introduction

Olfa Boubaker

University of Carthage, National Institute of Applied Sciences and Technology, Tunis, Tunisia

Medical robots are machines used in healthcare applications. They include all devices and technologies that can assist doctors, technical staff, and patients with health difficulties. Medical robots can involve surgical robots, machines, and technologies used for diagnosis, exploration, and therapy, artificial organs, laboratory devices, drug delivery systems, rehabilitation devices, and all assistive technologies designed for medical applications. Examples of these devices are shown in Fig. 1.

Compared to manual devices used in the medicinal field, medical robots offer a broad variety of benefits. They are flexible and can be trained to perform several tasks. They reduce human fatigue and improve the precision and capabilities of physicians and technical people in medical and biomedical fields. Furthermore, they offer many benefits for patients such as a decrease of trauma, faster recuperation, scar control, cost reduction, and ease of use (Boubaker, 2020).

According to (Market and Market, Inc, n.d.), the global medical robotic systems market (GMRSM) is anticipated to reach USD 12.7 billion by 2025 from an estimated USD 5.9 billion in 2020 at a compound annual growth rate (CAGR) of 16.5% for the forecast period 2020–25. Furthermore, according to (Grand View Research, Inc, n.d.), the GMRSM size was estimated at USD 16.1 billion in 2021 and is projected to grow at a CAGR of 17.4% from 2022 to 2030. It is claimed that the end-users of these machines are principally hospitals, ambulatory surgery centers, and rehabilitation centers, and that the geriatric population is the most affected by the evolution of these products. It is also asserted that the increasing demand for laparoscopic surgeries and the growing percentage of trauma injuries are key factors driving market growth (Grand View Research, Inc, n.d.). According to Market and Market, Inc (n.d.), the top five companies working in medical robotics in the world are Intuitive Surgical, Inc., Stryker, Hocoma AG, Mazor Robotics Ltd., Hansen Medical, Inc, and Auris Surgical Robotics, Inc.

Medical robots can be classified using different criteria (Boubaker, 2020). For example, in Chapter 1 of this book, medical robots are categorized according to their applications. They are presented in four groups: rehabilitation robots; surgical robots; micro and nano robots; and service robots. A summary and a comparative analysis of commercially available robots of each class are provided.

In many references (see Boubaker (2020) and references therein), it is claimed that at least four requirements are essential for medical robots. These needs are depicted and hierarchized in Fig. 2. In all situations, safety considerations remain of higher importance. Compared to robotic systems used in other fields of application, medical robots should be

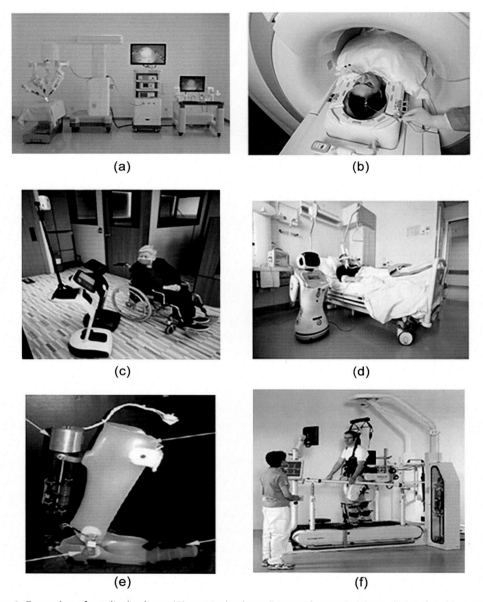

Fig. 1 Examples of medical robots: (A) surgical robots (Liu et al., 2022); (B) machines for diagnosis, exploration, and therapy (Gajjar et al., 2023); (C) companion robots and assistive technologies for elderly and disabled patients' assistance, therapy, and psychological support (Aymerich-Franch and Ferrer, 2022); (D) service robots and drug delivery systems (Yoganandhan et al., 2021); (E) prostheses, artificial organs, and wearable devices (Meng et al., 2015); (F) rehabilitation devices (Meng et al., 2015).

Fig. 2 Fundamental requirements of a medical robot.

more precise and also equipped with sensors and actuators guaranteeing higher levels of safety for users as well as high compliance with their environment. As medical robots can operate inside a human body by substituting a human organ, or interacting with a patient's limb, or even be used as a drug delivery system, it is essential that these devices be compatible with the human body by avoiding all possibly allergies and causes of rejection. Fig. 3 details the different levels of biomimicry and biocompatibility related to these machines.

Readers needing to explore more about soft robotics in the medical field can refer to Chapter 2 of this book. In this chapter, a comprehensive review of these robots is expanded. Soft robots are depicted in three groups: rehabilitation robots including gloves and wearable devices; cardiac assistive devices; and surgical robots. Recent advances and challenges are also presented. Various aspects of modeling, actuation, sensing, control, simulation, and implementation of these machines are considered.

Rehabilitation robots are machines used to support dynamically the movement of patients with functional difficulties and to reduce the cost of the rehabilitation process by decreasing the amount of person-to-person time that a therapist needs to spend with a patient. Robot-assisted rehabilitation has been extensively demonstrated to be effective in functional recovery for patients with stroke, spinal cord injury, traumatic brain injury, cerebral palsy, multiple sclerosis, and Parkinson's disease. Therapeutic robots can also be used to compile data to quantify the recovery patient's improvement in order to optimize the treatment process (Boubaker, 2020).

Rehabilitation robots can be categorized using different classifications (Boubaker, 2020). The most natural one is based on the anatomy of the human body and considers two groups of rehabilitation robots: lower extremity rehabilitation devices and upper extremity rehabilitation devices. Another categorization, often considered, divides these devices into three groups: grounded rehabilitation devices; end-effectors devices; and wearable exoskeletons (Boubaker, 2020). Fig. 4 explains such classifications. Readers

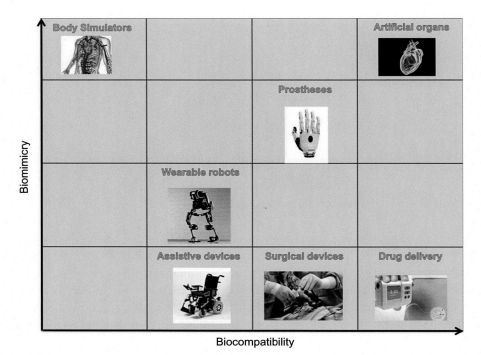

Fig. 3 Levels of compatibility of medical robots. *(Reproduced and adapted with permission from Olfa Boubaker, 2020. Copyright Elsevier, 2023 (Boubaker, O., 2020. Medical robotic. In: Boubaker, O. (Ed.), Control Theory in Biomedical Engineering: Applications in Physiology and Medical Robotics. Academic Press. https://doi.org/10.1016/B978-0-12-821350-6.00007-X).)*

needing to familiarize themselves with the field of rehabilitation robotics can refer to Chapter 3, in which the most recent innovations in this field are presented. A brief history of these machines is also given, along with a comprehensive survey of this area. In the same chapter, the reader can discover different design criteria of a rehabilitation robot involving control techniques, assessment criteria, and safety considerations. Future directions and related challenges are also established.

In the same framework, Chapters 4 and 5 are devoted to robotic devices applied to the rehabilitation of lower and upper limbs, respectively. In Chapter 4, the reader can find a state of the art of devices devoted for gait recovery, a safety analysis of these machines, a brief review of control techniques of these devices, and a discussion of the open challenges and new directions in this research area. In Chapter 5, two classes of upper limb assisted-therapy robots are considered: exoskeleton and end-effector robots. The chapter provides a survey of this research domain, and sets out the design, control, and clinical studies results of some notable available devices.

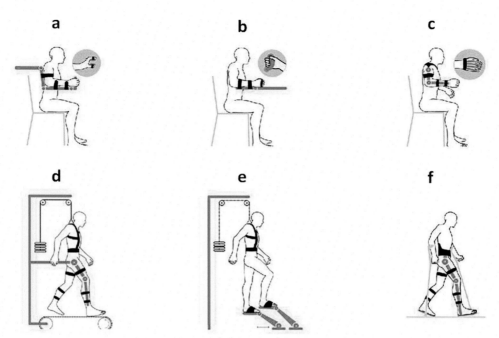

Fig. 4 Robot-aided rehabilitation classification. (A–C) Upper body rehabilitation-aided devices therapy. (D–F) Lower body rehabilitation devices. (A and D) Grounded rehabilitation devices. (B and E) End-effectors devices. (C and F) Wearable exoskeletons. *(Reproduced and adapted with permission from Olfa Boubaker, 2020. Copyright Elsevier, 2023 (Boubaker, O., 2020. Medical robotic. In: Boubaker, O. (Ed.), Control Theory in Biomedical Engineering: Applications in Physiology and Medical Robotics. Academic Press. https://doi.org/10.1016/B978-0-12-821350-6.00007-X).)*

From another point of view, rehabilitation devices can be seen as a subclass of all medical assistive technologies. These include prosthetic devices and artificial limbs, orthotic devices and exoskeletons, functional electro-stimulation, robotic aids, smart houses, robotics aids, and robot personal assistants. Examples of these devices and technologies are shown in Fig. 5.

An orthosis is a mechanism applied outside the human body to control the structural and functional characteristics of the neuromuscular and skeletal system. It is used to support ineffective joints, muscles, or limbs. Orthotics frequently take the shape of an exoskeleton. An exoskeleton is a powered anthropomorphic fit that is worn by the patient and has links and joints that correspond to those of the human body.

A prosthetic is a robotic device that replaces a lost part of the human body supplying mobility or manipulation abilities. Prosthetic limbs are then artificial limbs. They generally need the body energy to move. In Chapter 6, readers can find a state of the art of hip joint prostheses. As the hip joint can be subject to trauma, vascular necrosis, rheumatoid arthritis, posttrauma, and congenital deformities, replacement of the failing joint with an

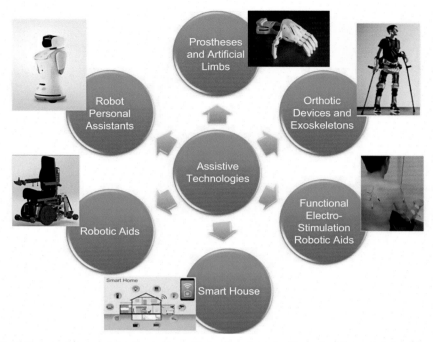

Fig. 5 Assistive devices and technologies. *(Reproduced and adapted with permission from Olfa Boubaker, 2020. Copyright Elsevier, 2023 (Boubaker, O., 2020. Medical robotic. In: Boubaker, O. (Ed.), Control Theory in Biomedical Engineering: Applications in Physiology and Medical Robotics. Academic Press. https://doi.org/10.1016/B978-0-12-821350-6.00007-X).)*

artificial joint is becoming more and more frequent. The limitations of such solutions are discussed in this chapter.

Alternatively, as key features impacting the GMRSM are increasing demand on laparoscopic surgeries, expanding demand for surgical procedures for geriatric population, and cumulative cases of trauma injuries, surgical robotics constitutes one of the most important classes of medical robots (Grand View Research, Inc, n.d.). Robotic surgery, also called robot-assisted surgery, includes any type of surgical procedures performed using robotic machines. Applications may include orthopedic surgery, throat surgery, gynecology, urology, neurosurgery, ophthalmology, abdominal surgery/laparoscopy, etc.

The main advantages of surgical robots are overcoming the weaknesses of traditional surgical procedures and enhancing the surgeon's abilities. These capabilities include improving precision, flexibility, and control. Robotic surgery is usually associated with minimally invasive surgery procedures made through small incisions. Robot-assisted surgeries are more expensive than conventional surgery procedures. The most used surgical robotic system is the da Vinci system (http://www.intuitivesurgical.com). It costs between USD 1 million and 3 million (Grand View Research, Inc, n.d.). The price of the CyberKnife radiosurgery robotic system (https://cyberknife.com) approved for treating a variety of cancers ranges from USD 4.2 million to 5.2 million per unit. Furthermore, additional fees are always required for maintenance (Grand View Research, Inc, n.d.).

Robotic surgery has experienced a great revolution in the past 50 years, profoundly marked by the three critical stages described by Fig. 6. A surgical system is generally constituted by three components as described by Fig. 7 for the case study of the da Vinci surgical system: a surgeon's console cart; a patient side cart; and the vision system cart.

Encouraged by the necessity of standardized medical procedures for the training of students in surgery and in order to minimize animal and patient tests, several useful body-part simulators have been proposed. We should acknowledge that combinations of soft materials, contractile actuators, and flexible sensors have contributed effectively to the design of body-part simulators for producing realistic healthcare scenarios (Boubaker,

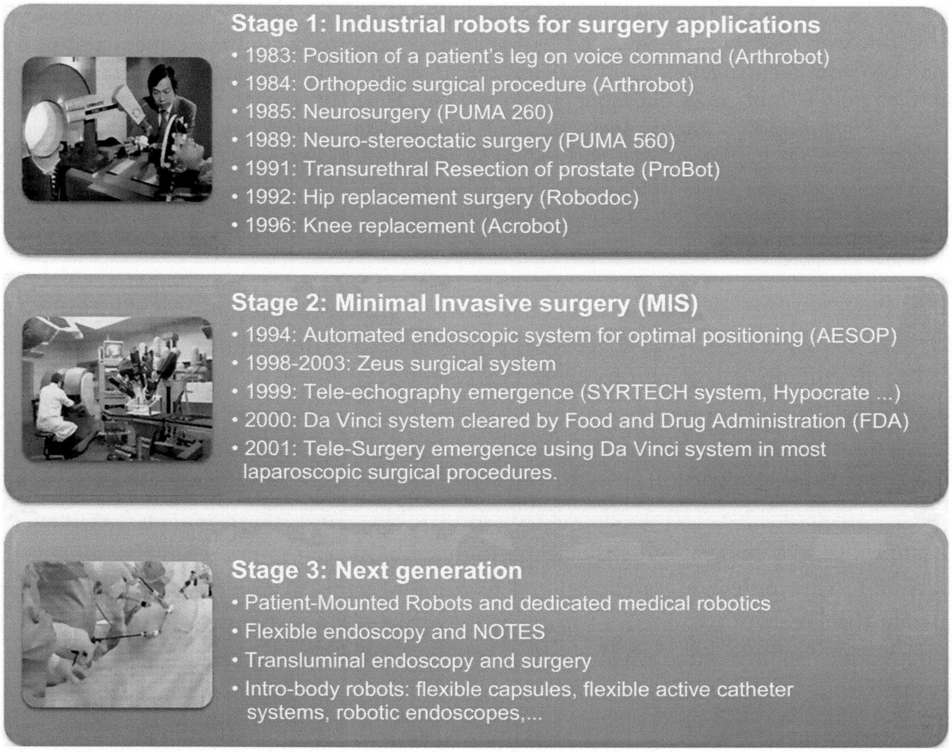

Stage 1: Industrial robots for surgery applications
- 1983: Position of a patient's leg on voice command (Arthrobot)
- 1984: Orthopedic surgical procedure (Arthrobot)
- 1985: Neurosurgery (PUMA 260)
- 1989: Neuro-stereoctatic surgery (PUMA 560)
- 1991: Transurethral Resection of prostate (ProBot)
- 1992: Hip replacement surgery (Robodoc)
- 1996: Knee replacement (Acrobot)

Stage 2: Minimal Invasive surgery (MIS)
- 1994: Automated endoscopic system for optimal positioning (AESOP)
- 1998-2003: Zeus surgical system
- 1999: Tele-echography emergence (SYRTECH system, Hypocrate ...)
- 2000: Da Vinci system cleared by Food and Drug Administration (FDA)
- 2001: Tele-Surgery emergence using Da Vinci system in most laparoscopic surgical procedures.

Stage 3: Next generation
- Patient-Mounted Robots and dedicated medical robotics
- Flexible endoscopy and NOTES
- Transluminal endoscopy and surgery
- Intro-body robots: flexible capsules, flexible active catheter systems, robotic endoscopes,...

Fig. 6 Timetable of surgical robotics depicted by three stages. *(Reproduced with permission from Olfa Boubaker, 2020. Copyright Elsevier, 2023 (Boubaker, O., 2020. Medical robotic. In: Boubaker, O. (Ed.), Control Theory in Biomedical Engineering: Applications in Physiology and Medical Robotics. Academic Press. https://doi.org/10.1016/B978-0-12-821350-6.00007-X).)*

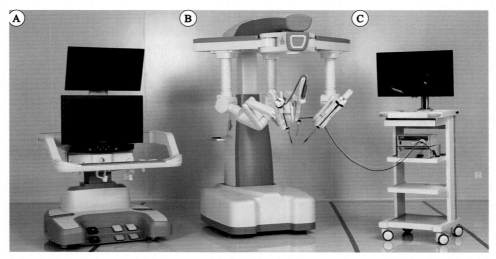

Fig. 7 The da Vinci surgical system composed of three parts: (A) a surgeon's console, (B) a patient side robotic cart with four arms manipulated by the surgeon (one to control the camera and three to manipulate instruments), and (C) a high-definition 3D vision system. *(Reproduced and adapted with permission from Li et al., 2023. Copyright Elsevier 2023 (Li, X., Xu, W., Fan, S., Xiong, S., Dong, J., Wang, J., Dai, X., Yang, K., Xie, Y., Liu, G., Meng, C., 2023. Robot-assisted partial nephrectomy with the newly developed KangDuo surgical robot versus the da Vinci Si surgical system: a double-center prospective Randomized Controlled Noninferiority Trial. Eur. Urol. Focus 9(1), 133–140. https://doi.org/10.1016/j.euf.2022.07.008).)*

2020). Furthermore, artificial intelligence integrating machine-learning algorithms and analytical software have been of great support to improve the training procedures. Thus, training of students has helped to propel the growth of the market of surgical and diagnosis robots. As safety and high-quality education are very important in the learning procedure of surgical trainees before operation on real patients, Chapters 7 and 8 are devoted to learning more about such applications. In Chapter 7, artificial intelligence in robot-assisted surgery for surgical skills assessment and transfer is considered. A comprehensive review of artificial intelligence-powered approaches that detect and incorporate the underlying skills-related features of surgical trajectories to classify and improve the levels of expertise of users in surgical training platforms are proposed. To this end, the functionality, advantages, and drawbacks of current skills evaluation and transfer methods are investigated. In Chapter 8, a virtual-based haptic endoscopic sinus surgery training system is proposed from design to validation. The simulated anatomy and the trainee haptic feedback are improved in comparison with a number of current simulators. A quantitative and qualitative analysis is proposed.

According to Grand View Research, Inc (n.d.), the GMRSM segmentation by region (respectively by country) shows that North America (the USA and Canada), Europe

(Germany, the UK, Spain, France, and Italy), Asia-Pacific (Japan, China, India, South Korea, Australia, and Singapore), Latin America (Brazil, Mexico, and Argentina), and the Middle East and Africa (South Africa, Saudi Arabia, and the United Arab Emirates) show the most revenue growth in descending order, whereas Asia-Pacific detained the principal revenue share of more than 50.0% in 2021.

In order to illustrate and to understand the growth of revenue in Asia-Pacific, we have considered the evolution of medical robotics field in India in Chapter 9, as this country is one of the leading ones in this field over the region. This chapter considers the state of the art, the challenges, and the opportunities in robotic advancements in India. Four specific areas in the field are identified: micro and nano robotics in medicine; minimally invasive surgical robotics; image-guided surgical procedures; and rehabilitation robotics. Modeling, sensing, and control of these robots have also been summarized for a medical Indian platforms case study.

From another angle, and according to (Grand View Research, Inc, n.d.), the GMRSM segmentation based on type of outlook shows that surgical robots and service robotics are growing at the fastest rate among all medical robotics applications. Nevertheless, medical service robots, which is still a relatively new field of robotics in the world, led the medical market with a share of 65.8% in 2022. This situation can be explained by the COVID-19 pandemic and concerns about disinfecting hospitals. Fig. 8 shows examples of area disinfection units and disinfection robots used in hospitals during the COVID-19 pandemic.

Readers needing to discover more about this new field can refer to Chapter 10, which presents a comprehensive review about the trends and implementation of service robots in medical institutions during the Covid-19 pandemic. The chapter deals with the classification of service robots, their development, distribution in the last 10 years, and related applications. Particular attention is devoted to the contribution of artificial intelligence for the improvement of this field. Applications of these intelligent robots may include robots for remote treatment of patients, visiting and serving patients with food and medicines, disinfecting surgical rooms, robots for children, and elderly therapy and psychological support. This feature, known as social robotics, is important from social and emotional viewpoints, and it encompasses two broad domains: human–robot interaction and cognitive robotics. Fig. 9 shows some well-known companion robots used for such objectives.

Finally, the book closes with Chapter 11, which summarizes the main issues of all achievements in medical robotics and addresses related perspectives, challenges, and open issues.

(a) Xenex Lightstrike

(b) TRU-D Smart UVC

(c) UVD robot by Blue Ocean Robotics

(d) Sterilray Far-UVC robot

Fig. 8 Disinfection robots used in hospitals during the COVID-19 pandemic. *(Reproduced with permission from Taghipour et al., 2023. Copyright Elsevier 2023 (Mehta, I., Hsueh, H.Y., Taghipour, S., Li, W., Saeedi, S., 2022. UV disinfection robots: a review. Robot. Autonom. Syst. 104332. https://doi.org/10.1016/j.robot.2022.104332).)*

Fig. 9 Companion robots. *(Reproduced with permission from Shishehgar, Kerr and Blake, 2018. Copyright Elsevier, 2023 (Shishehgar, M., Kerr, D., Blake, J., 2018. A systematic review of research into how robotic technology can help older people. Smart Health 7–8, 1–18. https://doi.org/10.1016/j.smhl.2018.03.002).)*

References

Aymerich-Franch, L., Ferrer, I., 2022. Liaison, safeguard, and well-being: analyzing the role of social robots during the COVID-19 pandemic. Technol. Soc. 70, 101993. https://doi.org/10.1016/j.techsoc.2022.101993.

Boubaker, O., 2020. Medical robotic. In: Boubaker, O. (Ed.), Control Theory in Biomedical Engineering: Applications in Physiology and Medical Robotics. Academic Press, https://doi.org/10.1016/B978-0-12-821350-6.00007-X.

Gajjar, A.A., Le, A.H.D., Lavada, R.S., Boddeti, U., Barpujari, A., Agarwal, N., 2023. Evolution of robotics in neurosurgery: a historical perspective. Interdiscipl. Neurosurg., 101721. https://doi.org/10.1016/j.inat.2023.101721.

Grand View Research, Inc, Medical Robotic Systems Market Size, Share & Trends Analysis Report By Type (Surgical Robots, Exo-robots, Pharma Robots, Cleanroom Robots, Robotic Prosthetics, Medical Service Robots), By Region, and Segment Forecasts, 2022–2030. https://www.grandviewresearch.com/

industry-analysis/medical-robotic-systems-market#:~:text=The%20global%20medical%20robotic%20systems,17.4%25%20from%202022%20to%202030. (Accessed 12 March 2023).

Liu, Y., Liu, M., Lei, Y., Zhang, H., Xie, J., Zhu, S., Jiang, J., Li, J., Yi, B., 2022. Evaluation of effect of robotic versus laparoscopic surgical technology on genitourinary function after total mesorectal excision for rectal cancer. Int. J. Surg. 104, 106800. https://doi.org/10.1016/j.ijsu.2022.106800.

Market and Market, Inc, Medical Robots Market Report [Instrument & Accessories, Robotic Systems (Surgical Robots, Rehabilitation Robots)], Application (Laparoscopy, Radiation Therapy, Pharmacy), End User (Hospital, Ambulatory Surgery)—Global Forecasts To 2025. https://www.marketsandmarkets.com/Market-Reports/medical-robotic-systems-market-2916860.html. (Accessed 12 March 2023).

Meng, W., Liu, Q., Zhou, Z., Ai, Q., Sheng, B., Xie, S.S., 2015. Recent development of mechanisms and control strategies for robot-assisted lower limb rehabilitation. Mechatronics 31, 132–145. https://doi.org/10.1016/j.mechatronics.2015.04.005.

Yoganandhan, A., Kanna, G.R., Subhash, S.D., Jothi, J.H., 2021. Retrospective and prospective application of robots and artificial intelligence in global pandemic and epidemic diseases. Vacunas (Engl. Ed.) 22 (2), 98–105. https://doi.org/10.1016/j.vacune.2020.12.002.

CHAPTER 1

Assistive robotic technologies: An overview of recent advances in medical applications

Md Samiul Haque Sunny[a], Md Mahbubur Rahman[b], Md Enamul Haque[b], Nayan Banik[a], Helal Uddin Ahmed[c], and Mohammad Habibur Rahman[a,b,c]

[a]Department of Computer Science, University of Wisconsin-Milwaukee, Milwaukee, WI, United States
[b]Department of Mechanical Engineering, Milwaukee, WI, United States
[c]Biorobotics Laboratory, University of Wisconsin-Milwaukee, Milwaukee, WI, United States

1.1 Introduction

Medical robots have come a long way since the 1980s. Robotics has benefits more for medical professionals and patients compared to other scientific fields, which was unthinkable decades ago. Examples include skilled treatment, effective surgery, and excellent patient care in a secure setting (Riek, 2017). They assist in providing patients with intelligent and more comprehensive care, which results in a faster healing process (Khan et al., 2020). Robotic technologies realize augmented virtual telepresence that allows users to visit distant locations and interact with nearby people, things, and surroundings without physically traveling there (Zhang, 2021). Similarly, surgeries advance a lot with the help of robot technologies, for example, the application of robots in neurosurgery and cutting bones. However, the blessing of robotics has still not reached the marginal people of the world due to the high cost and task-specific requirements of rigorous training of surgeons, which are the major drawbacks.

Moreover, one billion people, or 15% of the world's population, are experiencing some form of disability; this is worst in developing countries (WorldBank, 2022). People with disabilities rely on their relatives and friends for support. Although these individuals attempt to cover care gaps, they have jobs and other responsibilities. Therefore, they cannot provide services that meet the needs of impaired individuals. On the other side, healthcare staffs are overburdened due to the labor shortage, where many more patients require care than workers to provide it (Smaling et al., 2022). In addition, the issue has been exacerbated due to the vulnerability of COVID-19 to affect them. Therefore, robotics technology has a big possibility of addressing care gaps and supporting healthcare professionals. Assistive robots such as exoskeletons are trained to restore gripping, gaits, and other bodily movements. Besides, exoskeletons are used to rehabilitate upper/lower extremities, such as to aid chronically immobile individuals (de Andrade et al., 2023).

Medical and Healthcare Robotics
https://doi.org/10.1016/B978-0-443-18460-4.00004-4

Robots have been used in various settings, including hospitals, clinics, homes, schools, and nursing homes, as well as in urban and rural settings. In 2020, hospitals held the largest global medical robots market share and projected 12.7 billion by 2025 (Market, 2022). Technologists, researchers, providers, and end users must collaborate to ensure robots' successful adoption in medical applications.

1.2 Current state of healthcare robotics

Microrobotics, surgical robotics, and interventional robotics are three areas where internal robots have seen recent progress. Microrobotics are miniature mobile robots that can be used for various medical applications, including cancer detection and targeted therapy. In surgery, robotics facilitates fewer scars, less pain/discomfort, less bleeding, and faster recovery (Chen et al., 2022). In addition, robots' remote surgery and precision expedite surgeons' productiveness to a large extent. Similarly, therapeutic interventions have been speeded up with frequent monitoring, increased repetition, and systematic data acquisition. For instance, gait rehabilitative and assistive robots are being used as great complementary tools to therapists. Currently, they are providing more frequent therapy to more patients than before with the help of remote therapy and robotics. They are being privileged by faster recovery, and shorter hospital stays on the patient end. Therefore, all the robotic applications have made robotics inevitable parts of the modern healthcare system. According to the Agency for Healthcare Research and Quality (ARHQ, 2022), five major factors for deploying robots in healthcare applications are presented in Fig. 1.1. The scope of the robot must be well defined with clear goals, key stakeholders are involved, clinical success matrices should be determined, and quality improvement should be an ongoing process to avoid unintended consequences.

1.3 Robots in medical applications

In the human body, concentric tube robots are used as steering needles, making procedures inaccessible to traditional instruments. On the body, robotic prostheses and exoskeletons provide fine-grained dexterity, reach, and strength to people with upper and lower-limb amputation (Lassoued and Boubaker, 2020). On the other hand, outside of the human body, mobile manipulators are helping to treat highly infectious diseases such as Ebola (Gao et al., 2021). High-fidelity robotic patient simulators are used for training clinicians to simulate physiological cues (Pourebadi and Riek, 2018).

1.3.1 Rehabilitation robots

When it comes to daily activities, a person's own upper and lower extremities are the most important tools at their disposal. A shockingly high number of people live with the loss or impairment of one or more of their limbs. Individuals with neurologic-related

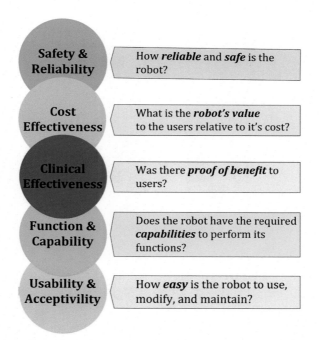

Fig. 1.1 Aspects of medical robot adoption that needs to be taken into consideration.

gait impairment can benefit from appropriate physical therapy to restore and recover their lower extremity function. There is still room for improvement when it comes to promoting physical activity and capitalizing on existing community-based recreation programs. Physiotherapists play an integral role in the lives of many people with physical disabilities and chronic conditions. Traditional rehabilitation therapies, however, call for a significant time commitment from the therapist or clinician. Physical therapists have a difficult and physically taxing job. In the last few decades, interventions for neuromotor rehabilitation have come a long way, especially in light of the financial strain that therapist/clinician shortages cause. Repetitive exercise of coordinated motor activities has improved patients' gait, arm function, and quality of life. Treatments based on exercise can be taxing on therapists and expensive for healthcare systems. They may become less expensive thanks to technological advancements like robotics and virtual reality.

Rehabilitation robots are divided into two types based on their treatment approaches. The first strategy involves continuous passive movement (CPM). CPM necessitates no volunteer effort from the patient because the robot controls and moves the limb. The second strategy is active-assisted movement, in which the patient sends a signal to the robot before it can carry out the action. Based on the structure, rehabilitation robots are end effectors and exoskeletons. End effectors are simple robots with a distal movable handle that patients attach to and follow. As rehabilitation requires various movement

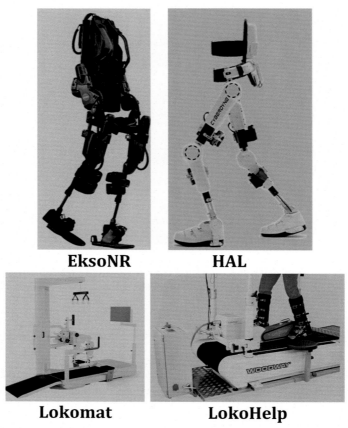

EksoNR   **HAL**

Lokomat   **LokoHelp**

Fig. 1.2 Therapeutic robots for rehabilitation are now in the market and are widely used in clinical settings.

sizes and shapes, this robot can adapt. Exoskeleton robots encase limbs in splints or bionic structures. Exoskeleton robots determine joint torque and control limb movement. Exoskeletons need less space than end-effector robots. Based on application, rehabilitation robots can be classified into two types: explicitly designed for use with the upper extremities (UE) and lower extremities (LE). Each scenario allows for a distinct categorization of the rehabilitative robots' core functions. This chapter focuses on LE rehabilitation robots. Fig. 1.2 shows some commercially available rehabilitation robots that support gait motion.

Many researchers have worked to improve the lower-limb exoskeleton presented in Table 1.1. SLEX used individualized gait pattern generation (IGPG) to allow patients with different anthropometric measurements to share a lower-limb exoskeleton (Wu et al., 2018). It can adapt to different body sizes, gait patterns, and walking speeds.

Table 1.1 Research aimed at improving the performance of the lower extremity rehabilitation robot (LERR).

Research	Purposes	Technology	Remarks
SLEX, Wu et al. (2018)	Efficiency improvement of LERR by adapting gait pattern	Gaussian regression with automatic relevance determination	The model can predict individuals gait
VALOR, Liu et al. (2019)	Vision-assisted LERR gait pattern planning	RGB-D camera extracts ground object, then controller moves it	The method can adapt to indoor environments with few obstacles
Yang et al. (2020)	Adopting uncertainties and disturbances	Uniform boundedness in disturbances, fuzzy set theory	Optimal control gains have been theorized
Li et al. (2018)	Estimation of human-joint muscular torques	Inverse dynamic approach	Accuracy with a variety of subjects and motion types
Ma et al. (2018)	Switching the gait of the exoskeleton robot	BP neural network was used to predict knee joint angles in real time using sEMG signals	The method predicts human intent and changes gait in real time
Chen et al. (2019)	Robust adaptive control; rehabilitative LERR	Leakage-type and Udwadia-Kalaba theory to compensate for uncertainty	Modeling error, initial condition deviation, friction force, and other external disturbances are handled
Zhang et al. (2021)	It provided a framework for the real-time continuous gait phase-variable estimation; LERR	It used phase-variable-based control	It introduced an online learning mechanism for extracting and learning gait features from previous strides
Chang et al. (2017)	It generates the exoskeleton's constraints by processing the received signal from the patient	The finite state machine controls double stance, early swing, late swing, and weight acceptance	The system could adapt to individual needs to coordinate biomechanical and muscle activation

VALOR (Liu et al., 2019), a vision-assisted lower-limb exoskeleton robot that helps patients make independent decisions by parameterizing step length and height. The controller extracts ground object features that may affect movement from RGB-D camera data. Another fuzzy set theory-based controller for the 2DOF lower-limb exoskeleton is proposed in Yang et al. (2020), where the method optimizes gain. Simulations showed high precision tracking control, and good uncertainty robustness (Yang et al., 2020). A new method of estimating lower-limb human-joint muscular torque (JMT) based on the body's inverse dynamics is proposed in Li et al. (2018). In addition to that, a gait–switching method for stairs and obstacles is also developed (Ma et al., 2018). Their exoskeleton switches gait across obstacles in real time using surface electromyography (sEMG).

In clinical settings, some commercial rehabilitation devices based on treadmills have been used extensively for lower–limb rehabilitation. The Lokomat (Hacoma, 2022), LokoHelp (Woodway, 2019), and ReoAmbulator (Neurorehabdirectory, 2022) are just a few examples. Functional exercise therapy is organically integrated with the assessment and feedback system (Voloshkin et al., 2022) thanks to the new Lokomat robot that can simulate physiological gait trajectory, drive the patient's unilateral or bilateral lower limbs, and accurately control the speed of the treadmill to match the patient's gait consistent.

Exoskeleton–based lower extremity rehab robots (EBLERRs) combine AI and mechanical power devices to help patients with lower extremity movement and magnify muscle energy and activities, which are a direct and effective way to compensate for or regain physical dysfunction (Liu et al., 2016). Strong and durable EBLERRs enable people with lower–limb impairments to reintegrate into society (Shi et al., 2019).

In a rehabilitation treatment program, it is required to provide challenging rehabilitation exercises to the patients. By adjusting the intensity and duration of the exercises based on the patient's progress, therapists can increase the likelihood that their patients will achieve their objectives. Game-like exercises are used to motivate patients to complete their therapy. As a result of the augmented performance feedback (APF) shown in Fig. 1.3, the benefits of Lokomat training are amplified.

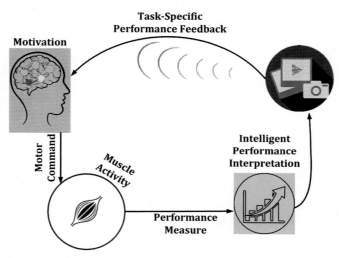

Fig. 1.3 Intelligent performance interpretation of Lokomat based on task-specific performance feedback.

1.3.2 Surgical robots

Robotics has brought a changing era in the field of surgery. This modern technology facilitates fewer scars, less pain, less bleeding, and quicker recovery (Chen et al., 2022). In addition, robotics surgery has included remote and isolated areas in the world, where experts are sending their services seating in big cities (Lokhande and Patil, 2022). Moreover, during the COVID-19 pandemic, direct contact of healthcare professionals with patients posed a major risk, which resulted in the cancelation of surgeries, that is, massive economic loss for the healthcare system and depriving patients of treatment. The robots partly solve this problem as an intermediary shield against pathogen contamination (Zemmar et al., 2020). This section discusses some robotic applications in surgeries. Fig. 1.4 listed frequent surgical operations and procedures carried out with robotic assistance, whereas Fig. 1.5 presents some of the surgical robots used in various settings. To advance the standard of surgical care, there is an unprecedented demand for the provision of an accurate, objective, and automated evaluation of the surgical skills of surgical trainees. A convolutional neural network (CNN) was developed by Ismail Fawaz et al. (2019) to classify surgical skills. This was accomplished by identifying latent patterns in trainees' motions during robotic surgery. Now that data on surgical procedures are becoming more widely available, sophisticated deep learning techniques can be used to evaluate a seasoned doctor's proficiency and impart that knowledge to a novice surgeon (Soleymani et al., 2021a). The DESK dataset provides examples of surgical procedures across a variety of robotic platforms with significant variation in setup, and Rahman et al. (2019) proposed a method that makes use of compact image representation with

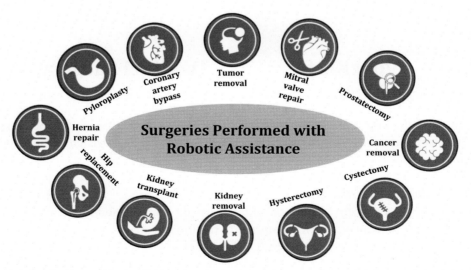

Fig. 1.4 Surgical operations and other procedures are typically performed with robotic assistance.

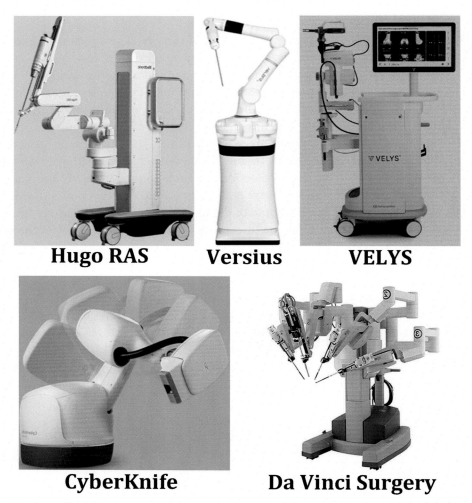

Fig. 1.5 As an illustration, here are some types of surgical robots currently in use in hospitals.

kinematic features for surgeme recognition. A comparative study of several commercially available surgical robots in healthcare with their purpose and associated technology is given in Table 1.2.

1.3.2.1 Laparoscopic robots

A laparoscopy is a modern surgical procedure carried out in the abdomen or pelvis utilizing a camera (Thigpen, 2022). Compared to traditional surgery, the laparoscopic procedure results in a smaller scar, less pain, and a faster recovery, because it is done with smaller incisions with less pain. However, this laparoscopy has some limitations, which have been resolved largely by robotics laparoscopy. These are mentioned in the subsequent paragraph.

Table 1.2 Characteristics of some surgery robots.

Research	Purposes	Technology	Remarks
Hugo RAS (HugoTM RAS System, 2022)	Robotic-assisted surgery technology across both open and laparoscopy	Collaborative robots; subsystems for vision, tissue manipulation, anatomical accessibility, and training in operator technology (Desai, 2018)	Flexible configurations to fit OR space
Versius (CMR Surgical, 2022)	Itfacilitates minimal access surgery	Remote control with 3D vision	Virtual access to OR, enhanced with 3D HD vision, and ergonomic seating and standing of surgeons
Da Vinci Surgery (Davincisurgery, 2022)	It also focuses on minimal invasive option	Vinci vision system delivers 3D high-definition views, giving your surgeon a crystal clear view of the surgical area	Vinci system because it extends the capabilities of their eyes and hands what the human eye sees
VELYS (J&J MedTech, 2022)	Robotic-assisted solution simplifies knee replacement surgery	*NATURAL CONTROL* Technology to maintain the saw cut, *ADAPTIVE TRACKING* Technology, high-speed camera for triple-drive motion, and *PURESIGHT* Hydrophobic optical Reflectors to accurate plan of resection	It providing valuable insights, versatile execution, and verified performance designed to deliver efficiency for surgeons and optimize patient outcomes
CyberKnife (2022)	It facilitates precision for personalized radiotherapy	Stereotactic body radiation therapy (SRS SBRT) to deliver precise doses of radiation with extreme accuracy	CyberKnife is approved for treating a variety of cancers, including brain tumors, breast, liver, lung, pancreatic, and prostate cancers

In conventional laparoscopy, the movement of the laparoscopic instrument to the right by the surgeon causes the instrument's tip to move to the left inside the patient's body due to the fulcrum movement effect (Fulmer, 2020), which is counterintuitive and takes much time for surgeons to be hands-on. In robotic laparoscopy, this problem is solved by a left-right reverse algorithm (Oesterreich et al., 2022).Usual laparoscopy provides mechanical scaling of surgeons' hand movement to instrument tips. This procedure transfers hand tremor effects to the tips, which result in imprecise movements (Breedveld et al., 1999). Fortunately, robotic systems can minimize or remove the hand tremor effects by filtering those noises (Fulmer, 2020), which are accomplished by filtering out tremor motion from the voluntary motion of hand using low-pass filter or advanced algorithm, for example, Kumar et al. (2020) and Pan et al. (2022). Further, a laparoscopic instrument tip has four degrees of freedom (DoFs), which cannot mimic human hand movement. However, robotic laparoscopy has articulated tips, which have seven DoFs. This mimics human wrist movements, including rotation (Wu et al., 2022). Further, laparoscopy also facilitates palpation by attaching force sensors at end effectors (Sušić et al., 2019).

1.3.2.2 Nonlaparoscopic robots

The opposite of laparoscopy is the conventional exploratory laparotomy, where the abdomen is opened and subsequent testing and treatment are applied. This includes electrode implants, eye surgery, continuum robots, etc. Some of these are discussed here.

Electrode implant helps collect cortical signals regarding joint movement for people with severe disabilities who have few other control options, for example, those who have advanced amyotrophic lateral sclerosis, brain stem strokes, or extreme cervical tetraplegia. This implanting procedure requires open surgeries. Previously, implanting procedures often suffered from tremor, drift, and accurate force feedback (De Seta et al., 2021). Now robotic electrode implanting facilitates minimal drilling and proper alignment of the electrode array in a faster operation.

Robotics has fascinated us in eye surgery with precise and small incisions, a higher degree of freedom, and tremble-lessness and fatigue-free operation (Li et al., 2021) through overcoming human borders, and physiological limits (Urias et al., 2019). Now, surgeons can manipulate their tools on a micrometer scale which caters to a lot of control in multi DoFs. Since the Da Vinci Robotic Surgical System (Intuitive) was certified for laparoscopic surgery in 2000, robotic surgery has been progressing toward becoming a significant component of ophthalmic surgical procedures (Singh and Krishna, 2014).

Contrary to rigid-link theories, continuum robots aim to resemble organic trunks, snakes, and tentacles (Hansen, 2022). Thus, a continuum robot is a particular kind of robot with endless joints and degrees of freedom. Numerous designs of continuum robots have achieved commercial success and have diverse applications, ranging from medicine to underwater exploration (Singh and Krishna, 2014). This new technology overwhelmed us with better access, safer interactions, and new procedures such as early-stage diagnoses of lunch cancer.

Robotic medical capsules are now being widely used in a renaissance in surgeries, medicine delivery, and diagnosis. Previously, a thin and flexible tube with a camera and light was inserted into the stomach and duodenum during the operation, known as gastroscopy, to look for any abnormalities (Martincek et al., 2020), which were painful for the patients. Then, the capsule robot originated from the science fiction concept of robots entering the body to identify and treat diseases. A capsule endoscope designed to take pictures of the digestive system was the first capsule robot that was commercially available (Mapara and Patravale, 2017).

1.3.3 Soft, micro, and nanorobots

Soft robots are made of flexible materials rather than rigid linkages to ensure safety when working near people (Rus and Tolley, 2015). Their flexibility enables them to fit in spaces that stiff bodies cannot, which may be advantageous in many situations. Soft robots are difficult to handle because of their flexibility and compliance, which makes them useful. Evolutionary algorithms and other automated design tools are frequently used to create soft robots because conventional equations to control rigid links typically cannot be applied in soft robots (Bongard, 2013). Soft materials, such as the wide variety of elastomers used in much current research, are inexpensive, widely available, and simple to work with. While this improves patient safety, it comes at the cost of other features (such as low force exertion, poor controllability, and a lack of sensing capabilities) (Hughes et al., 2016). The various micro- and nanorobotic prototypes have come a long way since their early iterations in terms of power, motion control, functional versatility, and capabilities, thanks to the tireless efforts of the nanorobotics community. As their sophistication increases, nano- and microscale robots open up exciting new possibilities in biomedicine. A comparative study of several available soft robots in healthcare with their purpose, associated technology, and application is given in Table 1.3.

Numerous studies have shown these robots to be capable of navigating through complex biological media or narrow capillaries, allowing them to autonomously release their payloads at predetermined destinations after performing localized diagnosis, removing biopsy samples, and imaging the affected area. No cords, batteries, or other external power sources are necessary to power these wireless robots (Li et al., 2017). Many of them are constructed from biocompatible materials that will degrade or even vanish after their task is done. Fig. 1.6 depicts some soft, micro, and nanorobots applicable to medical fields. The components of invendoscopy (Peters et al., 2018) are a single-use processing unit, a reusable hand-held controller, and a disposable colonoscope. Continuum (Hansen, 2022) has the potential to enhance minimally invasive surgical procedures in the medical field and provides brand new opportunities for inspections and repairs in the field of engine maintenance. Using semiconductor components, researchers at Cornell have developed the first microrobot Laser Jolts (Miskin et al., 2020) capable of being controlled and made for walking using standard electronic signals. Laser light can independently bend each of their legs. The robot moves forward by toggling the laser

Table 1.3 Summary of commercial soft robots in healthcare.

Research	Design principle	Technology and controllable DOF	Application
Khadem et al. (2017)	Steerable needle with notch	Ultrasound guided with 3-DOF	Brachytherapy
Su et al. (2016)	Concentric tube	3 Piezoelectrically actuated tubes with 6-DOF	MRI-guided surgery
Gafford et al. (2020)	Concentric tube	2 Motor-driven dexterous arms with 3-DOF	Central airway obstruction
Graetzel et al. (2019)	Monarch: Concentric tube	Cable driven with 10-DOF	Bronchoscopy
Amanov et al. (2018)	Backbone	2 Tendon-driven segments with 2-DOF	Single-site partial nephrectomy
Zhang et al. (2018)	Backbone	Tendon-driven extensible backbone with multi-DOF	Neurosurgery
Ahmad et al. (2019)	McKibben's fabricated muscles	Pneumatic with 2-DOF	Fetal surgery
Garbin et al. (2018)	Bellows in parallel	3 Pneumatic bellows at the tip with 3-DOF	Gastroscopy
Inoue and Ikuta (2016)	Serial bellows	2 Hydraulic segments with multi-DOF	Endovascular

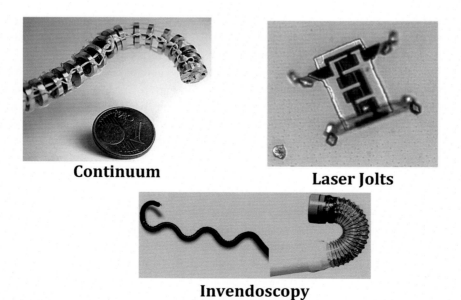

Continuum

Laser Jolts

Invendoscopy

Fig. 1.6 Examples of soft and nanorobots for minimally invasive surgical procedures or diagnostics.

between its front and back legs. Small enough to be injected into the body, tiny robots could 1 day be programmed to hunt down and destroy cancer cells. Magnetic actuation systems that can remotely command precise five DoFs control (three translational DoFs and two rotational DoFs) of a single magnetic device have a fascinating contribution to microrobot (Ryan and Diller, 2017). Examples can be found in targeted neural cell delivery, and selective connection of neural networks, where operation is done remotely by applying magnetic field (Kim et al., 2020).

1.3.4 Service robots

Thanks to service robots, healthcare workers can focus on patient care instead of mundane administrative tasks. Many of these robots can carry out their functions independently, and some can even report back to their human masters when they finish a job. In hospitals, these robots are used for various tasks, including patient room preparation, inventory management, the replenishment of medical supply cabinets, and the transportation of soiled bed linens to the laundry. Service robots allow healthcare workers to devote more time to meet the immediate needs of patients, which have been shown to boost morale. Moreover, those devices can provide passive support in maintaining safe distances and curing infectious diseases (Soleymani et al., 2021b). A comparative study of several commercially available service robots in health care with their purpose and associated technology is given in Table 1.4.

Among several commercially available service robots presented in Fig. 1.7, LD-UVC from Omron (2022) is an automated cleaning and disinfecting robot that can fight against viruses and bacteria with a great success ratio. This efficient robot can overcome the burden of manual cleansing in healthcare centers amid the emergencies like the COVID-19 pandemic. Moreover, the sensors can navigate the robot to a predefined path, shut down the UV in human presence, and provide 360-degree coverage to the operating region. Similar to LD-UVC, LightStrike from Xenex (2022) is another cleaning and disinfecting robot that is easy to use, operationally fast, effective, and customizable. It uses high-intensity UV light to disinfect hospital premises and permits the service staff to better serve their responsibilities, providing they can concentrate on other tasks without overlapping. TUG (Aethon, 2022) is another high-performance hospital-based food and supply robot carrying up to 453 kg of racks, carts, and bins containing food, medications, and specimens in hospitals. It provides healthcare providers, including doctors and nurses, to concentrate on their work and increase the quality of service in the center. Notable features of TUG include low noise, predefined maps, secure medicine handling, a cleaner environment, and many more.

Veebot (2022) is another robotic device that helps phlebotomists to accurately scan for veins in the arm using a camera, ensure the selected vein has proper blood circulation through ultrasound, and upon confirmation, insert the onboard needle to extract the blood. However, the robot is incapable of further sample testing leaving it for manual

Table 1.4 Summary of commercial service robots in healthcare.

Research	Purposes	Technology	Remarks
LD-UVC (Omron, 2022)	Automated cleaning and disinfecting of virus and bacteria	UVC lamps, PIR motion sensors, Lidar sensors, steel guard, audio and visual buzzers	Touch-based control interface, capable of avoiding obstacles, maneuver in narrow spaces, follow mapped routes, automatic human detection to avoid radiation exposure
LightStrike (Xenex, 2022)	Automated cleaning and disinfecting of virus and bacteria	Pulsed xenon ultraviolet light (PX-UV), sensor-based motion detection, cloud reporting, auto-updated software, disinfection Pod	PX-UV includes both UV-B and UV-C in its range of germicidal UV which are fast and effective to clean patient room within 20 min without damaging medical apparatus
TUG (Aethon, 2022)	Providing hospital-based food and supplying	Map in memory, scanning lasers, ultrasonic and infrared sensors, wireless network, intermittent charging with 10 h backup	Automated transport of medicines, lab samples, food items, waste and trash to enhance workers productivity and safety, improve patient care and satisfaction
Veebot (2022)	Helping phlebotomists to extract blood from veins	Inflatable cuff pumps, infrared and ultrasound sensors, computer vision-based machine learning	Increase safe, faster, and successful needle insertion rate on suitable veins for different age groups, easy to use the machine to minimize contamination and needle stick injuries
YuMi (2022)	Collaboration in different laboratory-based tasks	Integrated IRC5 controller, different I/O interfaces, support HMI devices like ABB's teach pendant and commercial tablets	Effective deployment with laboratory staff, perceive surrounding with cameras, real-time algorithms for collision avoidance

human intervention. On the other side, YuMi (2022) is a full-fledged dual-arm collaborative robot assisting the human workforce safely and effectively in many fields. YuMi is an appropriate addition to the medical domain, where efficiency is crucial. YuMi is now deployed in laboratories and healthcare facilities to perform diagnostics, routine testing, blood cultures, etc.

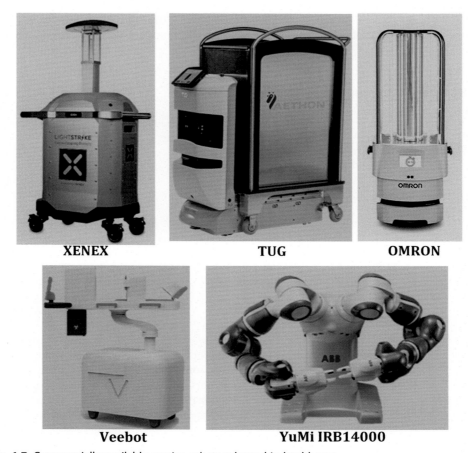

XENEX   **TUG**   **OMRON**

Veebot   **YuMi IRB14000**

Fig. 1.7 Commercially available service robots adopted in healthcare.

1.3.5 Social robots

A social robot is an autonomous robot that can connect and communicate with humans and other social robots by adhering to the social behaviors and rules associated with its role in a group. Authors identified commonalities, such as that social robots are agents with some degree of physical embodiment and engage in social interactions with humans through verbal exchange, cooperative problem-solving, and independent decision-making (Henschel et al., 2021). Social robots live up to their promise when used in settings suited to their design and purpose. Robots have many potential applications in society, including but not limited to serving as companions for the elderly and the cognitively impaired, working as teachers, and supporting cognitive and behavioral change interventions. Further, social robots facilitate educational entertainment, for example, video games, with an educational aspect (Sequeira, 2020). A comparative study of several commercially available social robots in health care with their purpose and associated technology is given in Table 1.5.

Table 1.5 Summary of commercial social robots in healthcare.

Research	Purposes	Technology	Remarks
Myon (2022)	Improve social interactions among elderly and patients through humanoid robot	6 Body parts, 48 actuators of Dynamixel RX-28, 32 DoFs	Robust and maintainable system with runtime attachable and detachable body parts each having separate power systems and sensors
Robear (Thestron, 2022)	Assist in mobility tasks like moving and transporting	Actuators with low gear ratio, three types of sensors including torque sensors, tactile sensors	Comparatively lighter than its predecessor with backdrivability, gentle movements, retractable legs avoid falling over and to move through the narrow space
Pepper (2022)	Enhance emotional interactions of patients through sensory activities	Touch display, RGB camera, microphone, touch, gyro, sonar, laser, bumper sensors, omnidirectional wheels, WiFi, and Ethernet connectivity	Smart emotion detection using voice and facial expression analyzer, machine learning module to extract behavior based on past communications with patients
PARO (2022)	Impacts social interactions through seal-shaped robotic pet	Dual 32-bit processors, 3 microphones, 12 tactile sensors, touch-sensitive whiskers, a system of motors and actuators	Responds to touch by petting tail or blinking eyes, memorize names, making baby seal sounds, programmed for day and nighttime-specific activities
Aibo (2021)	Reduce loneliness and provide interactive activities through dog-shaped robotic pet	64-Bit quad-core CPU, speakers, microphones, OLED eyes, range, motion, touch, gyro and light sensors, SIM card slots, LTE, WiFi connectivity	Functionalities as a virtual pet including virtual feeding, voice and touch commands, human identification, special features like making friends with other aibo robots, automatic placement in charging dock, etc.

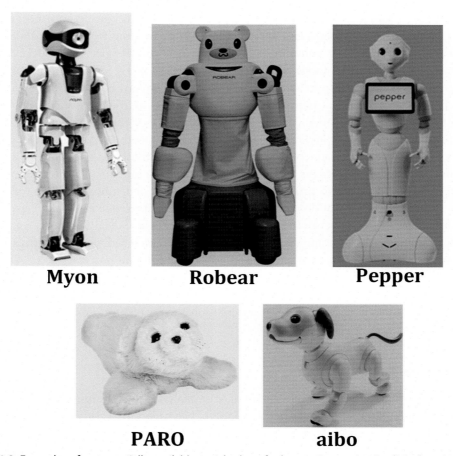

Myon   **Robear**   **Pepper**

PARO   **aibo**

Fig. 1.8 Examples of commercially available social robots for human interaction in clinical settings or at home.

As a social robot, Myon (2022) depicted in Fig. 1.8 has an independent modular design resembling a young child. The interactive nature of the robot helps the elderly and patients requiring continuous company a chance to practice day-to-day social activities, including singing, dancing, and playing games. However, the locomotive power of Myon requires further improvement to ensure agility for a human. The stressful transfer of patients from bed to wheelchair or assisting them to stand up, Robear (Thestron, 2022) is deployed to address the shortage of trained caregivers or professional therapists in many countries. Although this humanoid robot can lift and carry patients avoiding lower-back pain for the nurses, further research is ongoing to make it lighter and maintain proper balance on its feet. With its multitude of use in diverse domains, Pepper (2022) achieves a place in healthcare facilities due to its capabilities of learning from repetitive interaction with patients and showing emotional attitudes through facial expressions, gestures, and tone of voice. The intelligent humanoid can be operated with voice commands, touch, and an onboard tablet interface. The machine learning module to learn from past dialogues with a patient is a continuous field of research for Pepper. As a social robot, PARO (2022) impacts patients' well-being by reducing stress, improving interaction, and providing mutual respect among them and their caregivers. This award-winning seal-shaped robotic pet can mimic repetitive interactions by the carrier and provide an acoustic signal as a real animal. However, the immobility of Paro and its nonhumanoid nature confines its applicability to a versatile social robot like others. Aibo (2021) is widely known as a virtual pet because of its shape and in-home activities as a pet dog. This companion robot keeps the patients occupied and reduces their loneliness. Aibo incorporates AI-enabled technologies to provide interactive activities depending on different situations. Moreover, Aibo's affordable price and portable nature make it a perfect choice to consider as a must-have social robot in every household. Social robots are generally restricted in their programming and have no emotions, which badly impact human interaction. Although they can work nonstop, they require constant power to do so. Despite initial apprehension, studies show that exposure to a social robot can reduce anxiety and increase the desire to engage with it. People may be more open to future interactions with social robots if their initial encounters with them are humorous or lighthearted rather than serious or task-oriented.

1.4 Future trends

With the development of machine learning, data analytics, and computer vision, among other technologies, medical robotics will evolve, becoming increasingly capable of performing tasks autonomously and effectively. Future engineering research efforts should be focused on the most promising areas to accelerate this development. It is essential to learn the value added by robots and the technologies they use in the medical field. To date, medical robots have been designed to improve outcomes in ways that are not directly related to lowering the cost of human labor. In contrast, robots are used as autonomous agents in almost every other industry to reduce human labor costs.

The minimum degree of autonomy required for practical clinical use is usually provided. Efforts should be made to streamline research conducted independently on designing, developing, and evaluating medical robots. It aids in the development of a common interface through which various types of robots can interact with one another. The level of cooperation and output will rise as a result. In response to a diagnostic robot's vitals report, a robot could transport a patient from an ambulance to an emergency room. Furthermore, a surgical robot waiting in the emergency room may initiate surgery without minimal human supervision. For example, radiotherapy robots function at a high level of conditional autonomy by computing and executing a radiation exposure trajectory to deliver the prescribed radiation dose to a patient while protecting surrounding tissues to an acceptable level. Orthopedic robots can independently mill out a prescribed cavity for knee and hip implants.

On the contrary, laparoscopic surgical robots are effective while under the constant control of an operator, even though they currently provide only minimal robotic assistance. Some clinical applications that could benefit from robotic solutions, such as transcatheter mechanical thrombectomy and heart valve repair, still lack such solutions. The current level of autonomy of robotic systems is expected to increase. By far, the most exciting area of software development is autonomous systems. Highly autonomous systems for remotely performing emergency mechanical thrombectomies to treat stroke, for instance, would significantly expand access to this treatment and reduce the time to treatment. Bionic implants that enhance or restore bodily functions, for instance, will eventually become so naturally integrated with their host that they will not need constant conscious control.

Technical fields of healthcare robotics depend on cutting-edge computer vision, intelligent manipulation, accurate navigation, high precision sensors, and robust communication interfaces. So, development in the aforementioned domains directly contributes to the better modeling of human health care and needs to be supported by robots. Moreover, the integration of the Internet of things (IoT) and 5G connectivity could make healthcare robots a great resource for gathering up-to-the-minute health information. Complex machine learning models are being created to skim through this mountain of data in search of the impetus for its creation. It will aid in developing smart robots for use in the medical field.

Even for experienced medical staff, the learning curve associated with healthcare robotics is steep. Setting up the robot for deployment calls for a sizable workforce. Expertise in operating a medical robot does not need to be acquired overnight, and researchers should work toward making robots that are both lightweight and simple to set up.

1.5 Conclusions

Humans and robots have always had a complicated relationship. Despite continued human skepticism, medical robotics research continues to flourish. Robots have become so advanced in intelligence that they are increasingly escaping their laboratories and manufacturing environments. They are now widely used in hospitals, and society, aiding people in everything from space travel to medical procedures. Even though robots have gotten smarter and more complex, current AI technology is still far from human intelligence. Humans and machines will likely collaborate in the workplace of the future. Robots will assist in relieving overworked healthcare workers, surgeons, housekeepers, caregivers, and delivery personnel, as well as keep factories' assembly lines running smoothly. However, the greater impact of AI and medical robotics will be enhancing and expanding human capabilities rather than restricting them.

References

Aethon, 2022. Mobile Robots for Healthcare—Pharmacy, Laboratory, Nutrition and EVS. Available from: https://aethon.com/mobile-robots-for-healthcare. (Accessed 29 August 2022).

Ahmad, M.A., Ourak, M., Gruijthuijsen, C., Legrand, J., Vercauteren, T., Deprest, J., Ourselin, S., Vander Poorten, E., 2019. Design and shared control of a flexible endoscope with autonomous distal tip alignment. In: 2019 19th International Conference on Advanced Robotics (ICAR), pp. 647–653.

Aibo, 2021. Aibo. Available from: https://us.aibo.com. (Accessed 28 August 2022).

Amanov, E., Nguyen, T.-D., Markmann, S., Imkamp, F., Burgner-Kahrs, J., 2018. Toward a flexible variable stiffness endoport for single-site partial nephrectomy. Ann. Biomed. Eng. 46 (10), 1498–1510.

ARHQ, 2022. Agency for Healthcare Research and Quality (AHRQ). Available from: https://www.ahrq.gov. (Accessed 29 August 2022) Available from:.

Bongard, J.C., 2013. Evolutionary robotics. Commun. ACM 56 (8), 74–83.

Breedveld, P., Stassen, H.G., Meijer, D.W., Jakimowicz, J.J., 1999. Manipulation in laparoscopic surgery: overview of impeding effects and supporting aids. J. Laparoendosc. Adv. Surg. Tech. A 9 (6), 469–480.

Chang, S.R., Nandor, M.J., Li, L., Kobetic, R., Foglyano, K.M., Schnellenberger, J.R., Audu, M.L., Pinault, G., Quinn, R.D., Triolo, R.J., 2017. A muscle-driven approach to restore stepping with an exoskeleton for individuals with paraplegia. J. Neuroeng. Rehabil. 14 (1), 1–12.

Chen, X., Zhao, H., Zhen, S., Sun, H., 2019. Adaptive robust control for a lower limbs rehabilitation robot running under passive training mode. IEEE/CAA J. Autom. Sin. 6 (2), 493–502.

Chen, Z.Z., Xu, S.Z., Ding, Z.J., Zhang, S.F., Yuan, S.S., Yan, F., Wang, Z.F., Liu, G.Y., Qiu, X.F., Cai, J.C., 2022. Comparison between laparoscopic-assisted natural orifice specimen extraction surgery and conventional laparoscopic surgery for left colorectal cancer: a randomized controlled study with 3-year follow-up results. Zhonghua Wei Chang Wai Ke Za Zhi 25 (7), 604–611.

Surgical, C.M.R., 2022. Versius. Available from: https://cmrsurgical.com/versius. (Accessed 29 October 2022).

CyberKnife, 2022. Home-CyberKnife. Available from: https://cyberknife.com/. (Accessed 29 October 2022).

Davincisurgery, 2022. Da Vinci Surgery—Robotic Assisted Surgery for Patients. Available from: https://www.davincisurgery.com. (Accessed 29 October 2022).

Desai, J.P., 2018. Encyclopedia of Medical Robotics, The (In 4 Volumes). World Scientific.

de Andrade, R.M., Fabriz Ulhoa, P.H., Fragoso Dias, E.A., Filho, A.B., Vimieiro, C.B.S., 2023. Design and testing a highly backdrivable and kinematic compatible magneto-rheological knee exoskeleton. J. Intell. Mater. Syst. Struct. 34 (6), 653–663.

De Seta, D., Daoudi, H., Torres, R., Ferrary, E., Sterkers, O., Nguyen, Y., 2021. Robotics, automation, active electrode arrays, and new devices for cochlear implantation: a contemporary review. Hear. Res. 414, 108425.

Fulmer, B.R., 2020. Laparoscopic and robotic radical prostatectomy: practice essentials, preparation, technique. Available from: (Accessed 28 August 2022) https://emedicine.medscape.com/article/458677overview.

Gafford, J.B., Webster, S., Dillon, N., Blum, E., Hendrick, R., Maldonado, F., Gillaspie, E.A., Rickman, O.B., Herrell, S.D., Webster, R.J., 2020. A concentric tube robot system for rigid bronchoscopy: a feasibility study on central airway obstruction removal. Ann. Biomed. Eng. 48 (1), 181–191.

Gao, A., Murphy, R.R., Chen, W., Dagnino, G., Fischer, P., Gutierrez, M.G., Kundrat, D., Nelson, B.J., Shamsudhin, N., Su, H., et al., 2021. Progress in robotics for combating infectious diseases. Sci. Robot. 6 (52), eabf1462.

Garbin, N., Wang, L., Chandler, J.H., Obstein, K.L., Simaan, N., Valdastri, P., 2018. Dual-continuum design approach for intuitive and low-cost upper gastrointestinal endoscopy. IEEE Trans. Biomed. Eng. 66 (7), 1963–1974.

Graetzel, C.F., Sheehy, A., Noonan, D.P., 2019. Robotic bronchoscopy drive mode of the Auris Monarch platform. In: 2019 International Conference on Robotics and Automation (ICRA), pp. 3895–3901.

Hacoma, 2022. Lokomat®—Hocoma. Available from: https://www.hocoma.com/us/solutions/lokomat. (Accessed 29 August 2022).

Hansen, S., 2022. Continuum Robot Offers Brand New Possibilities. MTU AEROREPORT. Available from: https://aeroreport.de/en/innovation/continuum-robot-offers-brand-new-possibilities. (Accessed 29 August 2022).

Henschel, A., Laban, G., Cross, E.S., 2021. What makes a robot social? A review of social robots from science fiction to a home or hospital near you. Curr. Robot. Rep. 2 (1), 9–19.

Hughes, J., Culha, U., Giardina, F., Guenther, F., Rosendo, A., Iida, F., 2016. Soft manipulators and grippers: a review. Front. Robot. AI 3, 69.

Hugo™ RAS System, 2022. Medtronic. Available from: https://www.medtronic.com/covidien/en-us/robotic-assisted-surgery/hugo-ras-system.html. (Accessed 29 October 2022).

Inoue, Y., Ikuta, K., 2016. Hydraulic driven active catheters with optical bending sensor. In: 2016 IEEE 29th International Conference on Micro Electro Mechanical Systems (MEMS), pp. 383–386.

Ismail Fawaz, H., Forestier, G., Weber, J., Idoumghar, L., Muller, P.-A., 2019. Accurate and interpretable evaluation of surgical skills from kinematic data using fully convolutional neural networks. Int. J. Comput. Assist. Radiol. Surg. 14 (9), 1611–1617.

MedTech, J&J, 2022. Robotic-Assisted Solution—DePuy Synthes. Available from: https://www.jnjmedtech.com/en-US. (Accessed 29 October 2022).

Khadem, M., Rossa, C., Usmani, N., Sloboda, R.S., Tavakoli, M., 2017. Robotic-assisted needle steering around anatomical obstacles using notched steerable needles. IEEE J. Biomed. Health Inform. 22 (6), 1917–1928.

Khan, Z.H., Siddique, A., Lee, C.W., 2020. Robotics utilization for healthcare digitization in global COVID-19 management. Int. J. Environ. Res. Publ. Health 17 (11). https://doi.org/10.3390/ijerph17113819.

Kim, E., Jeon, S., An, H.-K., Kianpour, M., Yu, S.-W., Kim, J.-Y., Rah, J.-C., Choi, H., 2020. A magnetically actuated microrobot for targeted neural cell delivery and selective connection of neural networks. Sci. Adv. 6 (39), eabb5696.

Kumar, A., Kumar, S., Kaushik, A., Kumar, A., Saini, J.S., 2020. Real time estimation and suppression of hand tremor for surgical robotic applications. Microsyst. Technol. 28 (1), 1–7.

Lassoued, A., Boubaker, O., 2020. Modeling and control in physiology. In: Control Theory in Biomedical Engineering, Elsevier, pp. 3–42.

Li, J., de Ávila, B.E.-F., Gao, W., Zhang, L., Wang, J., 2017. Micro/nanorobots for biomedicine: delivery, surgery, sensing, and detoxification. Sci. Robot. 2 (4), eaam6431.

Li, M., Deng, J., Zha, F., Qiu, S., Wang, X., Chen, F., 2018. Towards online estimation of human joint muscular torque with a lower limb exoskeleton robot. Appl. Sci. 8 (9), 1610.

Li, Y., Wolf, M.D., Kulkarni, A.D., Bell, J., Chang, J.S., Nimunkar, A., Radwin, R.G., 2021. In situ tremor in vitreoretinal surgery. Hum. Factors 63 (7), 1169–1181.

Liu, W., Yin, B., Yan, B., 2016. A survey on the exoskeleton rehabilitation robot for the lower limbs. In: 2016 2nd International Conference on Control, Automation and Robotics (ICCAR), pp. 90–94.

Liu, D.-X., Xu, J., Chen, C., Long, X., Tao, D., Wu, X., 2019. Vision-assisted autonomous lower-limb exoskeleton robot. IEEE Trans. Syst. Man Cybern. Syst. 51 (6), 3759–3770.

Lokhande, M.P., Patil, D.D., 2022. Object identification in remotely-assisted robotic surgery using Fuzzy inference system. In: Demystifying Federated Learning for Blockchain and Industrial Internet of Things, IGI Global, pp. 58–73.

Ma, X., Wang, C., Zhang, R., Wu, X., 2018. A real-time gait switching method for lower-limb exoskeleton robot based on sEMG signals. In: International Conference on Cognitive Systems and Signal Processing, pp. 511–523.

Mapara, S.S., Patravale, V.B., 2017. Medical capsule robots: a renaissance for diagnostics, drug delivery and surgical treatment. J. Control. Release 261, 337–351.

Market, M.R., 2022. Medical Robots Market worth $12.7 billion by 2025. Available from: https://www.marketsandmarkets.com/PressReleases/medical-robotic-systems.asp. (Accessed 28 October 2022).

Martincek, I., Banovcin, P., Goraus, M., Duricek, M., 2020. USB capsule endoscope for retrograde imaging of the esophagus. J. Biomed. Opt. 25 (10), 106002.

Miskin, M.Z., Cortese, A.J., Dorsey, K., Esposito, E.P., Reynolds, M.F., Liu, Q., Cao, M., Muller, D.A., McEuen, P.L., Cohen, I., 2020. Electronically integrated, mass-manufactured, microscopic robots. Nature 584 (7822), 557–561.

Myon, 2022. Neurorobotics Research Laboratory (NRL)—Myon. Available from: http://www.neurorobotik.de/robots/myon_en.php. (Accessed 28 August 2022).

Neurorehabdirectory, 2022. ReoAmbulator. Available from: https://www.neurorehabdirectory.com/rehabproducts/reoambulator. (Accessed 29 August 2022).

Oesterreich, R., Varela, M.F., Moldes, J., Lobos, P., 2022. Laparoscopic approach of pediatric adrenal tumors. Pediatr. Surg. Int. 38 (10), 1–10.

Omron, 2022. LD-UVC—Omron. Available from: https://web.omron-ap.com/th/ld-uvc. (Accessed 29 August 2022).

Pan, M., Yang, Q., Su, T., Geng, K., Liang, K., 2022. An effective tremor-filtering model in teleoperation: three-domain wavelet least square support vector machine. Appl. Soft Comput. 130, 109702.

PARO, 2022. PARO Therapeutic Robot. Available from: http://www.parorobots.com. (Accessed 28 August 2022).

Pepper, 2022. Pepper the Humanoid and Programmable Robot—SoftBank Robotics. Available from: https://www.softbankrobotics.com/emea/en/pepper. (Accessed 28 August 2022).

Peters, B.S., Armijo, P.R., Krause, C., Choudhury, S.A., Oleynikov, D., 2018. Review of emerging surgical robotic technology. Surg. Endosc. 32 (4), 1636–1655.

Pourebadi, M., Riek, L.D., 2018. Expressive robotic patient simulators for clinical education. In: Proceedings of the R4L Workshop on Robots for Learning—Inclusive Learning at the 13th Annual ACM/IEEE International Conference on Human-Robot Interaction (HRI'18).

Rahman, M.M., Sanchez-Tamayo, N., Gonzalez, G., Agarwal, M., Aggarwal, V., Voyles, R.M., Xue, Y., Wachs, J., 2019. Transferring dexterous surgical skill knowledge between robots for semi-autonomous teleoperation. In: 2019 28th IEEE International Conference on Robot and Human Interactive Communication (RO-MAN), pp. 1–6.

Riek, L.D., 2017. Healthcare robotics. Commun. ACM 60 (11), 68–78. https://doi.org/10.1145/3127874.

Rus, D., Tolley, M.T., 2015. Design, fabrication and control of soft robots. Nature 521 (7553), 467–475.

Ryan, P., Diller, E., 2017. Magnetic actuation for full dexterity microrobotic control using rotating permanent magnets. IEEE Trans. Robot. 33 (6), 1398–1409.

Sequeira, J.S., 2020. Robotics in Healthcare: Field Examples and Challenges. vol. 1170 Springer.

Shi, D., Zhang, W., Zhang, W., Ding, X., 2019. A review on lower limb rehabilitation exoskeleton robots. Chin. J. Mech. Eng. 32 (1), 1–11.

Singh, P.K., Krishna, C.M., 2014. Continuum arm robotic manipulator: a review. Univers. J. Mech. Eng. 2 (6), 193–198.

Smaling, H.J., Tilburgs, B., Achterberg, W.P., Visser, M., 2022. The impact of social distancing due to the COVID-19 pandemic on people with dementia, family carers and healthcare professionals: a qualitative study. Int. J. Environ. Res. Public Health 19 (1), 519.

Soleymani, A., Li, X., Tavakoli, M., 2021a. Deep neural skill assessment and transfer: application to robotic surgery training. In: 2021 IEEE/RSJ International Conference on Intelligent Robots and Systems (IROS), pp. 8822–8829.

Soleymani, A., Torabi, A., Tavakoli, M., 2021b. A low-cost intrinsically safe mechanism for physical distancing between clinicians and patients. In: 2021 IEEE International Conference on Robotics and Automation (ICRA), pp. 3677–3683.

Su, H., Li, G., Rucker, D.C., Webster III, R.J., Fischer, G.S., 2016. A concentric tube continuum robot with piezoelectric actuation for MRI-guided closed-loop targeting. Ann. Biomed. Eng. 44 (10), 2863–2873.

Sušić, I., Zam, A., Cattin, P.C., Rauter, G., 2019. Enabling minimal invasive palpation in flexible robotic endoscopes. In: New Trends in Medical and Service Robotics, Springer, pp. 70–77.

Thestron, 2022. The Strong Robot With the Gentle Touch—RIKEN. Available from: https://www.riken.jp/en/news_pubs/research_news/pr/2015/20150223_2. (Accessed 28 August 2022).

Thigpen, B., 2022. Single incision laparoscopic myomectomy in pregnancy. Gynecol. Obstet. Clin. Med. https://doi.org/10.1016/j.gocm.2022.04.001.

Urias, M.G., Patel, N., He, C., Ebrahimi, A., Kim, J.W., Iordachita, I., Gehlbach, P.L., 2019. Artificial intelligence, robotics and eye surgery: are we overfitted? Int. J. Retin. Vitr. 5 (1), 1–4. https://doi.org/10.1186/s40942-019-0202-y.

Veebot, 2022. VEEBOT SYSTEMS INC. Available from: https://www.veebot.com. (Accessed 29 August 2022).

Voloshkin, A., Tereshchenko, A., Carbone, G., Rybak, L., Nozdracheva, A., 2022. Design of a suspension lever mechanism in biomedical robotic system. Front. Robot. AI 9, 906691.

Woodway, 2019. LokoHelp—Electromechanical Gait Trainer—Woodway. Available from: https://www.woodway.com/products/loko-help. (Accessed 29 August 2022).

WorldBank, 2022. Disability Inclusion Overview. World Bank. Available from: https://www.worldbank.org/en/topic/disability. (Accessed 29 August 2022).

Wu, X., Liu, D.-X., Liu, M., Chen, C., Guo, H., 2018. Individualized gait pattern generation for sharing lower limb exoskeleton robot. IEEE Trans. Autom. Sci. Eng. 15 (4), 1459–1470.

Wu, J., Chen, W., Guo, D., Ma, G., Wang, Z., He, Y., Zhong, F., Lu, B., Wang, Y., Cheung, T.H., et al., 2022. Robot-enabled uterus manipulator for laparoscopic hysterectomy with soft RCM constraints: design, control and evaluation. IEEE Trans. Med. Robot. Bionics 4 (3), 656–666.

Xenex, 2022. LightStrike™ Robot—Xenex® UV Disinfection. Available from: https://xenex.com/lightstrike. (Accessed 29 August 2022).

Yang, S., Han, J., Xia, L., Chen, Y.-H., 2020. An optimal fuzzy-theoretic setting of adaptive robust control design for a lower limb exoskeleton robot system. Mech. Syst. Signal Process. 141, 106706.

YuMi, 2022. ABB's Collaborative Robot—YuMi. Available from: https://new.abb.com/products/robotics/collaborative-robots/irb-14000-yumi. (Accessed 29 August 2022).

Zemmar, A., Lozano, A.M., Nelson, B.J., 2020. The rise of robots in surgical environments during COVID-19. Nat. Mach. Intell. 2, 566–572. https://doi.org/10.1038/s42256-020-00238-2.

Zhang, J., 2021. Human-Robot Interaction in Augmented Virtuality: Perception, Cognition and Action in 360° Video-Based Robotic Telepresence Systems (Ph.D. thesis). Staats-und Universitätsbibliothek Hamburg Carl von Ossietzky.

Zhang, Y., Sun, H., Jia, Y., Huang, D., Li, R., Mao, Z., Hu, Y., Chen, J., Kuang, S., Tang, J., et al., 2018. A continuum robot with contractible and extensible length for neurosurgery. In: 2018 IEEE 14th International Conference on Control and Automation (ICCA), pp. 1150–1155.

Zhang, B., Zhou, M., Xu, W., et al., 2021. An adaptive framework of real-time continuous gait phase variable estimation for lower-limb wearable robots. Robot. Auton. Syst. 143, 103842.

CHAPTER 2

Soft robotics in medical applications: State of the art, challenges, and recent advances

Mostafa Kaviri, Ali Jafari Fesharaki, and Soroush Sadeghnejad
Bio-Inspired System Design Laboratory, Department of Biomedical Engineering, Amirkabir University of Technology (Tehran Polytechnic), Tehran, Iran

2.1 What is soft robotics?

Robotic applications have been increasing in recent decades. Traditionally, robotic research assumes that robots are built from rigid links and joints, which require a strict kinematic chain. This assumption necessitates dynamic models and the kinematics of robots and robot controlling methods to be developed with a wide range of techniques (Baltes et al., 2017; Gerndt et al., 2015; Mirmohammadi et al., 2021) in order to reach a high level of reliability, accuracy, and efficiency (Siciliano et al., 2016). As a result of the rigidity of typical robotic links, most rigid-link robots have an actuator at every joint (Bauer et al., 2014) and the applied loads and environmental properties of each link and joint are geometrically independent. In "soft robots," force is distributed and integrated throughout the robot. The word "soft" refers to the body of the robot (Rus and Tolley, 2015).

Soft robots are typically used for the mechanical compliance of the components of a machine, with this compliance affected by both material and mechanical structural properties (Polygerinos et al., 2017). In recent years, the soft robotic field has rapidly expanded due to its extensive practical applications, with the emergence of many new journals, research groups, and worldwide open-access resources focused on this field. The primary differences between soft and hard robots are demonstrated in Fig. 2.1. Young's modulus (Rus and Tolley, 2015) is a useful measure of the tensile properties of materials used for fabricating robotic structures. Rigid robotic structures typically have high Young's modulus values in the range of 10^9 to 10^{12} Pa, whereas soft tissue structures are constructed from much lower tensile materials and therefore have much lower Young's modulus values in the order of 10^4 to 10^5 Pa. When a soft robot faces an obstacle, this absence

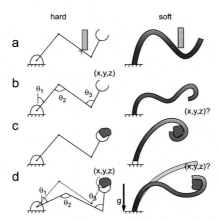

Fig. 2.1 A comparison of the function of a rigid (left) and soft (right) robot in terms of (A) dexterity, (B) control, (C) manipulation, and (D) applied force (Bauer et al., 2014).

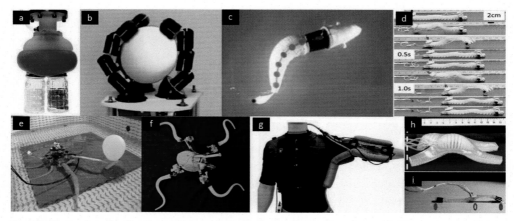

Fig. 2.2 Examples of soft robots: (A) a universal gripper (Amend et al., 2012), (B) a Pisa/IIT SoftHand (Catalano et al., 2014), (C) an autonomous soft robotic fish (Marchese et al., 2014), (D) a soft worm (Umedachi et al., 2016), (E) a soft octopus (Cianchetti et al., 2015), (F) a soft-bodied underwater remotely operated vehicle (Arienti et al., 2013), (G) a multijoint soft wearable robot for upper limb assistance (Proietti et al., 2021), (H) a multigait soft robot (Umedachi et al., 2016), and (I) a soft robot inspired by caterpillars (Lin et al., 2011).

of rigidity allows flexibility and geometric adaptability to each obstacle. The compliance of soft robots is comparable to the soft tissues of a body or skin; however, this is a disadvantage in modeling high-order compliance scenarios for soft robots, as dynamic models to reliably predict and observe the position of every specific point on the robot do not exist. Fig. 2.2 reveals some types of soft robots.

2.2 Why soft robotics?

Softness and compliance are two essential properties of biological systems, which facilitate interaction with the environment and increase safety. Human interaction with robots is constantly increasing due to improvements in science and technology, but the safety of this interaction is a major concern. While techniques (control measures) exist to reduce the likelihood of a collision, contact remains too dangerous in many cases as robots are typically heavy and a likely collision could result in serious injury. The intrinsic ductility of soft robots provides the necessary safety controls for robot–human interaction as stretchable materials can deform and absorb much of the energy generated in a collision (Rus and Tolley, 2015). The infinite features of soft robots improve dexterity and interaction speed with biological organs. The continued deformation of soft robots also expands the working range, resulting in more dexterity. Considering this improved ability, a soft robot could work well in unstructured environments. For example, the force generated between the gripper and the object must be measured to accurately control the robot. The fabrication of bioinspired robots was highly complex before the introduction of soft robots because they have many joints and connections, which increases complexity and the likelihood of errors. With the expansion of the soft manufacturing paradigm, some mechanical systems have been replaced by flexible and continuous bodies, making them faster and easier to develop and implement. The primary advantages of soft robotics are:

○ Demonstrated applications include implantable and wearable devices (Polygerinos et al., 2017).
○ Robots can move through unpredictable terrain (Polygerinos et al., 2017).
○ For grasping an object, compliant materials adapt more readily to a variety of shapes (Rus and Tolley, 2015).
○ It provides interaction safety with humans and biological organs (Polygerinos et al., 2017).

The most common applications of soft robots are:

○ Soft mobile robots
○ Exosuits, wearable, and soft medical robots
○ Soft manipulators and grippers

Some disadvantages of soft robotics include:

○ Low reliability and repeatability
○ Can be challenging in control
○ Less powerful than rigid robotics

The following sections discuss the state of the art in rehabilitation devices, cardiac devices, and surgical robots, followed by a discussion of some present challenges and the current status of soft robotics.

2.3 Rehabilitation

Physical impairments and disabilities can drastically reduce a person's quality of life. For example, neurological diseases, posttraumatic arthritis, cerebral palsy, and many other conditions may affect the function of the hands or limbs. Imagine trying to cook or put on clothes without a fully functional hand; if you list all the daily activities for which you use your hands, you will realize their importance and comprehend the annoyance that arises when you have a hand impairment, and, when it is a permanent imperfection, it is almost unbearable. The hand is just one of the many organs that play a role in activities of daily living (ADLs), any of which limit the functionality of the human body if an impairment exists. Fortunately, developments in science and technology offer new approaches to this problem. Specifically, this chapter argues that assistive robotics can meet the needs of an aging and increasingly disabled society. When the body's ability to perform physical tasks or movements is impaired by an injury, a disease, or a congenital defect, assistive devices can be designed to restore the impaired ability. People who have limited mobility need to do rehabilitation exercises, namely, actively assisted shoulder flexion exercises and wrist flexion exercises with resistance. A number of robotic devices that can help patients with mobility problems currently exist. However, most of these devices are rigid robots. As demonstrated in Fig. 2.1, rigid robotics increases the force applied to a patient, placing strain on the other joints and parts of the body. Interacting with rigid robots also carries other risks due to their size and weight, posing additional danger to the patient and potentially leading to further injury. Adaptive soft robots increase the safety of interaction and largely solve these problems. Robots can actively adapt in two ways: impedance control and compliance. In the 1970s, active adaptive systems were found to be preferable to passive systems for lower limb support or replacement, and, in 1958, Joseph McKibben first highlighted the usefulness of adaptive systems for the upper limbs. The challenges of rehabilitative and assistive devices are similar and are discussed further.

2.3.1 Soft rehabilitation gloves

As human hands are kinematically complicated, it is challenging to develop technologies that smoothly interact with them. Grasping and handling movements of the hand can involve up to 19 joints and 29 muscles. In combination, these joints and muscles control the grasping strength and precise movements of the hand.

Examples of soft robotic gloves designed to help people with hand disabilities in daily living and home therapy are presented in Fig. 2.3A–G. In these examples, elastomers with fiber reinforcements were used for the design and development of hydraulically actuated multisegment soft actuators. The ability of soft actuators to mimic thumb and finger motions suitable for a variety of common grasping activities was demonstrated in a range of practical scenarios (references from the following section are

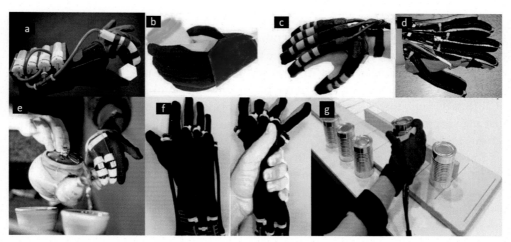

Fig. 2.3 (A) A valve-less soft rehabilitation glove (Ahmadjou et al., 2021). (B) A soft robotic glove (Yap et al., 2017). (C) A soft robotic glove that is pneumatically operated for use in conjunction with home rehabilitation (Polygerinos et al., 2015a, b). (D) A power assistive and rehabilitation wearable robot based on pneumatic soft actuators (Al-Fahaam et al., 2016a, b). (E) An Exo-Glove with a tendon routing system instead of pneumatic soft actuators (In et al., 2015). (F) A wearable exoskeleton robot for wrist rehabilitation (Al-Fahaam et al., 2016a, b). (G) An EMG-controlled soft robotic glove (Polygerinos et al., 2015a).

fine). In addition, the actuators were attached to the back of the hand, creating an open palm interface that facilitates object grasping. The complete device was designed as a wearable hip belt with a battery life of several hours per charge. The coarse and fine grasping capabilities of the robotic glove were qualitatively assessed on healthy volunteers through demonstrations in open space and interaction with various objects encountered in ADLs (Polygerinos et al., 2015a, b). In the control of these devices, open–loop SEMG logic (Fig. 2.3G) was introduced to discriminate muscle contractions then relay information to a low–level fluid pressure controller. This controller regulates pressure in preselected groups of glove actuators that were previously trialed (Polygerinos et al., 2015a, b).

Nine volunteers with C4–C7 spinal cord injuries completed hand function testing at the Toronto Rehabilitation Institute in 2018 to assess the effectiveness of the soft robotic glove. The soft robotic glove improved object manipulation in ADLs. All volunteers showed a statistically significant difference in the mean scores for all modified objects between baseline and assisted conditions. The results of this study support the effectiveness of the Toronto soft fabric-based robotic glove as an assistive device to help people with upper limb paralysis due to spinal cord damage use their hands (Cappello et al., 2018).

Another study discusses a soft robotic glove with textile actuators, soft sensors, and state machine control for intention identification. Through bench testing, the authors demonstrate that the actuators can produce forces and movements that are similar to those of a typical human finger. Capacitive sensors made of textile elastomers are integrated into the glove to track the flexion of the finger through strain and to detect the contact of objects through force. Intuitive user control is made possible by a state machine control that analyzes signals from the sensors to identify relative changes in the hand-object interaction (Laoutid et al., 2009).

Another preview describes a soft wearable robot called Exo-Glove (Fig. 2.3E) that used a soft tendon guidance system for under control (In et al., 2015) and could help patients complete ADLs that required grasping. The experimental results showed that tissue-reinforced actuators with a corrugated tissue layer and a reinforced elastic layer can maintain finger movements with the required force output at lower operating air pressure than the required pressure of previous actuators. A wearable hip belt incorporating a control system that allowed independent control of the individual actuators was constructed. Compared to previous glove systems, the glove was lighter. Five healthy volunteers tested the supportive effect of the glove's ROM and grip strength.

The results showed that the glove provided sufficient range of motion (ROM) when there was a lack of voluntary muscular control, and it could facilitate ADLs. The glove also enhanced the capacity of two stroke patients to execute functional grasping activities, according to the preliminary research conducted on their hands (Yap et al., 2017).

In another case, the research describes the design and construction of a small and soft hand rehabilitation glove (Fig. 2.3A). To provide an input with the correct values for the bending actuators, the design used low-cost, accessible microcompressors for pneumatic circuits. Two configurations, one with a larger compressor than the other, were used to compare the performance of the proposed pneumatic circuit. The results showed that both circuits can adequately operate a single bending transducer. The pressure of a closed circuit can be increased to a maximum of 160 kPa using two LPM microcompressors. The circuit's average power usage was 3.4 W, whereas the pressure was at its highest level. To demonstrate the application of their concept in a pneumatic hand rehabilitation glove, the researchers presented a two-stage controller to adjust the angle of the end tip of the actuator. When a steady state is reached, the actuator can track a 3-Hz sinusoidal curve with a root minimum square error of 22.5 kPa (Ahmadjou et al., 2021).

2.4 Soft wearable robots

Another rapidly growing area of soft robotics is assistive soft wearable ones. In recent decades, several devices to assist movement in the upper and lower limbs have become available. These assistive devices may be cable-controlled or hydraulically operated devices and utilize textiles.

2.4.1 Upper limb assistive device

The preliminary design and evaluation of a soft, wearable robot to assist shoulder movement is shown in Fig. 2.4A–C (O'Neill et al., 2017). This design includes small, lightweight pneumatic actuators: one controlling shoulder abduction and two controlling horizontal flexion and arm extension in both directions. When tested and characterized individually on a test bench, the actuators performed sufficiently to justify their use in assisting a range of patients performing ADLs. Unpowered, the device weighs 0.48 kg, folds easily, and does not restrict the user's movement. New sensing and control algorithms were developed for the device, allowing the user to reach most of their natural working range without noticeably increasing muscular effort. Evaluation involved comparing the user's movements to those made while wearing the system but not turned on. Intrinsic mechanical transparency allows improved coupling between the robot's and users' movements, which subsequently allows the device to be used as an assessment tool, even when unpowered (Proietti et al., 2021).

When the accuracy of the IMU-based sensing technology was compared to data from an optical motion capture system, minimal RMSE results were achieved. This indicates that kinematic metrics may be estimated using this IMU-based sensing approach without the use of external optical motion capture equipment. This example is one demonstration of the potential of multijoint soft wearable robots (Fig. 2.4D) intended for upper limb support and treatment to assist patients performing ADLs. The design, fabrication, and initial evaluation of a robotic forearm orthosis in human subjects are shown in Fig. 2.4D. An antagonistic pair of soft tissue-based spiral actuators, which actively pronate and supinate the forearm, is incorporated into the robotic forearm orthosis. The knitted elastic band, stretch-resistant nylon fabric, and internal bladder are used to create the pneumatic, tissue-based helical actuators. When inflated to an internal pressure of

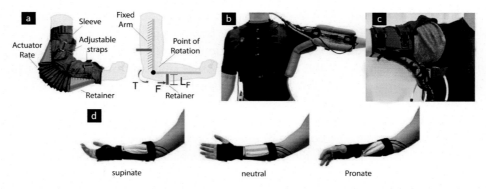

Fig. 2.4 (A) A soft elbow exosuit to augment bicep lifting capability (Thalman et al., 2018). (B) A soft wearable robot for upper limb support (Proietti et al., 2021). (C) A soft wearable robot for shoulder support (O'Neill et al., 2017). (D) Robotic forearm orthosis (Realmuto and Sanger, 2019).

690 kPa, empirical isometric loading characteristics show that the actuators can deliver up to 100 psi (1.7 Nm) of torque. During a preliminary evaluation ($N = 6$) on human subjects, the efficiency of torque support provided by the robotic forearm orthosis was assessed (Realmuto and Sanger, 2019). This research by Thalman et al. (2018) presents the design, characterization, and evaluation of a soft, lightweight elbow exosuit (Fig. 2.4A) designed to support elbow flexion in warehouse workers. This would be valuable for reducing muscle strain and injury in warehouse workers.

In order to maintain higher pressure and thus generate greater force, the suit consists of a series of overlapping TPU-based actuators housed in an inelastic fabric pocket. The base of each actuator is bent to conform to the curvature of the arm. During the course of a static test involving a volunteer in an exosuit carrying loads of 1.5 and 2.5 kg in assisted and unassisted conditions, respectively, it was found that the exosuit was an effective aid in isometric contractions of the biceps. When using the device, muscle activity decreased by 43% for the 1.5-kg weight and by 63% for the 2.5-kg weight, according to the EMG data. In addition, the exosuit was shown to maintain 93% of the original ROM when the elbow was fully extended to fully flexed. Moreover, it was shown to provide support over the entire ROM without loading and allowing an average reduction in muscle activity of 47% when the exosuit was used.

2.4.2 Lower limb assistive device

In another recent study (Lee et al., 2017a, b), a powerful off-board activation platform and an improved control strategy have been utilized to enable accurate monitoring of the force profile of various shapes, sizes, and timings of soft exosuits (Fig. 2.5) at a normal walking and running pace. The bandwidth and tracking performance of the controller were greatly improved using FFMs, which account for disturbances due to thigh motion, nonlinear stiffness of the human and exosuit hips, and the transmission model of the actuator. The actuator system provides the opportunity to optimize support profiles to increase metabolic reduction for wearers during walking and jogging activities.

The design and methodological characterization of a soft sensor suit are discussed in order to record the angles of the hip, knee, and ankle joints in the sagittal plane (Fig. 2.5C). The idea used hyperelastic strain sensors based on microchannels of liquid metal in an elastomer and discretized stiffness gradients to increase mechanical endurance. These reliable sensors were demonstrated in tests to be able to stretch up to 396% of their original length, constrain the beam with only 0.17% of the torque of any joint, have a sensitivity of more than 2.2%, and change electromechanical specifications by less than 2% over 1500 loading and unloading cycles. By comparing the data with joint angle data collected with optical motion capture, the accuracy and variability of the soft-sensing tightening was also evaluated. At speeds of 0.89 m/s for walking and 2.7 m/s for running, the root mean square error (RMSE) of the sensor suit were less than 5% and 15%,

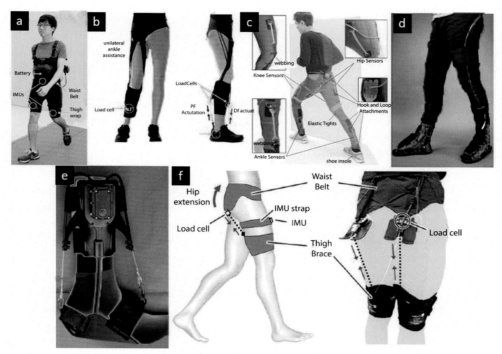

Fig. 2.5 (A) Reducing metabolic rate during walking and running with a versatile wearable exosuit (Kim et al., 2019). (B) A soft robotic exosuit improves walking in poststroke patients (Awad et al., 2017). (C) Wearable soft sensor suits to measure human gait (Mengüç et al., 2014). (D) ExoBoot, an inflatable robotic foot wear for ankle support (Chung et al., 2018). (E) A soft wearable robotic ankle-foot orthosis (Kwon et al., 2019). (F) Improved assistive profile tracking of soft exosuits for walking and jogging with off-board propulsion (Lee et al., 2017a, b).

respectively. Regardless of the difference in absolute readings, the repeatability of the sensor suit's joint angle measurements was statistically comparable to that of optical motion capture at all speeds. It is anticipated that wearable soft sensing devices will be used for a wide range of purposes other than wearable robots, including the facilitation of human–computer interactions and medical diagnosis (Mengüç et al., 2014).

As an extension of the ExoBoot, a soft, inflatable robotic shoe (Fig. 2.5D) to support plantar flexion of the ankle during walking has been proposed (Chung et al., 2018). This addition creates a flat, lightweight textile shoe with an IMU sensor and a soft textile-based actuator. When inflated by flexing the top of the shoe, the inflatable actuator produces supportive plantar flexion torque. The largest amount of torque produced by the ExoBoot at different pressures and ankle angles was 39 Nm at 483 kPa (70 psi) and 60°, respectively. In a pilot research, a healthy person's performance using the ExoBoot was assessed while walking. An open pressure regulator based on the inertial measurement unit's assessment of the ankle's rotational velocity initiated the actuation. The actuator's pressure reached 75% of its supply pressure at the height of active support, and the maximum torque that could be delivered to the ankle was calculated to be 23 Nm. These findings indicate that the ExoBoot may be able to lower the metabolic cost of walking.

Veale et al.'s (2021) technique illustrates the possibility of a soft scaffold for providing the stable and comfortable knee extension motion needed to stand up. In this case, a pleated pneumatic interference actuator (PPIA) is a structure that generates torque by bulging a confined fabric-reinforced rubber tube. It is lightweight, foldable, and can be packed into clothing. Based on this method, the soft lift assister for the knee (SLAK) and a soft orthosis with multiple PPIAs were developed. At a flexion angle of 82°, the SLAK produced a maximum torque of 324 Nm while inflated at a pressure of 320 kPa. This exceeds the 180-Nm peak torque required for STS and other routine activities. The torque requirement for STS, which is greater than 93% of the STS motion when worn on a test leg, was satisfied by the SLAK. During testing, the SLAK showed the potential to fully support more than 100% of the required motion. In this case, the theoretical PPIA model underestimated the PPIA torque at high flexion angles and overestimated the PPIA torque at low-to-moderate flexion angles. The PPIA is being further developed with human testing of the SLAK, to be faster and more flexible, to be smaller and lighter, and to function on increasingly accurate models.

Another soft robotic assistive device, described by Kim et al. (2018), is a hip-only soft exosuit designed to improve walking and running motions by providing the wearer with peak forces of 300 N at the hips. This device can be fully automated and controls movements based on an online classification method, with different fixed assist profiles for running and walking. This control method is based on the biomechanical insight that changes in the potential energy of the center of mass are out of phase during running and walking. Specifically, they used an IMU, positioned on the abdomen, to track vertical acceleration at the peak of hip extension. With the help of eight subjects outside and six people on a treadmill, this robot's validation was carried out. On the treadmill and off-road, with (13.6 kg) and without load, with exo-on and exo-off settings, and with different shoe kinds, the average accuracy of the 14 participants was 99.99%. The findings of the outdoor energetic evaluation for eight subjects revealed a substantial drop in walking (3.9%) when the exo was taken off. The exo increased running speeds by 3.9% when compared to running with and without it (12.2% and 8.2%, respectively). This study is the first to show how operational expenses may be decreased by an autonomous wearable robot.

Elsewhere, a hip-only soft exosuit (Fig. 2.5A) was invented by Kim et al. (2019) to move autonomously and improve human walking and running by providing the wearer with peak forces of 300 N at the hips. Based on an online classification method, various fixed assist profiles for running and walking were used. The method is based on the biomechanical insight that the changes in the potential energy of the center of mass are out of phase during running and walking. Specifically, they used an IMU positioned on the abdomen to track vertical acceleration at the peak of hip extension. For poststroke patients, a soft ankle–foot orthosis (Fig. 2.5E), which is affordable, lightweight, comfortable to wear, and can support gait during rehabilitation both in and out of the hospital, was proposed (Kwon et al., 2019).

The suggested solution comprises an adjustable ankle support and flexible 3D-printed splint that enable normal ankle flexion and extension while providing vertical support to keep the structure from buckling. A bidirectional tendon-driven actuator supports both dorsiflexion and plantar flexion. The system also includes a wearable gait measurement module for real-time feedback control, which measures leg trajectory and pressure on the feet. As it is powered by a rechargeable battery and wirelessly connects to the main controller, it is entirely untethered, portable, and practical. The recorded sensor data and the leg biomechanics are utilized to determine the gait phase in real time and to construct a gait aid algorithm, even though each person's gait patterns are unique. Both dorsiflexion and plantar flexion were assisted by a bidirectional tendon-powered drive. The system also includes a wearable gait measurement module for real-time feedback control that measures leg trajectory and foot pressure. The system is completely untethered, portable, and convenient, as it is powered by a rechargeable battery and can be wirelessly connected to the primary control unit. A gait aid algorithm for dorsiflexion and plantar flexion thus provides an accurate prediction of control phase and timing, despite the fact that each person's gait trajectories are unique. Gait phase is identified in real time using measured sensor data and leg biomechanics. An improvement in both foot fall prevention and gait propulsion was seen in the walking experiment conducted with a poststroke patient as a feasibility study.

2.5 Cardiac assist devices

When the heart cannot deliver enough blood to sustain tissue perfusion, heart failure (HF) occurs. Fluid overload, swelling of the lower limbs, and pulmonary edema can all result from decreased cardiac output. Shortness of breath, general fatigue, and decreased exercise capacity are typical signs of HF. Although patients with the illness may want for a heart transplant, ventricular assist devices (VADs) and direct cardiac compression (DCC) devices are more frequently employed as clinical therapies due to a lack of donors. VADs enhance the left and right ventricles' ability to contract by drawing blood out of the ventricles and pushing it back into the heart (Lee et al., 2017a, b). In contrast to DCC devices, which are positioned around the heart to guarantee proper blood flow but are not in direct touch with the blood, VADs are in direct contact with the blood and the metal body of the device may promote clotting and blood clot formation (Oz et al., 2002; Wamala et al., 2017).

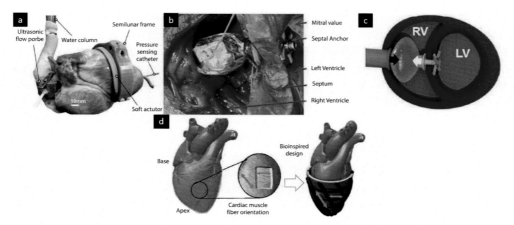

Fig. 2.6 (A) A soft robotic ventricular assist device (Payne et al., 2017). (B and C) A soft robotic device placed within the heart to help the failing right ventricle expel more blood (Horvath et al., 2017). (D) A soft robotic sleeve supports cardiac function (Roche et al., 2017).

Implanted soft robotic devices (Fig. 2.6A) have been suggested by Payne et al. (2017) to enhance cardiac function in isolated cases of left or right HF by inducing rhythmic loading of both the ventricles. The devices attach to the interventricular septum (IVS) and exert stresses on the ventricular free wall in order to produce convergence of the IVS and the ventricular free wall in systole and to aid recoil in diastole. Physiological monitoring of native hemodynamics enables organ-in-the-loop regulation of these robotic implants for totally autonomous enhancement of heart function. The devices are implanted into the beating heart under echocardiographic control. The idea is illustrated by in vivo experiments in a porcine model of both the right and left ventricles. To show how the device worked in various hemodynamic scenarios related to right and left HF, the authors employed a variety of HF models. These short-term in vivo experiments demonstrate that blood pressure and blood flow can be restored to prefailure levels with this assistive device. Sustained cardiac output is enabled by better filling of the ventricles during diastole, which was also demonstrated by a significant decrease in diastolic ventricular pressure. An implanted intracardiac robotic right ventricular ejection device (RVED) is suggested in order to dynamically approach the right ventricular (RV) free wall and IVS synchronously with the cardiac cycle to improve blood ejection in right heart failure (RHF). The RVED consists of a pneumatic linear actuator with an artificial muscle spanning the RV chamber between the two anchors, an RV free-wall anchor, and an anchoring system deployed through the IVS for safe and efficient intracardiac surgery. Horvath et al. (2017) described ventricular volume output, linear approach to various loads, and the effects of altered device actuation times on volume output using a ventricular simulator and a custom controller. Five adult pigs were used for the in vivo research on the RVED. First, utilizing 3D echo-cardiography, the device was successfully implanted into a beating heart ($n = 4$). An experimental model of RHF ($n = 1$) was used in a feasibility study to assess the device's capacity to improve RV ejection. Additional trials in chronic animals will provide additional information on the effectiveness of this support device. Activation of the RVED increased RV ejection performance in RHF. These results demonstrate that the use of the RVED inside the beating heart was successfully designed, developed, and implemented. This soft robotic ejection device can be used to rapidly install a mechanical circulatory support system in RHF.

Soft robotic or form-fitting, low-modulus implantable devices that can mimic or assist complicated biological processes such as myocardial contraction are of great interest. One recently trialed device (Roche et al., 2017) has been a soft robotic sleeve (Fig. 2.6D) implanted around the heart, which functions as a VAD by actively compressing and twisting the heart. The sleeve prevents the blood from coming into contact with the skin, eliminating the need for blood thinners or anticoagulants and avoiding problems with conventional VADs such as blood clotting and infection.

Roche et al.'s (2017) method mimics the alignment of the two outer mammalian heart muscle layers, using a biologically inspired design to arrange individual contracting parts, or actuators, in a layered spiral and circumferential pattern. With a stiffness equivalent to the heart tissue, the resulting implantable soft robot mimics the shape and functionality of the natural heart. In a pig model of acute HF, the viability of this soft sleeve device to maintain cardiac function was demonstrated. The soft robotic sleeve could be used for patients with HF as a bridge to transplantation as it can be tailored to individual needs (Roche et al., 2017).

2.6 Surgical robots

From open surgery to minimally invasive surgery (MIS), surgical methods have undergone a tremendous transformation in recent years. In MIS, long, rigid, or flexible surgical instruments are inserted into the body through tiny incisions (about 10–15 mm in diameter) or natural openings. In modern robotic MIS techniques, a surgeon uses a stiff robotic device to control the movement of customized surgical tools (Sadeghnejad et al., 2016; Sadeghnejad et al., 2019a, b, 2020). The practice of inserting an optical device directly into the area to be examined to observe the inside of the body is called endoscopy and is at the heart of MIS. There are numerous variations of the optical device known as an endoscope. Today, an endoscope is a long, flexible tube with a high-resolution camera and a light source at the tip that is 1.5–2 m long. Some of the main advantages of MIS are listed as follows:

○ Lower risk of infection (Bar-Cohen, 2000)
○ Improved cosmetics (Carrozza et al., 2006)

○ Shorter recovery time (Bethea et al., 2004)
○ Decreased trauma (Araromi et al., 2015)
○ Fewer postoperative complications and pain reduction (Runciman et al., 2019)

By restoring intuitiveness to the technique and adding maneuverability to the tip, robotic technology could help surgeons improve their precision, predictability, and repeatability, as demonstrated by the success of the da Vinci surgical robot. However, with the exception of the wrist, this method still requires hard and inflexible devices. Flexible tools, on the other hand, can be used in conjunction with MIS techniques and have high intrinsic flexibility. However, given that they are frequently lengthy, flexible, and have a rotating tip, these instruments may not be very dexterous once they are at the surgical site. These instruments are unstable and have a weak force output (Loeve et al., 2011). Traditional colonoscopies must be conducted by highly skilled medical professionals. Additionally, modern technologies run the danger of damaging the intestinal wall and causing pain. Some researchers designed, modeled, controlled, and tested a soft three-section modular robot. Three degrees of freedom, one translation, and two rotations were available for each robotic segment. The peristaltic motion used by the robot to translate was modeled after the motion produced by the intestine. The actuators of the robot were nine shape memory alloy (SMA) springs that could be adjusted separately. A brand-new silicone rubber skin provided the passive recovery force needed to extend the springs back to their initial position. Three air tubes were also included, one for each section, to provide forced convection and shorten the time that the SMA springs needed to cool. To optimize traction while yet providing adequate recovery force, a parametric analysis of the skin's thickness and curvature using finite element analysis (FEA) was carried out. The robot could attain any orientation between 90 and +90 in both pitch and roll in less than 4s with almost no steady-state error owing to a multiple input, multiple output (MIMO) controller based on fuzzy control that was devised and implemented for each of the parts. The robot's orientability and peristaltic motion were both put to the test. With a maximum speed of 4mm/s (24cm/min) and an average speed of 2.2cm/min, the robot could move in a peristaltic manner. Each section could also track periodic multiple input squared signals with an amplitude of 25 with less than 2% overshoot and almost no steady-state inaccuracy.

Manfredi et al. (2019) describes an SPID, or soft pneumatic inchworm double balloon, for colonoscopy. The Ecofex 00-30, used in the construction of mini-robots, has great compliance and flexibility for intrinsic movement, as shown in tests on a plastic deformable colon phantom. The extremely compliant twin balloon construction easily adapts to comply with the shape of various colonic sections and their diameters owing to the dexterous and compliant behavior of the 3-DOF soft pneumatic actuator linking the two balloons. Physical benefits of construction made of soft materials include their inherent flexibility, their mild, nontraumatic contact with the colonic wall, and their cheap manufacturing costs, which are necessary for the creation of disposable devices. This can help in sterilization, cross-contamination, and maintenance difficulties. A passive compliant interaction between the low Young's modulus and the intestinal wall lowers the anchoring pressure. The balloons never had any air leaks during the testing, which is crucial for guaranteeing the device's dependability. For colon examination, a camera that can be installed in the distal balloon's hollow is required. Traditional biopsy tools can be employed for both therapeutic and diagnostic purposes. A conduit connecting the camera to an external console can be incorporated into the SPID in order to introduce tools for treating questionable lesions. The main focus of the current investigation was on the transit rate and design of locomotion in a plastic colon phantom. The cable being inserted via the anus will result in extra drag and friction forces that were not taken into account in the current investigation. However, using a dedicated access port or an external device to feed the tether during movement, this friction may be all but eliminated. By employing manually controlled external piston cylinders, the SPA chambers and balloons were successfully activated, proving their functions. An active control will speed up movement, increase maneuverability, and simplify the process by speeding up activation and deactivation. Such control can be carried out using a smart control that follows the colonic lumen on its own or by employing an external user console with a joystick. This should provide extremely accurate control of the mini-robot as well as help from a technician or nurse practitioner during SPID training or use. In such a case, a skilled colonoscopist may possibly direct numerous concurrently running processes, such as requesting a biopsy or reversing locomotion to get a second opinion, while also causing a disruption in cost. Finally, an SPID may raise the compliance rate for CRC screening of the asymptomatic population, toward mass screening campaigns for early detection, by lowering the pressure against the colon wall (Manfredi et al., 2019).

Lengthy procedures undertaken with a high risk of unintentional patient harm result from instruments that are difficult to handle, and this situation is far from ideal. In addition, years of training are often required to master their use. Highly rigid endoscopic instruments pose a risk of deformation or perforation of the tissue surrounding the organ, often causing pain or other unintentional side effects. Wrapping the colon during a colonoscopy is an example of how the use of inflexible devices can harm or injure the patient. Unfortunately, there are still issues with placement, dexterity, force application, and visualization when using flexible instruments and endoscopes. The goal of soft robotic research is to combine the safety of soft materials with the better controllability of rigid robotics and accessibility of flexible tools (Runciman et al., 2019).

The use of soft robotic technologies in this context holds much potential. Most importantly, they are considered intrinsically safe for use in MIS due to their soft nature. In contrast, surgical robots made of rigid components require complicated control methods to ensure that the forces applied to the soft tissue are minimized, thereby reducing the risk of patient harm. With traditional rigid-component robots, computer malfunctions, although rare, can result in uncontrollable movements of the robotic arm that can have catastrophic effects on the patient. Because of their ability to move like a tentacle through small spaces and over long distances toward distant surgical sites, soft robots also hold promise for novel or advanced surgical procedures (Kwok et al., 2022). For example, Kwok et al. (2022) present a proof of concept for a brand-new early-stage surgical instrument. This instrument is based on two fundamental design principles:

○ The use of only soft materials
○ An underpowered system that can apply a specific force and adapt the finger shape to the target object

The operation of the instrument has been statistically studied, and the results serve as the basis for novel applications with ideal properties for the medical industry. The soft-body grippers have self-limiting properties and intrinsic safety features, which ensure extremely safe interactions with the biological tissue. According to this perspective, the effect of sensing surgical instruments, which continues to be a barrier to force feedback in MIS procedures, is minimized, and the designed tool, which is characterized by a self-regulating mechanism, opens the door for a paradigm shift in safe surgical manipulation (Rateni et al., 2015).

An elastic actuator (LEA) guaranteeing an output behavior in a single direction and preventing the isometric expansion of its elastomeric structure is presented by Gerboni et al. (2016). This LEA uses only one part—its sidewall—to replicate the linear push-and-pull action of a conventional single-action piston cylinder actuator. This makes it possible to create fluid actuators that are incredibly lightweight and show promise for use in medical and surgical applications. The LEA functions similarly to traditional piston cylinder actuators but is made entirely of disposable materials (silicone and polyester filaments) and has a simple manufacturing process. The actuator presented by Gerboni et al. (2016) has been used to actuate medical devices that require a linear push/pull action, due to the comparatively strong force with compact size generated by the LEA. The tool, referred to as a gripper for MIS, combines conventional commercial laparoscopic surgical jaws with a unique fluid-powered elastic actuator. This approach has created a useful, locally actuated device that is also familiar to surgeons. The system was verified to meet the force requirements for performing typical tasks (MIS) by evaluating the full range of gripping forces applied by the gripper (in various configurations and degrees of jaw opening) using the mechanism's rigid transmission. This evaluation resulted in a maximum gripping force of approximately 5.6 N. A research study by Diodato et al. (2018) shows how soft robotics can be used in conjunction with conventional rigid tools to enhance the capabilities of the robotic system without compromising the usability of the platform.

This is the first time that a surgical robot has been equipped with an endoscopic tool made of soft materials. The da Vinci Research Kit master console enhances work space and dexterity without sacrificing usability and makes it easier to manage the soft endoscopic camera. The soft robotic technique proposed in this work (Abidi et al., 2017) uses flexible fluidic actuators (FFAs) that enable highly agile and fundamentally safe navigation. An idealized fluid chamber design inside the robot modules lends dexterity to the robots. The entire structure of the squeezable two-module robot is made of soft and compliant elastomers that ensure safe physical interactions. The flexible endoscopic tools are guided along the central axis through an internal free lumen or chamber. To further highlight the robot's capabilities, a constant curvature (CC)-based inverse kinematics model is developed. To evaluate the capabilities of the robot in comparison with those of traditional systems in a realistic setting, experimental testing using MIS on a corpse model was conducted.

2.6.1 Materials and methods

From a hardware and software perspective, the endoscopic camera manipulator arm of the da Vinci Research Kit was coupled with a bioinspired soft manipulator equipped with a miniature camera. The usability of the integrated system was tested using a standard procedure with inexperienced users to determine the challenges in controlling the soft manipulator.

2.7 Challenges and state-of-the-art methods

In this section, challenges and state-of-the-art methods for researchers interested in soft robotic studies are listed.

2.7.1 Modeling

The majority of the time, continuum robots take the form of a cone or a cylinder and are propelled by tendons or fluids. Their analytical modeling approaches can be characterized according to whether or not the geometrical approximation for the manipulator makes use of (piecewise) CC assumptions. In the (P)CC model, the bending body is thought of as an arc or arcs with curvature(s) but not torsion. This is because the name of the model suggests it. Common models that can be easily applied to soft continuum robots include the finite element model of deformation (Webster and Jones, 2010) (Grazioso et al., 2019), static-equilibrium models (including forces) (Camarillo et al., 2008), and even dynamic models (Tatlicioglu et al., 2007; Sadati et al., 2020). However, there are situations when pure (P)CC assumptions are unable to reflect the intricacies of the deformation or handle cases that involve external loadings (Wang et al., 2018b). Using physical mechanisms, more complicated methods have been proposed, such as the spring-mass model (Yekutieli et al., 2005) and the Cosserat rod model (Rucker and Webster Iii, 2011; Tunay, 2013). However, due to the fact that high computing efficiency and heuristic experimental calibration would be necessary, only a small amount of work has been validated on actual robots (Renda et al., 2014). Another difficulty is the necessity of utilizing several sensing devices (Wang et al., 2020b) in order to supply the models with the essential information and states that are associated with their configurations. As a result of this, learning-based (or data-driven) approaches have garnered an increasing amount of interest in relation to the control of continuum robots. These approaches appear to be a promising replacement for traditional analytical modeling.

2.7.2 Control

Control is the most important part of every robot, and soft robots are no exception; control algorithms, if implemented, highly depend on actuation and sensing, and, so, control challenge is divided into sensing challenges, actuation challenges, and algorithm complexity challenges (Esfandiari et al., 2015, 2017; Kolbari et al., 2015a, b, 2016; Ebrahimi et al., 2016; Khadivar et al., 2017; Sadeghnejad et al., 2019a, b; Jafari et al., 2022).

2.7.2.1 Actuation

It is possible to achieve robotic movement in soft manipulators through the utilization of pressurized fluids (Ikuta et al., 2002; Gandarias et al., 2020; Berthet-Rayne et al., 2021), tendon displacement (Shiva et al., 2016), or the activation of smart materials (Chautems et al., 2020; Peters et al., 2019; Shiva et al., 2016). In soft material structures, hollow chambers can be manipulated using liquid and gas fluids to inflate (Shiva et al., 2016; Berthet-Rayne et al., 2021; Peters et al., 2020) or deflate (Robertson and Paik, 2017) the chambers, causing the material to deform in the process (Bernth et al., 2017; Berthet-Rayne et al., 2021). The behavior of the material can be described as bending or elongation as a result of these deformations (Cianchetti et al., 2014). Pipes that are connected to devices located on the outside of the pressurization system, such as pressure regulators or compressors, are used to start the pressurization process. The robotic systems themselves may be shrunk and kept at a relatively low weight, which makes bending and elongation behaviors much easier to achieve as a result of this. This is one of the primary benefits of this approach. In tendon-driven systems, electromagnetic actuators are also located on the exterior of the device. Bending can be done by drawing wires that are fixed at the tips of the manipulators' down channels that run along the manipulators themselves (Wurdemann et al., 2015b). After being subjected to temperature change-induced deformation, these metal composites can revert back to their original shape when using smart materials such as SMAs. This deformation can generate driving forces, which makes it possible for soft robotic manipulators to act upon their environment (Sohn et al., 2018). By passing a current through the resistive SMA, it is possible to produce an increase in temperature, which results in an increase in the amount of thermal heat.

2.7.2.2 Sensing

One area that is extremely important in sensing is related to measure position and orientation of the end effector. Employing the proper kinematic models that relate the joint positions (measured by position sensors, such as shaft encoders in the case of rotary joints) to the instrument's tip via the chain of rigid links that make up the overall robotic structure from the base to the tip is an easy way to acquire the pose of a robot manipulator that is made up of rigid links and stiff joints. This can be accomplished using a robot that is made up of rigid links and stiff joints. The position and orientation of the instrument's tip can be computed with the help of inverse kinematics, which is frequently available in an analytical, closed form for rigid-component robots. Furthermore, with the help of an appropriate controller, the necessary joint motor commands can be generated in order to move the instrument's tip to the desired location inside the patient's abdomen and then to orient the end effector (Wang et al., 2020a).

On the basis of this, stiff connected robot systems are easily capable of attaining precision levels of less than 1 mm, particularly when used in conjunction with visual feedback in a teleoperational setup (Javadi et al., 2017; Su et al., 2019; Khadivar et al., 2020; Kolbari et al., 2018). This level of accuracy is comparable to that which may be accomplished with the use of laparoscopic equipment. The control architecture of the da Vinci surgical system provides a user-friendly interface that enables the surgeon to concentrate solely on moving the instrument's end effector into the desired location in a highly intuitive manner. This is analogous to how a user navigates the cursor across the screen of a computer using a mouse (Tewari et al., 2002). The input device that is provided by the system also enables the surgeon to move the end effector into the desired location. This method makes even the most difficult procedures, like suturing, much easier to carry out. This is a significant improvement over the traditional laparoscopic method, which requires all movements to be more complicated in order to account for the fulcrum point. This method also makes it much easier to carry out procedures like suturing. If a surgeon wants to safely perform a minimally invasive treatment using laparoscopic equipment, it may take them years of training. However, if the surgeon uses a robot like the da Vinci system, even a beginner can tie a knot into a suture after only a 5-min practice session. Although recent RAMIS systems have made it possible to perform intuitive navigation of surgical tools by utilizing a rigid-component structure, the maneuvering of a soft robot inside of an abdominal cavity presents its own unique set of difficulties. When those challenges extend beyond the abdominal cavity, as is the case when considering soft endoscopic systems for applications in the colon, and with NOTES (natural orifice transluminal endoscopic surgery) in general, and when there is a need to utilize considerably longer endoscopes, they become even more difficult. It is challenging to construct basic kinematic models due to the nonlinear mechanical characteristics of a device made of soft or extremely flexible materials (silicone, rubber, or cloth are some examples (Lee et al., 2017a, b)). However, research into the creation of real-time models to assist in controlling robots made of soft materials is still ongoing. For the purpose of computing the robot's stance and the nature of its physical interactions with its surroundings, these models rely on information from sensors that are implanted into the soft robot itself. The next sections present an overview of distributed touch and force sensors in soft robotic instruments as well as pose sensors. These sensor technologies are well-suited for integration with soft robot arms, which are utilized in abdominal surgery in addition to other endoluminal and transluminal surgical procedures. Although the addition of diagnostic sensors is acknowledged to be important, doing so would go beyond the parameters of the scope of this chapter.

The ability to retrieve data on the precise location of surgical tools and their physical interactions with their immediate environment is equally as vital as receiving diagnostic information from a surgical site. It is possible for tactile and force sensors that have been integrated into surgical gear to bring into action an essential sensing modality known as the provision of haptic feedback. As the surgical community gradually transitioned from open surgery to laparoscopic surgery, we effectively lost the primary advantage of direct haptic feedback. As a result, this problem became glaringly clear to everyone involved in the field. In point of fact, it is generally acknowledged that, in the past, surgeons made excellent use of their haptic capabilities, which enabled them to differentiate between healthy and diseased tissues. Their fingertips were equipped with tens of thousands of minute tactile sensors, which allowed them to carry out advanced diagnostics by simply palpating the organs that were in question. The execution of suturing, organ manipulation, and other tasks that might require coordination of two laparoscopic tools became considerably more complicated alongside the significant reduction in the sense of touch. Despite the fact that laparoscopy is being widely hailed as superior to open surgery because of the scarring and hospital inpatient duration, it comes with its own catalogue of disadvantages. The introduction of robot-assisted minimally invasive surgery (RAMIS) has brought this last argument to the forefront: the sense of touch is completely eliminated in today's robot-assisted surgical equipment like the da Vinci system from Intuitive Surgical. When determining how instruments interact with their surroundings, vision has historically been regarded as a suitable alternative to the senses of tactile and force feedback. This has been the case for a number of reasons. By viewing the deformations that are generated, users of da Vinci systems, which have an advanced stereo vision feedback system, claim that they are able to "see" the force that is being imparted by hard instruments onto the soft tissue. Even while it is generally agreed upon that surgeon-controlled robot-assisted surgery must have some form of visual feedback for it to be successful, there is a growing movement among surgeons to incorporate haptic feedback into the procedure. It is hypothesized that equipping surgeons with haptic sensation would result in an improvement in the quality of surgical procedures, a reduction in surgical margins, and a decrease in the number of occasions in which excessive force resulted in unintended tissue harm (Roberts et al., 2010). In spite of the fact that the surgical sector continues to heavily rely on visual feedback, significant advancements have been made in the creation of tactile and force sensors (Cianchetti et al., 2014; Fraś et al., 2015). Recent years have also seen the development of technologies that make it possible to incorporate soft sensors into soft robots (Ren et al., 2021; Dawood et al., 2021; Soter et al., 2018). The challenge here is to make the transition from existing sensor technologies, which are typically made from rigid components, to new sensing options that are as compliant as the robots into which they are to be integrated. This will be accomplished by taking the progressive step away from existing sensor technologies and moving toward new sensing options.

When one considers the malleable nature of soft robotics, it becomes obvious that the two sensing modalities of posture and kinaesthetic information are strongly intertwined. Kinaesthetic information refers to the pressures that are exerted upon the robot by its surroundings. Several examples are provided to illustrate this point: (1) an external force that is applied to the soft robotic structure will induce bending, which is a change in pose, even if no actuation signal is emitted and (2) even if a certain actuation signal is emitted, which would normally lead to a correspondent bending of the structure in free space, the resulting anticipated pose might not be attained if an intervening obstacle imparts a force onto the robot. Recognizing this aspect of soft robotic behavior makes it abundantly evident that we require an integrated sensor system that is capable of monitoring the robot's position and kinesthetics in a manner that is dissociated from one another in a distinct manner. It should be noted that this is a significantly more difficult task than the corresponding sensing task in stiff-joint, rigid-component robots, in which these two modalities are decoupled by default. Indeed, a rigid-link robot will assume its desired pose (as specified by the user) independently of external forces; it will not bend or adapt to the environment but rather move any obstacles out of its way (within the limits of the strength of its links and its joint actuator forces). In the context of robot-assisted surgery, it is obvious that a robot with rigid links has the potential to cause damage to the patient if it is not appropriately constrained by an intelligent control architecture. This risk can be mitigated, however, by ensuring that the robot is properly guided by the surgeon. In recent years, a number of force and pose sensors that have the potential to be integrated into soft surgical robotic instruments have been created (Wang et al., 2018a).

The incorporation of sensors into the thin shafts of soft robotic surgical or endoscopic instruments is the primary focus of this particular piece of research. Point sensors are not going to be discussed in this particular publication. It is also important to note that there are other methods that make use of external sensors, such as cameras and fluoroscopy (which is of considerable interest in an MIS setting), 3D vision sensors also called RGBD (red, green, blue, depth) sensors, and electromagnetic sensors in order to obtain information about a robot's shape/pose and/or the forces to which it is subjected (Runciman et al., 2019). Another consideration that must be taken into account is the fact that the overall rigidity of the robot must not be sacrificed in any way throughout the process of developing the "ideal" sensor. This prerequisite must be met regardless of the approach that is chosen. There have been significant developments in sensors that are appropriate for integration with slender continuous robots, such as endoscopes and catheters, in order to measure the posture of a structure. Examples of these types of robots are endoscopes. Methods that measure the shape of flexible structures have been developed, which make use of grated fiber optics, a technique that is more frequently referred to as fiber Bragg grating (FBG). A periodic change in the refractive index of the fiber is produced as a result of gratings being scribed onto it. This variation acts like a dielectric mirror, reflecting specific wavelengths. Bending the fiber in a different direction causes a change in the distance between the gratings, which, in turn, causes a change in the specific wavelength, which, in turn, causes a change in the light pattern that is received by the optical interrogator. The changes in the light pattern may have something to do with the degree of bending that the structure undergoes or, to put it another way, the shape that the structure takes at various places. Because of their tiny diameters, FBG sensors can be effortlessly incorporated into catheters and endoscopes for the purpose of pose estimation. As an illustration, it has been demonstrated that FBG-based shape sensors for catheters that are 1 m in length may offer accurate estimations of the location of the end effector in relation to the catheter base (Shi et al., 2016).

Despite the fact that they are quite accurate in measuring shape and are able to resist any impact from local forces as a result of their contact with the environment, they also have a number of important drawbacks. In order to process the signals coming from the grated fibers, they make use of an extremely pricey optical interrogation system. This just serves to make the cost issue that much more difficult to resolve. In point of fact, the cost of these systems is approximately one thousand times more than that of light intensity modulation systems. They also have a problem with temperature drift, which requires additional reference fibers to be used in order to adjust for it. Because FBG fibers cannot be stretched, incorporating them into soft, stretchable robots presents a challenge (they cannot stretch in soft robots).

An accurate model of the bending behavior of the structure is necessary for the computation of the structure's ultimate shape, which results from the bending process. Searle (Coad et al., 2020) presented a method that might be used to measure the shape of flexible structures at a relatively low cost. For the purpose of transferring light to the portion of the structure where the shape is going to be measured, this method makes use of ordinary optical fibers. In this instance, the only thing that causes the transmission of light is the fibers. A transmission fiber sends light to a mirror, which reflects the light back into a receiving fiber, which is then connected to optoelectronics, which converts the received light into electrical signals that can be processed by a computer. Fibers are typically used in pairs. The mirror is fastened to the structure in such a way that any bending of the structure will cause it to shift relative to the tips of both the transmitting and receiving fibers, thus causing a change in the amount of light that is reflected. After that, a correlation can be drawn between the amount of bend in the structure and the measured light intensity.

The FBG approach is more accurate than this strategy, which provides good estimations of the shape of the structure in which they are embedded, but this approach is less accurate. Temperature drift can also be an issue, and, in terms of performance, this light intensity-based technique is inferior to FBG alternatives (Coad et al., 2020). On the plus side, this light intensity-based approach is very much less expensive than FBG solutions. Other methods that make use of optical fibers rely on the fact that the fibers absorb light when they are bent in a particular way under specific circumstances. Abraded optical fibers, which are fibers in which the outer sheath has been removed in places, are known to leak light due to a disruption in the complete internal reflection that is intrinsic to commercial optical fibers. The amount of light that is lost increases in proportion to the degree to which the abraded fiber is bent. Furthermore, the light intensity that is measured at the receiving end of this sensor can be related to the amount of bending, which, in turn, can be related to the shape of the structure (Fig. 2.1B). Because the optical fibers that are used in this method cannot stretch, their incorporation into soft, stretchable robots is limited (Godaba et al., 2020; Lunni et al., 2018; Zhao et al., 2016). This is another low-cost method that can be used to acquire relatively accurate estimates of the shape of flexible structures, and it lends itself to doing so.

Light waveguides built from transparent silicone have been created by researchers (Zhao et al., 2016; Godaba et al., 2020; Lunni et al., 2018), expanding on the concept of lossy optic fibers. These waveguides behave in a manner that is analogous to that of abraded optical fibers; the higher the bending of the silicone waveguides, the greater the amount of light that is lost, and estimates of the structure's shape can be derived from this behavior. These sensors are advantageous in that they are mechanically as soft as the material that is used to make soft robots. As a result, they are well-suited to integration with soft robots and will not have any impact on the compliance and stretchability of the structure of the robot. Even though this strategy has been implemented with great success, an accurate estimation of the structure's shape cannot be achieved because of the intrinsic hysteresis and nonlinear behavior of the material. Additionally, they are sensitive to the forces that are imposed on them when they come into direct physical touch with their surroundings. As a result, it can be challenging to differentiate between the shape of a structure and the forces acting on it as a result of its interaction with its surroundings. Because these kinds of sensors are more cumbersome than regular optical fibers, there are constraints placed on the level of miniaturization that can be achieved (Godaba et al., 2020, Lunni et al., 2018, Zhao et al., 2016).

Stretchable and electro-resistive yarns provide the foundation for yet another fascinating method for assessing the shape of constructions made of soft materials (Wurdemann et al., 2015). These yarns have a resistance that changes when they are stretched, and, when they are incorporated into a soft robot, they will expand and contract in tandem with the body of the robot. The method is not expensive and may be simply incorporated into soft robotic devices. Furthermore, the compliance and stretchability of the underlying structure are not adversely affected in any significant way by its application. However, because of hysteresis as well as changes in resistance over time, the yarn estimates of shape are not nearly as accurate as they are when utilizing optical fibers. It has been determined that e-gain, often known as liquid metal, is a potentially useful sensing approach for soft robotic systems. Creating channels in the substrate and body of the robot that are impermeable to fluids makes it possible for this technology to be relatively simply integrated with soft silicone robots. The deformation in the e-gain-containing channels of the robot causes a change in the conductivity of the liquid-sensing medium, and this change can be generated either by external forces or by controlled bending in the robot itself.

The implementation of e-gain in soft robots makes it possible to sense force and tactile in addition to shape (White et al., 2016; Dickey, 2017). However, because the sensor inputs for force and shape sensing are coupled in this method, additional sensing modalities and more complex modeling approaches are necessary to disambiguate between the two. Despite the fact that this method generates measurements that are relatively precise, there have been reports of difficulties in connecting these sensors to the usual electronics that are required for reading out the measurements (Wang et al., 2018b).

Other researchers have concentrated their efforts on the development of sensor coverings that can supply the underlying structure with capabilities for tactile sensing. Those that are founded on the idea of capacitive sensing have proved themselves to be particularly well-suited for incorporation into soft robotic constructions (Dawood et al., 2020). The electronic skin that Dawood et al. developed (Xu et al., 2021) is comprised of numerous layers of elastomeric material, each of which has carbon grease electrodes encased inside it. The carbon electrodes perform the function of capacitor plates, whereas the elastomer fulfills the role of providing the necessary insulation between the electrodes. Within the sensor skin, the electrodes are arranged in a grid pattern, with many electrode strips running across the skin in parallel. These electrode strips are spaced at different levels and are separated by insulating silicone layers. Because each set of electrode strips is perpendicular to its neighboring set, a matrix is produced that enables the collection of capacitance measurement data at a high spatial resolution via terminals at the skin edges. This is made possible because the electrode strips are arranged in a manner that is perpendicular to one another. Local deformations brought on by contact with the surrounding environment cause a shift in the distance that exists between the electrodes, which, in turn, causes a shift in capacitance. This shift in capacitance can then be related to the force that is being applied by means of a calibration procedure. Sensor skins that are only a few millimeters in thickness are capable of achieving a spatial resolution of approximately 8.5 mm. Given the elastic nature of such skins, they are excellent for integration with soft robotic structures for the purpose of distributed tactile sensing. This is possible because such skins do not have any substantial impact on the overall stiffness of the robot. On the other hand, each shape change that the soft robot goes through would, unfortunately, also lead to changes in the capacitances that are measured on its skin. The primary focus of the work being conducted right now is on developing machine learning approaches that are able to differentiate between global bending and local deformation caused by external forces. In recent years, attempts have been made to estimate the shape of a soft robot in an indirect manner. One method that has been used to do this is by utilizing pressure sensors (and/or airflow sensors), which are typically integrated into regulators that are used to adjust pressure within pneumatically actuated soft robot arms. This strategy shows a lot of potential because it does not call for the structure of the soft robot to be altered in any way, which reduces the risk of there being any kind of trade-off with regard to its pliability. Despite the complexities of the nonlinear properties of the material used and the deformative effect of any physical interaction with the environment on the robot, there are, however, other significant challenges associated with this approach. One of these challenges is the requirement of a highly accurate model of the soft robot in order to precisely predict its mechanical behavior. This is the case despite the fact that the approach is being used. There has been a significant amount of development in the establishment of control architecture that is capable of providing a positional error that is sufficiently low when estimating the configuration of a robot. This method, which can be used to forecast the shape of the robot, can be assisted by sensors introduced remotely (such as force/torque sensors installed at the robot base to monitor forces transmitted through the robot's soft body). Table 2.1 summarizes the control, actuation, and sensing methods and actuator types of the articles examined in this chapter.

Most of the challenges in soft robotics are universal and similar in their subareas. Fig. 2.7 categorizes the typical challenges in soft robotics.

On the other hand, there are several methods for developing soft robots and soft actuators. These methods are detailed in Fig. 2.8.

In summary, a selection of current methods applied to soft robot modeling is listed in Table 2.2.

A selection of common control strategies for soft robots is detailed in Table 2.2. Due to the material compliance requirements and degree of freedom of soft robots, complex designs are often required. Modeling and simulation tools allow simulation of complex tasks, as is commonplace for rigid robots. However, most soft robot modeling tools are still under development. Modeling and simulation tools for soft robots are more complex than for rigid robots, as the motion of soft robots occurs through structural deformation rather than joints. Deformable mechanics should be used to model the behavior of soft robots. In the following section is a list of these tools that are currently available, most of which have tutorials available.

Table 2.1 Reviewed articles by control, actuation, and sensing methods and actuator types.

Authors	Control method	Actuation method	Sensing method	Actuator
Polygerinos, P. (Polygerinos et al., 2015a, b)	Open-loop SEMG logic, sliding mode control (SMC)	Pneumatic	Fluidic pressure sensors	Soft segmented fiber reinforcement
Cappello, L. (Cappello et al., 2018)	PID	Pneumatic	Textile strain and force sensors	Multiarticular textile actuators
Yap, H.K. (Yap et al., 2017)	PID	Pneumatic	Air pressure sensor	–
In, H. (In et al., 2015)	Feedback linearization	Cable-driven	–	Tendon-driven
Proietti, T. (Proietti et al., 2021)	Bang-bang control	Pneumatic	IMU	Inflatable
O'Neill, C. (O'Neill et al., 2017)	–	Pneumatic	Encoder	Inflatable, textile base
Realmuto, J. and T. Sanger (Realmuto and Sanger, 2019)	Bang-bang control	Pneumatic	Piezoresistive silicon pressure sensor	Fabric-based helical actuator
Thalman, C. (Thalman et al., 2018)	–	Pneumatic	–	Inflatable, textile base
Lee, G. (Lee et al., 2017a, b)	Switching admittance-position control	Cable-driven	IMU, loadcell, encoder	Tendon-driven, textile base
Chung, J. (Chung et al., 2018)	Open-loop gate control	Pneumatic	IMU, pressure sensor	Inflatable, textile base
Veale, A.J., K. Staman, and H. van der Kooij (Veale et al., 2021)	–	Pneumatic	–	PPIA
Roche, E.T. (Roche et al., 2017)	–	Pneumatic	–	Modified McKibben PAM

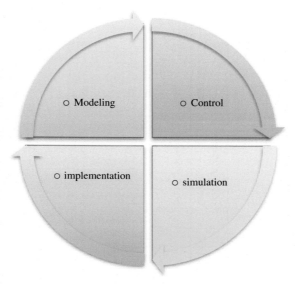

Fig. 2.7 The four most common challenges in soft robotics.

Fig. 2.8 The most common methods for developing soft robots and actuators.

2.7.3 Simulation

In the realm of robot control, simulation is a useful tool that, in general, can be implemented in one of two ways. The typical approach is to first perform a previous validation, which then serves as a virtual environment in which to test a suggested model or controller. This is the standard method. Examples can be found in challenging modeling methodologies, such as geometrical or dynamic modeling, as well as task evaluation. The success of the simulation, on the other hand, will be directly proportional to how we configure a soft robot simulator, particularly regarding the interactions that take place in an MIS scenario. It is also necessary to take into consideration the transfer of sophisticated factors as well as the feedback that is readily available from the actual environment. A sensing device known as FBG (Selvaggio et al., 2018) has a lot of potential because it can provide information on strain and shape in sophisticated models. The alternative method utilizes simulations as a parallel testing platform and an essential component in real-time control. This makes the simulations an extremely powerful tool. These simulators make it possible to collect data quickly for controller initialization or even for

Table 2.2 Common control modeling methods for soft robots (Sadati et al., 2017).

	Lumped system	CC and Euler-Bernoulli	CC and virtual energy	Modified CC	Cosserat/beam theory	Series ID
Specifications	Dynamic, forward/ inverse	Static, forward/ inverse	Static, inverse	Static, forward	Static, forward	Static, forward/ inverse
Accuracy (error, %)	Medium (22%)	Low (31%)	Low-medium (28%)	Low- medium (27%)	High (12%)	High (11%)
Applications	Simulation, control	Inaccurate control	Design	Inaccurate control	Simulation	Control

online modification of the control scheme. One of the applications that is most representative is called reinforcement learning. There has been research (Xu et al., 2021) that has looked into surgical robot learning using da Vinci Research Kit (dVRK)-compatible systems. Communication between the actuators, sensors, robot modules, and control platforms will be quite challenging in order to implement this in MIS. This will be a huge challenge. This is mostly due to the fact that the extraction of real-robot data from accessible robots and the transmission of signals both take a significant amount of time.

Reinforcement learning is becoming increasingly prominent within the robotic community alongside the utilization of soft robotics as a direct result of the remarkable achievements of (deep) reinforcement learning in AlphaGo (Wurdemann et al., 2015b). The stumbling block of analytical modeling of soft robots can be avoided by adopting control methods that are based on learning instead. This enables the robots to adapt to complicated situations and complete complex tasks, both of which may be constrained when using traditional control methods. It has been established that controllers and motion planners educated through learning can achieve high accuracy and satisfy stated requirements in vivo. This demonstration pertains to surgical robots used in MIS. However, because the quality (accuracy and distribution) of sampled sensory data is critical in techniques for machine learning based on regression or iterative processes, the stability of controllers could not be reliably guaranteed on each individual robot. The interaction time for controller training would also be an uncontrollable aspect in reinforcement learning. This is one of the reasons why more and more recent research has been focusing on transferring simulation-trained controls to actual robots.

That being said, the substantial amount of precollected or online interaction data continues to be the primary problem, which helps explain why learning in soft robots for MIS has not yet advanced beyond the research level (Wang et al., 2021). Recently, soft robots have attempted the concept of shared control and autonomy. This technique combines human engagement with machine intelligence to lessen the load on the human operator and the automatic controller (Selvaggio et al., 2021; Jitosho et al., 2021). It is possible that shared control and autonomy could serve as the foundation for a revolutionary approach to solving complex problems that occur in unstructured contexts. An excellent illustration of this would be MIS.

Various simulation solutions are proposed in the literature to improve insights into how and what to design for a soft robot. Table 2.3 presents the most common simulation solutions developed, utilized, or created by researchers in recent years.

2.8 Conclusions

We discussed rehabilitation in the first section, which means providing assistance to a limb of the body that involves ADLs. One of the most important of these activities is manipulation, which is divided into two sections: (i) grasping and (ii) positioning.

Table 2.3 Common soft robot simulation solutions developed, utilized, or created by researchers.

Simulation solutions	
	○ SOFA: Software toolbox for soft robot modeling, simulation, and control (Coevoet et al., 2017)
	○ PyElastica: A compliant mechanics environment for soft robotic control (Naughton et al., 2021)
	○ ChainQueen: A real-time differentiable physical simulator for soft robotics (Hu et al., 2019)
	○ TMTDyn: A MATLAB tool that uses discretized lumped systems and reduced-order models to model and control hybrid rigid-continuum robots (Sadati et al., 2021)
	○ SoRoSim: A soft robotics MATLAB toolbox using the geometric variable-strain approach (Mathew et al., 2021)
	○ S. AMBF (BASim): An open-source framework for rapid development of interactive soft-body simulations for real-time training (Munawar et al., 2020)
	○ SoMo: Simulation of continuum robots in complex environments: quick and accurate (Graule et al., 2021)
	○ DiffAqua: A differentiable computational design pipeline for soft underwater swimmers with shape interpolation (Ma et al., 2021)
	○ Soft IK: Soft IK with stiffness control (Bern and Rus, 2021)
	○ Viper: Volume invariant position-based elastic rods (Angles et al., 2019)
	○ Vine Simulator: A dynamics simulator for soft growing robots (Jitosho et al., 2021)
	○ IPC-sim: Incremental potential contact: large deformation dynamics without intersection and inversion (Li et al., 2020)
	○ SoGut: A soft robotic gastric simulator (Dang et al., 2021)

The upper limbs are responsible for object positioning, whereas the wrist and hand are responsible for grasping. Assistive devices that implement grasping mainly focus on hand gestures and grasping force. In these areas, we reviewed articles about "soft rehabilitation gloves," which are the most inclined solution for grasping failure. Positioning imperfections handled by upper limb assistive devices are reviewed in the second section. Most reviewed devices have various configuration actuators and sensors, which help patients perform better in various tasks.

The lower limb is in charge of human movement; stability, speed, and efficiency of movement depend on gait cycles. Perfect gait cycles need to exert proper force and provide enough velocity. In this section, we reviewed the studies that discussed lower limb problems and potential solutions. The heart is a vital part of the body, and cardiovascular disorders, including heart disease, are the leading cause of mortality globally. So, cardiac assist devices could be useful. Because of the soft tissue of the heart, devices that are in direct contact with it must have a similar stiffness as the heart tissue. As a result, researchers have recently become more interested in the soft robotic solutions discussed in the section. Since the advent of soft robotics roughly two decades ago, research interest in the discipline has grown rapidly. It is motivated by the industry's admiration for the large spectrum of soft materials accessible for use in creating extremely dexterous robots with adaptive properties much beyond what rigid-component devices can achieve. The potential of soft robots to compliantly adapt to their surroundings has piqued the interest of the surgical robotic community. In this review, an overview of the recent progress in the development of soft robotics has been provided.

Conflict of interest statement

The authors confirm that this book chapter and its corresponding research work involve no conflict of interest.

Acknowledgments

We are grateful to the Bio-Inspired System Design Laboratory at the Biomedical Engineering Department of Amirkabir University of Technology (Tehran Polytechnic) for supporting this research.

References

Abidi, S.H.J., Gerboni, G., Brancadoro, M., Diodato, A., Cianchetti, M., Wurdemann, H., Althoefer, K., Menciassi, A., 2017. Highly dexterous 2-module soft robot for intra-organ navigation in minimally invasive surgery. Int. J. Med. Robot. Comput. Assist. Surg. 14.

Ahmadjou, A., Sadeghi, S., Zareinejad, M., Talebi, H.A., 2021. A compact valveless pressure control source for soft rehabilitation glove. Int. J. Med. Robot. 17, e2298.

Al-Fahaam, H., Davis, S., Nefti-Meziani, S., 2016a. Wrist rehabilitation exoskeleton robot based on pneumatic soft actuators. In: 2016 International Conference for Students on Applied Engineering (ICSAE), IEEE, pp. 491–496.

Al-Fahaam, H., Davis, S., Nefti-Meziani, S., 2016b. Power assistive and rehabilitation wearable robot based on pneumatic soft actuators. In: 21st International Conference on Methods and Models in Automation and Robotics (MMAR).

Amend, J.R., Brown, E., Rodenberg, N., Jaeger, H.M., Lipson, H., 2012. A positive pressure universal gripper based on the jamming of granular material. IEEE Trans. Robot. 28, 341–350.

Angles, B., Rebain, D., Macklin, M., Wyvill, B., Barthe, L., Lewis, J.P., Von Der Pahlen, J., Izadi, S., Valentin, J., Bouaziz, S., 2019. VIPER: volume invariant position-based elastic rods. Proc. ACM Comput. Graph. Interact. Tech. 2, 1–26.

Araromi, O., Gavrilovich, I., Shintake, J., Rosset, S., Richard, M., Gass, V., Shea, H., 2015. Rollable multisegment dielectric elastomer minimum energy structures for a deployable microsatellite gripper. IEEE/ASME Trans. Mechatron. 20, 438–446.

Arienti, A., Calisti, M., Giorgio-Serchi, F., Laschi, C., 2013. PoseiDRONE: design of a soft-bodied Rov with crawling, swimming and manipulation ability. In: 2013 Oceans—San Diego, 23–27 Sept. 2013, pp. 1–7.

Awad, L., Bae, J., O'Donnell, K., De Rossi, S.M.M., Hendron, K., Sloot, L., Kudzia, P., Allen, S., Holt, K., Ellis, T., Walsh, C., 2017. A soft robotic exosuit improves walking in patients after stroke. Sci. Transl. Med. 9, eaai9084.

Baltes, J., Tu, K.Y., Sadeghnejad, S., Anderson, J., 2017. HuroCup: competition for multi-event humanoid robot athletes. Knowl. Eng. Rev. 32.

Bar-Cohen, Y., 2000. Electroactive Polymers as Artificial Muscles-Capabilities, Potentials and Challenges. Handbook on Biomimetics., p. 11.

Bauer, S., Bauer-Gogonea, S., Graz, I., Kaltenbrunner, M., Keplinger, C., Schwodiauer, R., 2014. 25th anniversary article: a soft future: from robots and sensor skin to energy harvesters. Adv. Mater. 26, 149–161.

Bern, J.M., Rus, D., 2021. Soft ik with stiffness control. In: 2021 IEEE 4th International Conference on Soft Robotics (RoboSoft), IEEE, pp. 465–471.

Bernth, J.E., Arezzo, A., Liu, H., 2017. A novel robotic meshworm with segment-bending anchoring for colonoscopy. IEEE Robot. Automat. Lett. 2, 1718–1724.

Berthet-Rayne, P., Sadati, S.H., Petrou, G., Patel, N., Giannarou, S., Leff, D.R., Bergeles, C., 2021. Mammobot: a miniature steerable soft growing robot for early breast cancer detection. IEEE Robot. Automat. Lett. 6, 5056–5063.

Bethea, B., Okamura, A., Kitagawa, M., Fitton, T., Cattaneo, S., Ameli, M., Baumgartner, W., Yuh, D., 2004. Application of haptic feedback to robotic surgery. J. Laparoendosc. Adv. Surg. Tech. A 14, 191–195.

Camarillo, D.B., Milne, C.F., Carlson, C.R., Zinn, M.R., Salisbury, J.K., 2008. Mechanics modeling of tendon-driven continuum manipulators. IEEE Trans. Robot. 24, 1262–1273.

Cappello, L., Meyer, J.T., Galloway, K.C., Peisner, J.D., Granberry, R., Wagner, D.A., Engelhardt, S., Paganoni, S., Walsh, C.J., 2018. Assisting hand function after spinal cord injury with a fabric-based soft robotic glove. J. Neuroeng. Rehabil. 15, 59.

Carrozza, M.C., Cappiello, G., Micera, S., Edin, B., Beccai, L., Cipriani, C., 2006. Design of a cybernetic hand for perception and action. Biol. Cybern. 95, 629–644.

Catalano, M.G., Grioli, G., Farnioli, E., Serio, A., Piazza, C., Bicchi, A., 2014. Adaptive synergies for the design and control of the Pisa/IIT SoftHand. Int. J. Robot. Res. 33, 768–782.

Chautems, C., Tonazzini, A., Boehler, Q., Jeong, S.H., Floreano, D., Nelson, B.J., 2020. Magnetic continuum device with variable stiffness for minimally invasive surgery. Adv. Intell. Syst. 2, 1900086.

Chung, J., Heimgartner, R., O'Neill, C.T., Phipps, N.S., Walsh, C.J., 2018. ExoBoot, a soft inflatable robotic boot to assist ankle during walking: design, characterization and preliminary tests. In: 2018 7th IEEE International Conference on Biomedical Robotics and Biomechatronics (Biorob), 26–29 Aug. 2018, pp. 509–516.

Cianchetti, M., Ranzani, T., Gerboni, G., Nanayakkara, T., Althoefer, K., Dasgupta, P., Menciassi, A., 2014. Soft robotics technologies to address shortcomings in today's minimally invasive surgery: the Stiff-Flop approach. Soft Robot. 1, 122–131.

Cianchetti, M., Calisti, M., Margheri, L., Kuba, M., Laschi, C., 2015. Bioinspired locomotion and grasping in water: the soft eight-arm Octopus robot. Bioinspir. Biomim. 10, 035003.

Coad, M.M., Blumenschein, L.H., Cutler, S., Zepeda, J.A.R., Naclerio, N.D., El-Hussieny, H., Mehmood, U., Ryu, J.H., Hawkes, E.W., Okamura, A.M., 2020. Vine robots. IEEE Robot. Automat. Mag. 27, 120–132.

Coevoet, E., Morales-Bieze, T., Largilliere, F., Zhang, Z., Thieffry, M., Sanz-Lopez, M., Carrez, B., Marchal, D., Goury, O., Dequidt, J., Duriez, C., 2017. Software toolkit for modeling, simulation, and control of soft robots. Adv. Robot. 31, 1208–1224.

Dang, Y., Liu, Y., Hashem, R., Bhattacharya, D., Allen, J., Stommel, M., Cheng, L.K., Xu, W., 2021. SoGut: a soft robotic gastric simulator. Soft Robot. 8, 273–283.

Dawood, A.B., Fras, J., Aljaber, F., Mintz, Y., Arezzo, A., Godaba, H., Althoefer, K., 2021. Fusing dexterity and perception for soft robot-assisted minimally invasive surgery: what we learnt from Stiff-Flop. Appl. Sci. 11, 6586.

Dawood, A.B., Godaba, H., Ataka, A., Althoefer, K., 2020. Silicone-based capacitive E-skin for exteroception and proprioception. In: 2020 IEEE/RSJ International Conference on Intelligent Robots and Systems (IROS), 24 Oct.–24 Jan. 2021, pp. 8951–8956.

Dickey, M., 2017. Stretchable and soft electronics using liquid metals. Adv. Mater. 29, 1606425.

Diodato, A., Brancadoro, M., De Rossi, G., Abidi, H., Dall'alba, D., Muradore, R., Ciuti, G., Fiorini, P., Menciassi, A., Cianchetti, M., 2018. Soft robotic manipulator for improving dexterity in minimally invasive surgery. Surg. Innov. 25, 69–76.

Ebrahimi, A., Sadeghnejad, S., Vossoughi, G., Moradi, H., Farahmand, F., 2016. Nonlinear adaptive imped-ance control of virtual tool-tissue interaction for use in endoscopic sinus surgery simulation system. In: 2016 4th International Conference on Robotics and Mechatronics (ICRoM), IEEE, pp. 66–71.

Esfandiari, M., Sadeghnejad, S., Farahmand, F., Vossoughi, G., 2015. Adaptive characterisation of a human hand model during intercations with a telemanipulation system. In: 2015 3rd RSI International Confer-ence on Robotics and Mechatronics (ICRoM), IEEE, pp. 688–693.

Esfandiari, M., Sadeghnejad, S., Farahmand, F., Vossoughi, G., 2017. Robust nonlinear neural network-based control of a haptic interaction with an admittance type virtual environment. In: 2017 5th RSI inter-national conference on robotics and mechatronics (ICRoM), IEEE, pp. 322–327.

Fraś, J., Czarnowski, J., Maciaś, M., Główka, J., Cianchetti, M., Menciassi, A., 2015. New Stiff-Flop module construction idea for improved actuation and sensing. In: 2015 IEEE International Conference on Robotics and Automation (ICRA), IEEE, pp. 2901–2906.

Gandarias, J.M., Wang, Y., Stilli, A., García-Cerezo, A.J., Gómez-De-Gabriel, J.M., Wurdemann, H.A., 2020. Open-loop position control in collaborative, modular variable-stiffness-link (VSL) robots. IEEE Robot. Automat. Lett. 5, 1772–1779.

Gerboni, G., Brancadoro, M., Tortora, G., Diodato, A., Cianchetti, M., Menciassi, A., 2016. A novel linear elastic actuator for minimally invasive surgery: development of a surgical gripper. Smart Mater. Struct. 25, 105025.

Gerndt, R., Seifert, D., Baltes, J.H., Sadeghnejad, S., Behnke, S., 2015. Humanoid robots in soccer: robots versus humans in RoboCup 2050. IEEE Robot. Automat. Mag. 22 (3), 147–154.

Godaba, H., Vitanov, I., Aljaber, F., Ataka, A., Althoefer, K., 2020. A bending sensor insensitive to pressure: soft proprioception based on abraded optical fibres. In: 2020 3rd IEEE International Conference on Soft Robotics (RoboSoft), IEEE, pp. 104–109.

Grazioso, S., Di Gironimo, G., Siciliano, B., 2019. A geometrically exact model for soft continuum robots: the finite element deformation space formulation. Soft Robot. 6, 790–811.

Graule, M.A., Teeple, C.B., Mccarthy, T.P., Kim, G.R., Louis, R.C.S., Wood, R.J., 2021. Somo: fast and accurate simulations of continuum robots in complex environments. In: 2021 IEEE/RSJ International Conference on Intelligent Robots and Systems (IROS), IEEE, pp. 3934–3941.

Horvath, M.A., Wamala, I., Rytkin, E., Doyle, E., Payne, C.J., Thalhofer, T., Berra, I., Solovyeva, A., Saeed, M., Hendren, S., Roche, E.T., Del Nido, P.J., Walsh, C.J., Vasilyev, N.V., 2017. An intracardiac soft robotic device for augmentation of blood ejection from the failing right ventricle. Ann. Biomed. Eng. 45, 2222–2233.

Hu, Y., Liu, J., Spielberg, A., Tenenbaum, J.B., Freeman, W.T., Wu, J., Rus, D., Matusik, W., 2019. Chainqueen: a real-time differentiable physical simulator for soft robotics. In: 2019 International Con-ference on Robotics and Automation (ICRA), IEEE, pp. 6265–6271.

Ikuta, K., Ichikawa, H., Suzuki, K., 2002. Safety-active catheter with multiple-segments driven by micro-hydraulic actuators. In: International Conference on Medical Image Computing and Computer-Assisted Intervention, Springer, pp. 182–191.

In, H., Kang, B.B., Sin, M., Cho, K.-J., 2015. Exo-glove: a wearable robot for the hand with a soft tendon routing system. IEEE Robot. Automat. Mag. 22, 97–105.

Jafari, B., Shams, V., Esfandiari, M., Sadeghnejad, S., 2022. Nonlinear contact modeling and haptic char-acterization of the ovine cervical intervertebral disc. In: 2022 9th IEEE RAS/EMBS International Con-ference for Biomedical Robotics and Biomechatronics (BioRob), IEEE, pp. 1–6.

Javadi, M., Azar, S.M., Azami, S., Ghidary, S.S., Sadeghnejad, S., Baltes, J., 2017. Humanoid robot detection using deep learning: a speed-accuracy tradeoff. In: Robot World Cup. Springer, Cham, pp. 338–349.

Jitosho, R., Agharese, N., Okamura, A., Manchester, Z., 2021. A dynamics simulator for soft growing robots. In: 2021 IEEE International Conference on Robotics and Automation (ICRA), IEEE, pp. 11775–11781.

Khadivar, F., Sadeghnejad, S., Moradi, H., Vossoughi, G., Farahmand, F., 2017. Dynamic characterization of a parallel haptic device for application as an actuator in a surgery simulator. In: 2017 5th RSI Inter-national Conference on Robotics and Mechatronics (ICRoM), IEEE, pp. 186–191.

Khadivar, F., Sadeghnejad, S., Moradi, H., Vossoughi, G., 2020. Dynamic characterization and control of a parallel haptic interaction with an admittance type virtual environment. Meccanica 55 (3), 435–452.

Kim, J., Heimgartner, R., Lee, G., Karavas, N., Perry, D., Ryan, D.L., Eckert-Erdheim, A., Murphy, P., Choe, D.K., Galiana, I., Walsh, C.J., 2018. Autonomous and portable soft exosuit for hip extension assistance with online walking and running detection algorithm. In: 2018 IEEE International Conference on Robotics and Automation (ICRA), 21–25 May 2018, pp. 5473–5480.

Kim, J., Lee, G., Heimgartner, R., Arumukhom Revi, D., Karavas, N., Nathanson, D., Galiana, I., Eckert-Erdheim, A., Murphy, P., Perry, D., Menard, N., Choe Dabin, K., Malcolm, P., Walsh Conor, J., 2019. Reducing the metabolic rate of walking and running with a versatile, portable exosuit. Science 365, 668–672.

Kolbari, H., Sadeghnejad, S., Bahrami, M., Kamali, A., 2015a. Bilateral adaptive control of a teleoperation system based on the hunt-crossley dynamic model. In: 2015 3rd RSI International Conference on Robotics and Mechatronics (ICRoM), IEEE, pp. 651–656.

Kolbari, H., Sadeghnejad, S., Bahrami, M., Kamali, A., 2015b. Nonlinear adaptive control for teleoperation systems transitioning between soft and hard tissues. In: 2015 3rd RSI international conference on robotics and mechatronics (ICRoM), IEEE, pp. 055–060.

Kolbari, H., Sadeghnejad, S., Parizi, A.T., Rashidi, S., Baltes, J.H., 2016. Extended fuzzy logic controller for uncertain teleoperation system. In: 2016 4th International Conference on Robotics and Mechatronics (ICRoM), IEEE, pp. 78–83.

Kolbari, H., Sadeghnejad, S., Bahrami, M., Kamali, A., 2018. Adaptive control of a robot-assisted telesurgery in interaction with hybrid tissues. J. Dyn. Syst. Meas. Control. 140.

Kwok, K.W., Wurdemann, H., Arezzo, A., Menciassi, A., Althoefer, K., 2022. Soft robot-assisted minimally invasive surgery and interventions: advances and outlook. Proc. IEEE 110, 871–892.

Kwon, J., Park, J.-H., Ku, S., Jeong, Y., Paik, N.-J., Park, Y.-L., 2019. A soft wearable robotic ankle-foot-orthosis for post-stroke patients. IEEE Robot. Automat. Lett., 1.

Laoutid, F., Bonnaud, L., Alexandre, M., Lopez-Cuesta, J.-M., Dubois, P., 2009. New prospects in flame retardant polymer materials: From fundamentals to nanocomposites. Mater. Sci. Eng. R Rep. 63 (3), 100–125. https://doi.org/10.1016/j.mser.2008.09.002.

Lee, C., Kim, M., Kim, Y.J., Hong, N., Ryu, S., Kim, H.J., Kim, S., 2017a. Soft robot review. Int. J. Control. Autom. Syst. 15, 3–15.

Lee, G., Ding, Y., Bujanda, I.G., Karavas, N., Zhou, Y.M., Walsh, C.J., 2017b. Improved assistive profile tracking of soft exosuits for walking and jogging with off-board actuation. In: 2017 IEEE/RSJ International Conference on Intelligent Robots and Systems (IROS), 24–28 Sept. 2017, pp. 1699–1706.

Li, M., Ferguson, Z., Schneider, T., Langlois, T., Zorin, D., Panozzo, D., Jiang, C., Kaufman, D.M., 2020. Incremental potential contact. ACM Trans. Graph. 39.

Lin, H.T., Leisk, G.G., Trimmer, B., 2011. GoQBot: a caterpillar-inspired soft-bodied rolling robot. Bioinspir. Biomim. 6, 026007.

Loeve, A., Breedveld, P., Dankelman, J., 2011. Scopes Too Flexible… and Too Stiff. Pulse. vol. 1 IEEE, pp. 26–41.

Lunni, D., Giordano, G., Sinibaldi, E., Cianchetti, M., Mazzolai, B., 2018. Shape estimation based on Kalman filtering: towards fully soft proprioception. In: 2018 IEEE International Conference on Soft Robotics (RoboSoft), IEEE, pp. 541–546.

Ma, P., Du, T., Zhang, J.Z., Wu, K., Spielberg, A., Katzschmann, R.K., Matusik, W., 2021. DiffAqua. ACM Trans. Graph. 40, 1–14.

Manfredi, L., Capoccia, E., Ciuti, G., Cuschieri, A., 2019. A soft pneumatic inchworm double balloon (SPID) for colonoscopy. Sci. Rep. 9, 11109.

Marchese, A.D., Onal, C.D., Rus, D., 2014. Autonomous soft robotic fish capable of escape maneuvers using fluidic elastomer actuators. Soft Robot. 1, 75–87.

Mathew, A.T., Hmida, I.B., Armanini, C., Boyer, F., Renda, F., 2021. Sorosim: A Matlab Toolbox for Soft Robotics Based on the Geometric Variable-Strain Approach. arXiv preprint arXiv:2107.05494.

Mengüç, Y., Park, Y.-L., Pei, H., Vogt, D., Aubin, P.M., Winchell, E., Fluke, L., Stirling, L., Wood, R.J., Walsh, C.J., 2014. Wearable soft sensing suit for human gait measurement. Int. J. Robot. Res. 33, 1748–1764.

Mirmohammadi, Y., Khorsandi, S., Shahsavari, M.N., Yazdankhoo, B., Sadeghnejad, S., 2021. Ball path prediction for humanoid robots: combination of k-Nn regression and autoregression methods. In: Robot World Cup. Springer, Cham, pp. 3–14.

Munawar, A., Srishankar, N., Fischer, G.S., 2020. An open-source framework for rapid development of interactive soft-body simulations for real-time training. In: 2020 IEEE International Conference on Robotics and Automation (ICRA), IEEE, pp. 6544–6550.

Naughton, N., Sun, J., Tekinalp, A., Parthasarathy, T., Chowdhary, G., Gazzola, M., 2021. Elastica: a compliant mechanics environment for soft robotic control. IEEE Robot. Automat. Lett. 6, 3389–3396.

O'Neill, C.T., Phipps, N.S., Cappello, L., Paganoni, S., Walsh, C.J., 2017. A soft wearable robot for the shoulder: design, characterization, and preliminary testing. In: 2017 International Conference on Rehabilitation Robotics (ICORR), 17–20 July 2017, pp. 1672–1678.

Oz, M., Artrip, J., Burkhoff, D., 2002. Direct cardiac compression devices. J. Heart Lung Transpl. 21, 1049–1055.

Payne, C., Wamala, I., Bautista-Salinas, D., Saeed, M., Story, D., Thalhofer, T., Horvath, M., Abah, C., Del Nido, P., Walsh, C., Vasilyev, N., 2017. Soft robotic ventricular assist device with septal bracing for therapy of heart failure. Sci. Robot. 2, eaan6736.

Peters, J., Nolan, E., Wiese, M., Miodownik, M., Spurgeon, S., Arezzo, A., Raatz, A., Wurdemann, H.A., 2019. Actuation and stiffening in fluid-driven soft robots using low-melting-point material. In: 2019 IEEE/RSJ International Conference on Intelligent Robots and Systems (IROS), IEEE, pp. 4692–4698.

Peters, J., Anvari, B., Chen, C., Lim, Z., Wurdemann, H.A., 2020. Hybrid fluidic actuation for a foam-based soft actuator. In: 2020 IEEE/RSJ International Conference on Intelligent Robots and Systems (IROS), IEEE, pp. 8701–8708.

Polygerinos, P., Correll, N., Morin, S.A., Mosadegh, B., Onal, C.D., Petersen, K., Cianchetti, M., Tolley, M.T., Shepherd, R.F., 2017. Soft robotics: review of fluid-driven intrinsically soft devices; manufacturing, sensing, control, and applications in human-robot interaction. Adv. Eng. Mater. 19.

Polygerinos, P., Galloway, K.C., Sannan, S., Heman, M., Walsh, C., 2015a. EMG controlled soft robotic glove for assistance during activities of daily living. In: IEEE International Conference on Rehabilitation Robotics (ICORR).

Polygerinos, P., Wang, Z., Galloway, K.C., Wood, R.J., Walsh, C.J., 2015b. Soft robotic glove for combined assistance and at-home rehabilitation. Robot. Auton. Syst. 73, 135–143.

Proietti, T., O'Neill, C., Hohimer, C.J., Nuckols, K., Clarke, M.E., Zhou, Y.M., Lin, D.J., Walsh, C.J., 2021. Sensing and control of a multi-joint soft wearable robot for upper-limb assistance and rehabilitation. IEEE Robot. Automat. Lett. 6, 2381–2388.

Rateni, G., Cianchetti, M., Ciuti, G., Menciassi, A., Laschi, C., 2015. Design and development of a soft robotic gripper for manipulation in minimally invasive surgery: a proof of concept. Meccanica 50, 2855–2863.

Realmuto, J., Sanger, T., 2019. A robotic forearm orthosis using soft fabric-based helical actuators. In: 2019 2nd IEEE International Conference on Soft Robotics (RoboSoft), 14–18 April 2019, pp. 591–596.

Ren, L., Li, B., Wei, G., Wang, K., Song, Z., Wei, Y., Ren, L., Liu, Q., 2021. Biology and bioinspiration of soft robotics: actuation, sensing, and system integration. Iscience 24, 103075.

Renda, F., Giorelli, M., Calisti, M., Cianchetti, M., Laschi, C., 2014. Dynamic model of a multibending soft robot arm driven by cables. IEEE Trans. Robot. 30, 1109–1122.

Roberts, W.B., Tseng, K., Walsh, P.C., Han, M., 2010. Critical appraisal of management of rectal injury during radical prostatectomy. Urology 76, 1088–1091.

Robertson, M.A., Paik, J., 2017. New soft robots really suck: vacuum-powered systems empower diverse capabilities. Sci. Robot. 2, eaan6357.

Roche, E.T., Horvath, M.A., Wamala, I., Alazmani, A., Song, S.E., Whyte, W., Machaidze, Z., Payne, C.J., Weaver, J.C., Fishbein, G., Kuebler, J., Vasilyev, N.V., Mooney, D.J., Pigula, F.A., Walsh, C.J., 2017. Soft robotic sleeve supports heart function. Sci. Transl. Med. 9.

Rucker, D.C., Webster Iii, R.J., 2011. Statics and dynamics of continuum robots with general tendon routing and external loading. IEEE Trans. Robot. 27, 1033–1044.

Runciman, M., Darzi, A., Mylonas, G.P., 2019. Soft robotics in minimally invasive surgery. Soft Robot. 6, 423–443.

Rus, D., Tolley, M.T., 2015. Design, fabrication and control of soft robots. Nature 521, 467–475.

Sadati, S., Naghibi, S.E., Shiva, A., Walker, I.D., Althoefer, K., Nanayakkara, T., 2017. Mechanics of continuum manipulators, a comparative study of five methods with experiments. In: Annual Conference Towards Autonomous Robotic Systems. Springer, pp. 686–702.

Sadati, S.M.H., Naghibi, S.E., Shiva, A., Michael, B., Renson, L., Howard, M., Rucker, C.D., Althoefer, K., Nanayakkara, T., Zschaler, S., Bergeles, C., Hauser, H., Walker, I.D., 2020. TMTDyn: a Matlab package for modeling and control of hybrid rigid-continuum robots based on discretized lumped systems and reduced-order models. Int. J. Robot. Res. 40, 296–347.

Sadati, S.M.H., Naghibi, S.E., Shiva, A., Michael, B., Renson, L., Howard, M., Rucker, C.D., Althoefer, K., Nanayakkara, T., Zschaler, S., 2021. TMTDyn: a Matlab package for modeling and control of hybrid rigid–continuum robots based on discretized lumped systems and reduced-order models. Int. J. Robot. Res. 40, 296–347.

Sadeghnejad, S., Elyasi, N., Farahmand, F., Vossoughi, G., Sadr Hosseini, S.M., 2020. Hyperelastic modeling of sino-nasal tissue for haptic neurosurgery simulation. Scientia Iran. 27 (3), 1266–1276.

Sadeghnejad, S., Khadivar, F., Abdollahi, E., Moradi, H., Farahmand, F., Sadr Hosseini, S.M., Vossoughi, G., 2019a. A validation study of a virtual-based haptic system for endoscopic sinus surgery training. Int. J. Med. Robot. Comput. Assist. Surg. 15 (6), e2039.

Sadeghnejad, S., Farahmand, F., Vossoughi, G., Moradi, F., Sadr Hosseini, S.M., 2019b. Phenomenological tissue fracture modeling for an endoscopic sinus and skull base surgery training system based on experimental data. Med. Eng. Phys. 68, 85–93.

Sadeghnejad, S., Esfandiari, M., Farahmand, F., Vossoughi, G., 2016. Phenomenological contact model characterization and haptic simulation of an endoscopic sinus and skull base surgery virtual system. In: 2016 4th international conference on robotics and mechatronics (ICRoM), IEEE, pp. 84–89.

Selvaggio, M., Cognetti, M., Nikolaidis, S., Ivaldi, S., Siciliano, B., 2021. Autonomy in physical human-robot interaction: a brief survey. IEEE Robot. Automat. Lett. 6, 7989–7996.

Selvaggio, M., Fontanelli, G.A., Ficuciello, F., Villani, L., Siciliano, B., 2018. Passive virtual fixtures adaptation in minimally invasive robotic surgery. IEEE Robot. Automat. Lett. 3, 3129–3136.

Shi, C., Luo, X., Qi, P., Li, T., Song, S., Najdovski, Z., Fukuda, T., Ren, H., 2016. Shape sensing techniques for continuum robots in minimally invasive surgery: a survey. IEEE Trans. Biomed. Eng. 64, 1665–1678.

Shiva, A., Stilli, A., Noh, Y., Faragasso, A., De Falco, I., Gerboni, G., Cianchetti, M., Menciassi, A., Althoefer, K., Wurdemann, H.A., 2016. Tendon-based stiffening for a pneumatically actuated soft manipulator. IEEE Robot. Automat. Lett. 1, 632–637.

Siciliano, B., Siciliano, B., Khatib, O., Khatib, O., Siciliano, B., 2016. Springer Handbook of Robotics. Springer, Berlin.

Sohn, J.W., Kim, G.-W., Choi, S.-B., 2018. A state-of-the-art review on robots and medical devices using smart fluids and shape memory alloys. Appl. Sci. 8, 1928.

Soter, G., Conn, A., Hauser, H., Rossiter, J., 2018. Bodily aware soft robots: integration of proprioceptive and exteroceptive sensors. In: 2018 IEEE International Conference on Robotics and Automation (ICRA), IEEE, pp. 2448–2453.

Su, H., Yang, C., Ferrigno, G., De Momi, E., 2019. Improved human-robot collaborative control of redundant robot for teleoperated minimally invasive surgery. IEEE Robot. Automat. Lett. 4, 1447–1453.

Tatlicioglu, E., Walker, I.D., Dawson, D.M., 2007. New dynamic models for planar extensible continuum robot manipulators. In: 2007 IEEE/RSJ International Conference on Intelligent Robots and Systems, IEEE, pp. 1485–1490.

Tewari, A., Peabody, J., Sarle, R., Balakrishnan, G., Hemal, A., Shrivastava, A., Menon, M., 2002. Technique of da Vinci robot-assisted anatomic radical prostatectomy. Urology 60, 569–572.

Thalman, C.M., Lam, Q.P., Nguyen, P.H., Sridar, S., Polygerinos, P., 2018. A novel soft elbow exosuit to supplement bicep lifting capacity. In: 2018 IEEE/RSJ International Conference on Intelligent Robots and Systems (IROS), 1–5 Oct. 2018, pp. 6965–6971.

Tunay, I., 2013. Spatial continuum models of rods undergoing large deformation and inflation. IEEE Trans. Robot. 29, 297–307.

Umedachi, T., Vikas, V., Trimmer, B.A., 2016. Softworms: the design and control of non-pneumatic, 3D-printed, deformable robots. Bioinspir. Biomim. 11, 025001.

Veale, A.J., Staman, K., Van Der Kooij, H., 2021. Soft, wearable, and pleated pneumatic interference actuator provides knee extension torque for sit-to-stand. Soft Robot. 8, 28–43.

Wamala, I., Roche, E., Pigula, F., 2017. The use of soft robotics in cardiovascular therapy. Expert. Rev. Cardiovasc. Ther. 15.

Wang, H., Totaro, M., Beccai, L., 2018a. Toward perceptive soft robots: progress and challenges. Adv. Sci. 5, 1800541.

Wang, W., Cao, Y., Wang, X., Yu, L., 2020a. Closed-form solution of inverse kinematics for a minimally invasive surgical robot slave manipulator similar to da Vinci robot system. J. Eng. Sci. Med. Diagnos. Ther. 3.

Wang, X., Fang, G., Wang, K., Xie, X., Lee, K.-H., Ho, J.D., Tang, W.L., Lam, J., Kwok, K.-W., 2020b. Eye-in-hand visual serving enhanced with sparse strain measurement for soft continuum robots. IEEE Robot. Automat. Lett. 5, 2161–2168.

Wang, X., Lee, K.H., Fu, D.K.C., Dong, Z., Wang, K., Fang, G., Lee, S.L., Lee, A.P.W., Kwok, K.W., 2018b. Experimental validation of robot-assisted cardiovascular catheterization: model-based versus model-free control. Int. J. Comput. Assist. Radiol. Surg. 13, 797–804.

Wang, X., Li, Y., Kwok, K.W., 2021. A survey for machine learning-based control of continuum robots. Front. Robot. AI 8, 730330.

Webster, R.J., Jones, B.A., 2010. Design and kinematic modeling of constant curvature continuum robots: a review. Int. J. Robot. Res. 29, 1661–1683.

White, E., Case, J., Kramer, R., 2016. Multi-mode strain and curvature sensors for soft robotic applications. Sens. Actuators A: Phys. 253.

Wurdemann, H.A., Sareh, S., Shafti, A., Noh, Y., Faragasso, A., Chathuranga, D.S., Liu, H., Hirai, S., Althoefer, K., 2015a. Embedded electro-conductive yarn for shape sensing of soft robotic manipulators. In: 2015 37th Annual International Conference of the IEEE Engineering in Medicine and Biology Society (EMBC), 25–29 Aug. 2015, pp. 8026–8029.

Wurdemann, H.A., Stilli, A., Althoefer, K., 2015b. Lecture notes in computer science: an antagonistic actuation technique for simultaneous stiffness and position control. In: Intelligent Robotics and Applications. Springer.

Xu, J., Li, B., Lu, B., Liu, Y.-H., Dou, Q., Heng, P.-A., 2021. Surrol: An open-source reinforcement learning centered and DVRK compatible platform for surgical robot learning. In: 2021 IEEE/RSJ International Conference on Intelligent Robots and Systems (IROS), IEEE, pp. 1821–1828.

Yap, H.K., Lim, J.H., Nasrallah, F., Yeow, C.H., 2017. Design and preliminary feasibility study of a soft robotic glove for hand function assistance in stroke survivors. Front. Neurosci. 11, 547.

Yekutieli, Y., Sagiv-Zohar, R., Aharonov, R., Engel, Y., Hochner, B., Flash, T., 2005. Dynamic model of the octopus arm. I. Biomechanics of the octopus reaching movement. J. Neurophysiol. 94, 1443–1458.

Zhao, H., O'Brien, K., Li, S., Shepherd, R.F., 2016. Optoelectronically innervated soft prosthetic hand via stretchable optical waveguides. Sci. Robot. 1, eaai7529.

Further reading

Kriegman, S., Blackiston, D., Levin, M., Bongard, J., 2020. A scalable pipeline for designing reconfigurable organisms. Proc. Natl. Acad. Sci. U. S. A. 117, 1853–1859.

Alcaide, J.O., Pearson, L., Rentschler, M.E., 2017. Design, modeling and control of a SMA-actuated bio-mimetic robot with novel functional skin. In: 2017 IEEE International Conference on Robotics and Automation (ICRA), 29 May–3 June 2017, pp. 4338–4345.

Wang, X., Guo, R., Liu, J., 2019. Liquid metal based soft robotics: materials, designs, and applications. Adv. Mater. Technol. 4, 1800549.

CHAPTER 3

Rehabilitation robotics: History, applications, and recent advances

Soroush Sadeghnejad, Vida Shams Esfand Abadi, and Bahram Jafari
Bio-Inspired System Design Laboratory, Department of Biomedical Engineering, Amirkabir University of Technology (Tehran Polytechnic), Tehran, Iran

3.1 Introduction

Despite the introduction of the realm of robotics to the industry and its dramatic development in manufacturing industrial products, construction, military, and many other fields, medical robots with rehabilitation purposes have been progressing at a slower rate due to the numerous clinical trials required. However, given the growing demand for rehabilitation by an aging society and those recovering from strokes or injuries, familiarity with the cutting-edge global developments made in rehabilitation robotics is of great importance.

3.1.1 Overview

We may be on the verge of a new era, when the PC will get up off the desktop and allow us to see, hear, touch and manipulate objects in places where we are not physically present—Bill Gates in 2007 (Gates, 2007).

In the past 50 years, computer systems and robotics have come in handy in many areas as a means to assist humans. Robots have helped or entirely replaced humans in a wide range of chores, some as easy as daily human tasks and some as severe as advanced healthcare systems, sports, dangerous industrial and agricultural processes, and life-threatening circumstances such as firefighting or military operations (Mirmohammad et al., 2021; Boubaker, 2020). With the development of technology, an extremely pivotal invention that has lately mushroomed and improved the progress of robotics is artificial intelligence (AI). As AI has advanced, automation technology and machine learning approaches have been added to machines (Zandieh et al., 2021), resulting in smart equipment and efficient production with minimal human interference. In addition, advanced AI in the field of detecting human actions, cognitions, and emotions has been able to understand and even predict human behavioral patterns and perform well in response to human communications (Wang and Siau, 2019).

With all the advancements mentioned earlier, it is perceived that robots are now capable of improved interactions with humans and enhanced performance in human

Medical and Healthcare Robotics
https://doi.org/10.1016/B978-0-443-18460-4.00008-1

communities. One way to make use of this innovation is to help improve public healthcare and overcome medical challenges. Therefore, attention has been centered on robotic systems in the realm of medicine for diagnosis, treatment, surgery, and disease management (Loh, 2018; Sadeghnejad et al., 2020). Robots can also aid the elderly or incapacitated in carrying out their daily activities and injured patients in recovering faster. The application of robots in assisting the elderly or providing rehabilitation services to recovering patients is studied in the field of rehabilitation robotics. Yet, industrial robots have significantly outnumbered rehabilitation robots. Rehabilitation robots, on the other hand, have developed at a slower pace. The popularization of rehabilitation robots has faced several challenges, including their cost, maintenance, safety, special design, interface adaptation, and applying the proper treatment to each patient (Li et al., 2021). Typically, several clinical trials are required to successfully deal with the mentioned challenges. In addition, technical familiarity with robotic systems and specific understanding and knowledge of the patient's requirements are necessary.

Therefore, we aim to familiarize the reader with the realm of rehabilitation robotics in the sections of this chapter. First, the necessary definitions and general history of rehabilitation robotics are introduced in Section 3.2, and we present several prominent examples of medical robots with assistive and therapeutic applications in the past 50 years. In addition, a summary of their design criteria, control strategies, and human–robot interfaces is partly described. In Section 3.3, the applications of rehabilitation robots for physical therapy, including the upper and lower extremities, with their leading technologies are evaluated. In addition, a number of novel robots introduced in the past decade and recent control strategies, evaluation, and safety assessments are presented. Finally, we will conclude the entire chapter with potential future paths and challenges in the way of rehabilitation robots.

3.1.2 Taxonomy and terminology

This part acquaints the reader with the classifications, terminologies, and several definitions used in this chapter. Familiarity with the terms and meanings helps to read about the background of rehabilitation robotics more easily, and it is the first step to understanding the new topics in this field.

The field of rehabilitation robotics is usually categorized into two general topics: robotics for assistance and aid (assistive robotics) and robotics for therapy and recovery (therapeutic robotics). Assistive robotics are systems and devices that help patients who have movement disabilities due to age, injury, or stroke, to move their bodies and perform their daily tasks easier and faster. However, these robots do not necessarily engage the patient's nervous system to regain control of the disabled or paretic limb nor do they provide any feedback on the patient's recovery. These robots should be designed carefully to be highly convenient since the application of such robots for many patients is long term or even permanent. A well-known example of this group is the wheelchair. In contrast, therapeutic systems play at least one role in facilitating the recovery of patients. These systems are usually used for a predetermined and limited time period to focus on improving a particular function of a patient. Therapeutic robots can work alongside a therapist or with the patient's caregivers, thus resulting in faster therapy for the patient and less exhaustion and strain for the third person.

Rehabilitation robotics can be applied to both upper and lower body extremities. A comprehensive description of these two groups is presented in Section 3.3. The following are brief definitions of several topics frequently mentioned in this chapter or in other similar contexts, and it is necessary to be familiar with them.

- **Prosthetics and orthoses**: The main goal of this field is to provide the patient with an artificial limb to replace and function like the natural one (prosthetics), such as artificial parts and implants to replace the fingers, hands, feet, breasts, eyeballs, and teeth (Jafari et al., 2022a), or to support a limb for protection or correction (orthoses), such as dental, knee, and ankle-foot braces (Boubaker, 2020). These words are often mentioned in topics pertaining to medical robots. They are related to but are not the same as rehabilitation robotics.
- **Emotional therapy by robotics**: While rehabilitation systems for physical assistance and therapy provide upper and lower limb movement promotion, this group of robots motivates a positive change in the behavior of people with cognitive or emotional disorders, such as patients with Alzheimer's disease and children with autism (Yousif, 2020). They also encourage user interaction and provide good companionship.
- **Human–robot interfaces**: This is an important design parameter to consider when developing rehabilitation robots since the people who use such robots have disabilities or disorders, which may influence human-robot interactions (Udupa et al., 2021). Therefore, the robotic system must be simple and friendly to interact with. Among methods of improving user interface are advanced robot sensors, controller design, visual and graphical assistance, and the ability to detect voices and expressions.
- **Spinal cord injury (SCI)**: This is a neurological disorder and the main cause of disability among the youth in the United States. This can lead to paralysis in several or all limbs (tetraplegia) (Kyrarini et al., 2021).
- **Cerebral palsy (CP)**: Affecting almost 3 out of 1000 children who are born in the United States (Krebs et al., 2008), this condition is referred to as a group of diseases damaging muscle coordination and impairing physical movements. Additionally, many infants are diagnosed later when developmental delays and symptoms become clear. CP is usually caused by brain damage through fetal growth or during childbirth.
- **Stroke**: As the major cause of long-lasting or permanent disability, the incidence of a cerebral vascular accident after age 55 doubles every 10 years. More than 50% of stroke patients survive and seek recovery (Krebs et al., 2008).

3.2 The history of rehabilitation robotics

Although presenting every historical detail and mentioning all rehabilitation robots are not practical in a single chapter, in this part, we will chronologically introduce several early rehabilitation robots and technologies, which have become clinically and commercially distinguished.

3.2.1 Assistive robotics

Today, assistive robots with advanced frameworks and even virtual reality controllers are used in many fields, including medical assistance. However, the field of rehabilitation robotics was not much developed five decades ago. The earliest ideas of rehabilitation robots in the assistive category before 1960 came from master/slave manipulators (mechanical arms). Manipulators were primarily designed and used to deal with dangerous and harsh environments such as radioactive materials. The bilateral master/slave manipulator developed by R.C. Goertz and the General Electric "Handi-Man" designed by Mosher and Wendel were the forerunners of this field in the 1960s and before that (Corker et al., 1979). The idea was to maintain a resemblance to a human's arm to employ an easier master/slave control system. The easier control that resulted from several articulations encouraged engineers toward anthropomorphic designs. Incidentally, the evolutionary invention of the integrated circuit had just taken place in the early 1960s, and computers were being miniaturized. Therefore, the next step was to add an advanced computer-controlled system with a powered orthosis design, which could make the paralyzed arm of a patient move in the desired pattern. The Case Institute of Technology team performed such manipulations successfully in the 1960s. However, using the paralyzed patient's limb attached to the manipulator came with safety issues, as many patients had lost the sensibility of the mentioned limb due to traumas, and, thus, they did not possess a warning feedback system. This, added to the structural complexity of a design capable of fitting an injured upper limb, has led engineers to come up with remote manipulators known as telemanipulators. The Rancho Los Amigos Remote Manipulator with 7 degrees of freedom (DOFs) is an example of a robotic arm with a set of bidirectional tongue switches to remotely control the manipulator (Fig. 3.1). Such telemetry-controlled manipulators have also been mounted on wheelchairs. Engineers gradually realized that there was no need for an anthropomorphic design when there was no manipulator limb attachment. The design of such stand-alone medical manipulators with no human–robot physical contact inspired producers to hire such robots for rehabilitation applications.

As previously mentioned, with the growth of telemetry-controlled manipulators, the application of manipulator arms on wheelchairs sought the attention of many designers at the time. The earliest attempts to mount a robot on a wheelchair were made in the 1970s with an acceptable mechanical structure (Mason and Peizer, 1978). However, the hook

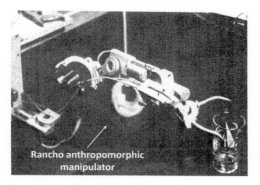
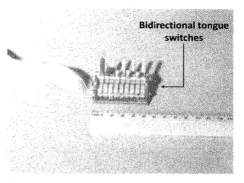

Fig. 3.1 The Rancho Los Amigos Remote Manipulator based on the study by Corker et al. (1979).

end effector was unsuccessful, and the control system was extremely basic. Later, an extremely simple user–robot interface was introduced as a push button. A patient suffering from muscular dystrophy could control the robot by pushing the button as an input (Zeelenberg, 1986). Such elementary designs were neither sophisticated nor the most convenient way to bear out every user's need. Yet, they are worth mentioning as they took the first steps toward the famous MANUS project (Fig. 3.2), initiated in the 1980s

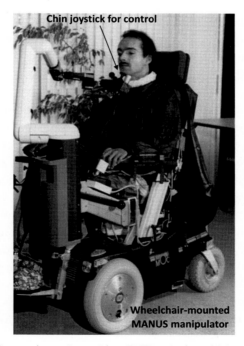

Fig. 3.2 The MANUS project and a patient with a C4/C5 spinal cord injury (Kwee, 1998).

by the efforts of Hok Kwee from the Institute for Rehabilitation Research in Hoens-broek, The Netherlands, and several other European institutes (Kwee, 1998). The initial design of the MANUS assistive robot came with a joystick and a small display for user control and feedback, respectively. Later, not only was it equipped with more advanced interfaces and control modes (at least 6 DOFs) but it also showed the way for several related research projects to follow. Raptor, developed by Mahoney from the Applied Resources Corporation, was a wheelchair-mounted arm that was commercially available in parallel with the MANUS project. Raptor designers tried to lower the costs and make the control system simpler with only 4 DOFs. The two wheelchair-mounted robotic arms were different in terms of functionality, structure, and cost. A comprehensive technical comparison of their kinematics and control systems for further review is available in the study by Alqasemi et al. (2005). Later designs focused on lightweight components, efficient control modes, and a high-quality graphical user interface.

Another key technology for wheelchair-mounted manipulators back in the 1980s was the navigation system, which then leaned toward being autonomous using the early methods of machine vision. An automated navigation system was particularly essential when it came to powered and smart wheelchairs. In this regard, ultrasonic sensors were able to calculate the distance of the objects within a specific visual field by sending a high-frequency sound wave. These sensor rings were attached to the wheelchair in great numbers and could provide an acceptable range of navigation. In the 2000s, advanced visual servo systems based on laser range finder sensors were developed and used for navigation purposes in wheelchairs. Although expensive, they were faster, more accurate, and more efficient in obstacle detection. Incidentally, Hephaestus Systems, using a sonar navigation system, became the pioneer of the 21st century smart wheelchairs (Simpson et al., 2002).

Aside from mobile and wheelchair-mounted arms, other assistive robots before the 2000s found their way to the market. A good example is the Oxford Intelligent Machines (OxIM), which produced several rehabilitation systems, many of which were followed and turned into other assistive robots in the 1980s and 1990s, such as the RT series, RAID project, and the MASTER system. They were stationary assistive systems designed to lay out a variety of human service tasks in a work environment. Perhaps, the most commercially notable assistive robot with a fixed workstation in the 1990s was the Handy-1 feeding robot (Boubaker, 2020), originally designed by the extensive research works of Mike Topping in the United Kingdom, thoroughly reviewed in the study by Topping (2001). Commercialized by Rehab-Robotics (UK), Handy-1 was affordable and well-received. It was first designed for the single purpose of feeding aid for people with disabilities such as children with CP. Nevertheless, later, extensions were developed to develop more multifunctional robots, able to perform or aid in painting, face shaving, and washing as well.

3.2.2 Therapeutic applications

The use of robotics in recovery and therapy has a shorter history than the application of assistive robots. There were very few research works to explore therapeutic robotics before 1990. However, a significant change of attention from assistive devices to therapeutic technologies is observed after the year 2000. From 1997 to 2007, the International Conference on Rehabilitation Robotics (ICORR) witnessed a 47% rise in submitted articles about therapeutic robots (Krebs et al., 2008). This is because the therapy provided by robots can be energy-, time-, and cost-efficient both for patients and clinicians. The remarkable developments through the history of therapeutic technologies and robots before the year 2000 are mentioned here, and Section 3.3 emphasizes the advancements of therapeutic robotics in the past two decades.

The beginning of robot application as a therapy tool for patients to regain their movement capabilities came with the idea of robots having actuators, force-controlled mechanisms, and programmable systems. In addition, any of the previously mentioned assistive robots, which later became capable of recording and calculating the level of recovery and progress of patients, could be categorized as early therapeutic systems. An RTX manipulator, for example, was equipped with an end effector with a haptic sensor to receive the patient's hit when the visual signal appeared and a software setup to record the response times and *ergo* assess the patient's recovery. In the 1980s, BioDex, developed by R. Krukowski, was a single-axis robot and could only exert resistance force against the patient's movement (passive motion). Afterward, another robotic system with two planar arms (2 DOFs for each) was introduced, featuring force sensors. Both of the mentioned robots were able to apply only passive motion (Senanayake and Senanayake, 2009; Khalili and Zomlefer, 1988).

In the 1990s, the famous MIT-MANUS project was started with Hogan and Krebs at the University of California, Berkeley, for stroke-suffering patients and neurological applications to improve the movement of the wrist, elbow, and shoulder (Krebs et al., 1998). MIT-MANUS featured a force-based impedance controller system, enabling it to detect any amount of movement by the patient and softly assist the movement as much as the patient needed but not more. It could also provide guidance when the patient deviated from the path of reaching the target (sensorimotor assistance). This therapy method was adapted and continued by later research groups. A subsequent work was a motion mirror image movement assistant, developed by Lum et al. (1999) in a collaborative effort between VA Palo Alto HCS and Stanford University. In addition to active-assisted and active-constrained modes, this robot offered a bilateral mode, by which the patient tries to make a mirror image movement using both hands. A digitizer measures the contralateral arm movement, and the robot assists the paretic arm to the mirror image position. The forces and torques applied to the robot are measured by several sensors. Another similar work in the late 1990s was known as the Assisted Rehabilitation and

Measurement Guide (ARM-GUIDE) developed in a joint effort between the Rehabilitation Institute of Chicago and the University of California, Irvine (Islam et al., 2020; Reinkensmeyer et al., 2014). This robot constrained the arm in a linear movement path but could be adjusted to different elevation angles. It was claimed to be useful for exact checking the arm injury and evaluating the effect of targeted therapy on the specific affliction.

The main goal of early lower extremity therapy was to design a system to support the body weight and maintain the patient's balance while the therapist encouraged or assisted the patient in moving the paretic limb. Later, such systems were able to obviate the need for subjecting patients to repetitive practice by overstrained and frustrated therapists. Two of the remarkable attempts made in this category were the Gait Trainer and the Lokomat project. The Gait Trainer was designed by Hesse's group to be a more efficient alternative to treadmill walking (Hesse and Uhlenbrock, 2000). It provided gait-like movement assistance, including swing and stance phases with a 40%–60% between them. The gait movement support could be complete or partial, and the patient's center of mass is vertically and horizontally controlled by the machine. Using this system, a hemiplegic person who was almost unable to move could perform the training with slight help from only one therapist, whereas, usually, at least two therapists are required to support such a patient in treadmill walking. In contrast to the Gait Trainer, the Lokomat system contained a treadmill, but a driven gait device was attached to the patient's legs that could be moved in a natural walking style on the moving treadmill. The earliest Lokomat prototype, designed by Colombo's group in the year 2000, had several actuators set on the knee and hip and was controlled by an adaptive position controller (Colombo et al., 2000). The driven gait device enabled the therapist to recognize the level of paresis and obtain the patient's gait pattern. The more recent designs for lower extremity therapy are focused on programmability, adjustability, being stand-alone, suitability for wearing, and ability to be used at home.

Many of these rehabilitation devices have been further modified and developed to provide the best treatment protocol to achieve better clinical outcomes. Other novel rehabilitation systems and more recent improvements of the already mentioned robots with further technical details are elaborated in Section 3.3.

3.3 Applications and recent advances

Today, rehabilitation is used in various fields in order to improve performance and restore ability in patients with movement disorders. Several problems such as the availability of therapists, the duration of treatment sessions, and the cost of rehabilitation tools have made researchers work on the development and improvement of robots that are able to perform the rehabilitation process, and, as a result, during the last decade, different robots have been developed to aid in this course. Yet, another important challenge to address remains the capability to provide sufficient support and practice for handling more complicated human activities involving multiple tasks simultaneously. Even in the assistive category, highly specific tasks are possible for patients to perform with the help of assistive robots, such as reaching with an arm or walking on a treadmill. In other words, such supportive devices provide help with only a portion of the damage while being incapable of assisting other functions with the impaired limb (arm, leg, etc.) at the same time. However, daily activities, such as going to the bathroom, climbing stairs, getting dressed, and making and having breakfast, undoubtedly require a vast range of tasks.

The design of rehabilitation robots faces different challenges. The lightness of a robot, its flexibility, and having a mechanism to perform normal muscular movement are among the essential parameters in the design of such robots (Li et al., 2021). In addition, the DOF of the robot and safety precautions are among the other important factors that should be considered. In this section, we will look at the recent advancements in the field of rehabilitation robotics in both upper and lower limb rehabilitation. Then, we will review notable control techniques and safety factors associated with rehabilitation robots.

Upper and lower limb rehabilitation robots play an important role in recovery and help patients with incapacitations and disorders. Several studies have shown that repetitive movements for injured limbs allow the brain to create new neural pathways and lead to regaining control of motion functions since repetitive exercises reactivate neural pathways in the damaged limb (Kyrarini et al., 2021). In addition, with the growth of the aging population, the need of people to have a nurse by their side has increased, and, therefore, upper/lower limb rehabilitation robots in the assistive category are suitable solutions to the problem of the insufficient number of available nurses in this field. In recent years, researchers have developed different types of upper limb rehabilitation exoskeleton robots (ULRERs) and lower limb rehabilitation exoskeleton robots (LLRERs), which can provide movement aids and therapy for different joints, assessment of upper limb function, gait, and daily activities.

3.3.1 Upper extremity

In recent years, many efforts have been made in the advancement of upper limb rehabilitation. One of the advances in ULRERs is a hand exoskeleton robot that was engineered by the researchers at the University of Berlin in Germany to improve hand movements. The control algorithm of this exoskeleton robot employs the principle of blind source separation to provide accurate movement for the joints of the fingers (DiCicco et al., 2004). In addition, a number of ULRERs employ artificial neural network control and are designed and used for patients with weak motion abilities by the interpretation of flexion and extension of the elbow and supination and pronation of the forearm. These robots amplify the EMG signals after extraction from seven arm muscles, and the average absolute value, the number of zero-crossings, and the length of the waveform are among the features used in the neural network control algorithm (Almassri et al., 2018).

Another advancement in the upper extremity rehabilitation category is the ARM-GUIDE robot, which was mentioned earlier in Section 3.2. This upper limb rehabilitation robot is used for training and evaluating the upper limb function, which employs the principle of reaching as a therapeutic technique in which the patient's arm is tied to a splint, and then the patient is told to reach for things (Reinkensmeyer et al., 2014).

Among other advances made in the field of upper limb rehabilitation robots, the design of the ARMin robot is a noteworthy one, which has 7 DOFs. These include three freedom levels for shoulder rotation: flexion and extension of an elbow, flexion and extension of the wrist, and supination and pronation of the forearm. None of the robot joints are beyond the limits of the human body. An ARMin robot contains a chair with a robotic arm to which the patients attach their hands and fit the length of the robotic arm to the desired level (Fig. 3.3). In addition, this robot contains many sensors, which has enabled it to be used as a system to identify movement defects (Keller et al., 2015; Islam et al., 2020).

The therapy Wilmington robotic exoskeleton (T-WREX) is another ULRER with 5 DOFs that was designed for patients with major weakness in the arm, and it provides concentrated training. This robot has two links secured to the upper arm and forearm. T-WREX possesses an antigravity feature, which can make the patient experience a floating arm in the air (Rehmat et al., 2018).

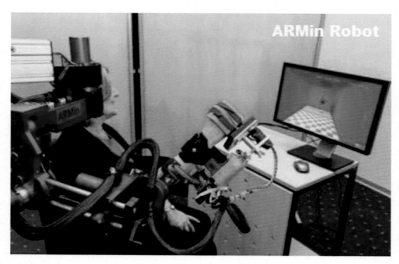

Fig. 3.3 An ARMin rehabilitation robot (Keller et al., 2015).

3.3.2 Lower extremity

LLRERs combine mechanical power devices and AI to help patients with lower limb dysfunction and improve muscle mobility, which causes the repair of physical disorders. The knee is one of the main and most vulnerable joints of the body, and its performance has a significant impact on a person's walking and the quality of daily activities. Knee exoskeleton robots can improve leg movements and provide the necessary ability to move the lower limb. One of the recent advancements in LLRERs is the development of a knee exoskeleton robot using a linkage mechanism that uses an inverse dynamic model to calculate the distribution of torque and force in the knee joint and initial angles of the joint as the inputs to predict sit-to-stand trajectory (Kamali et al., 2016; Kardan and Akbarzadeh, 2017) (Fig. 3.4). Experimental results and clinical studies have shown that this robot is remarkably effective and efficient for patients with knee injuries during sitting-to-standing activities. One of the latest developments in this field is the wearable soft knee exoskeleton. The IMU control system in this robot can measure joint angles of the knee and reduces the body's metabolic consumption by about 6.8% by helping the movement of the knee while walking. The remarkable point in the development of this robot is that the lower limb's actual movement and the patient's movement intention are not uniform; therefore, a combination of these two technologies can improve the function of the knee exoskeleton and make it safer for patients to use (Zhang et al., 2020).

Recent developments in ankle joint rehabilitation exoskeleton robots are the design and presentation of an ankle-foot orthosis (AFO), which allows users to adjust the stiffness of the orthosis. The patient can perform various activities of the lower limbs entirely (Li et al., 2020). Gait rehabilitation robots help patients with gait disorders to walk normally. An advancement of the gait rehabilitation exoskeleton is the design of novel force field control for walking, which measures the central position of the ankle joint trajectory by a motion capture system in order to guide the movement to the target path in a 3D

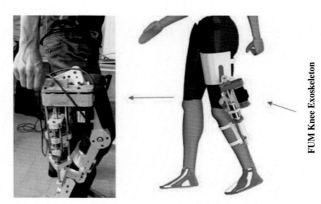

Fig. 3.4 An FUM knee exoskeleton (Akbarzadeh and Kardan, 2021; Kardan and Akbarzadeh, 2017).

space (Shi et al., 2019). Clinical studies and investigations have shown that the designed system can effectively realize motion control in different modes. In these robots, actuators can improve the adaptive control ability of robots, and, thus, robots can create an acceptable posture that is much more efficient and beneficial to patients (Li et al., 2017). The design of LLER robots with 3 DOFs is one of the remarkable developments in gait rehabilitation exoskeleton robots. These robots have a simple structure and are adjustable with the hip, knee, and ankle of patients of different heights (Wu et al., 2016). The design and presentation of a gait rehabilitation exoskeleton system can be used to customize and also modify gait and verify different modes in the rehabilitation process, which is one of the most unprecedented advancements. This prototype can also be employed to specify a personalized gait mode (Fig. 3.5).

A hip exoskeleton robot can increase the stability and strength of the trunk in patients and also decrease metabolic consumption (Zhou et al., 2021). Many researchers have worked on the development of hip exoskeleton robots in recent years. A hip rehabilitation exoskeleton uses a torque field control method to realize the proper state of the joint with 3 DOFs. It is extremely effective in the hip rehabilitation process and can be used as a reference control for rehabilitation robots. NREL is another hip exoskeleton robot controlled by a series of elastic actuators and can help in the balance of patients with muscle weakness and make them actively participate in walking (Fig. 3.6) (Zhang et al., 2017).

Lokomat is a new version of a rehabilitation robot, which can simulate physiological gait trajectory and adjust the speed of the treadmill accurately to adapt to the patient's gait and also combine functional therapy exercises with a thorough patient assessment and adjust its performance according to the patient's conditions and the assessment made

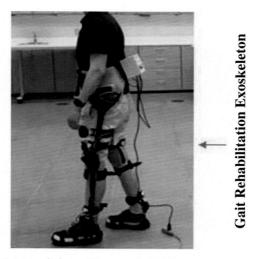

Fig. 3.5 A gait rehabilitation exoskeleton (Rose et al., 2020).

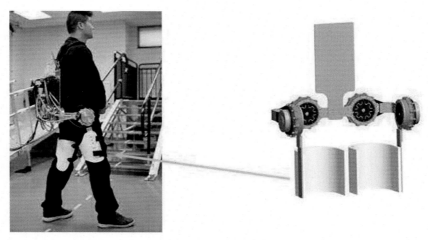

Fig. 3.6 An NREL rehabilitation exoskeleton (Zhang et al., 2017).

(Singh et al., 2016). Recently, in the United States, Lokomat has become accessible at 25 rehabilitation centers (Kyrarini et al., 2021).

Ekso Bionics and ReWalk are two renowned wearable exoskeletons that create the necessary force for patients' legs during rehabilitation practice (Naro et al., 2022). They have relatively similar designs to help the patient stand and walk during rehabilitation. A comparison of these two LLRERs is available in the study by Hong et al. (2020).

It is worth mentioning that apart from the physical assistive and therapeutic robots introduced here, socially assistive robots focus on patients' mental and psychological improvements. The target patient population is usually children with autism and people with depression, communication problems, and other psychological disorders (Yousif, 2020). Furthermore, mental and physical care was provided by specific robots during the COVID-19 pandemic. Their services included remotely detecting body temperature, nursing tasks, and disinfection (Mišeikis et al., 2020). Table 3.1 summarizes further therapeutic and assistive robots from recent publications, several of which were already mentioned in this chapter, and allows a comparison among these platforms. The reader is encouraged to look up the mentioned references for a more detailed elaboration of the listed systems.

3.3.3 Control techniques

In Section 3.2.1, master/slave control systems were briefly mentioned and described on early manipulators. These control systems force a patient to go on a predetermined path and motion regardless of the patient's desire and active voluntary determination. In such position controllers, the patient is kept passive, his/her active contribution will be disregarded, and any deviation from the preestablished pattern will not be permitted by the robot. This scenario is no longer recommended for the learning process since motor

Table 3.1 An outline of rehabilitation robots in the recent literature.

Device	Robot category	Tasks	Remarks	References
Soft knee wearable exoskeleton	Lower limb assistive	Helping the knee joint move during walking	Control system, including gait estimation model and knee torque model	Zhang et al. (2020)
Adaptive ankle exoskeleton	Lower limb assistive	Walking assistance	Adapting to walking tasks at different speeds and loads using novel series elastic actuators	Shao et al. (2019)
Lokomat	Lower limb therapeutic	Walking rehabilitation	Supporting full body weight and providing customizable walking patterns	Eguren and Contreras-Vidal (2020)
ARMin	Upper limb therapeutic	Helping shoulder rotations	Making natural shoulder movements possible and providing the CR mechanism	Islam et al. (2020)
ReWalk	Lower limb therapeutic	Gait rehabilitation	FDA-approved for home and community use/specialized in walking swing pattern	Hong et al. (2020)
Ekso	Lower limb therapeutic	Gait rehabilitation	Footplates equipped with sensors providing stepping patterns and foot trajectory	Hong et al. (2020)
Burt	Upper limb therapeutic	Increasing the motivation in repetitive therapy sessions	Game-based with an end effector	Kyrarini et al. (2021)
Barrett WAM	Upper limb therapeutic	Assisting upper limb movement	Game-based with 4- and 5-DOF configurations	Baritz (2020) and Kyrarini et al. (2021)

Table 3.1 An outline of rehabilitation robots in the recent literature—cont'd

Device	Robot category	Tasks	Remarks	References
Baxter	Assistive	Providing assistance with personalized dressing	Multitask hierarchical control strategy	Zhang et al. (2019)
JACO-based feeding robot	Assistive	Assisting with drinking and eating tasks	Providing improved bite acquisition and transfer	Gallenberger et al. (2019)
ARNA	Assistive in hospitals	Walking patients/ fetching objects/ monitoring patient's health	Novel controller frameworks	Das (2019)
Humanoid TUGs	Assistive in hospitals	Delivery of medications/ providing hospital supplies	Human body-shaped/easy human–robot interactions	Rivas et al. (2021) and Sowmiya et al. (2022)
Lio	Assistive	Delivery of items/ entertaining and monitoring patients	Interacting with humans with high safety	Mišeikis et al. (2020)
Nao	Socially assistive	Providing an assistant tutor for an autistic/ physiotherapeutic/ teaching assistant for children with autism	Improving the learning process	Yousif (2020)

learning progress occurs when deviations in movement and making errors are possible. Nowadays, applications using "assistance-only-as-needed" control systems are trending (Asl et al., 2020). These patient-cooperative controllers allow deviations in a limb movement from a predefined repetitive course, letting the patient actively participate in the movement and trajectory selection. Furthermore, the feeling of reward and success as a result of the patient's control over the robot will increase the motivation and take full advantage of the therapeutic effect. This strategy provides sufficient aid to let users perform the task while decreasing the aid according to the instantaneous performance measurement of the user in order to inspire the patients to pick up and complete the task on their own. When using "assistance-only-as-needed" control algorithms, there are time instants or longer periods during which the patient should not feel the existence of the robot, a behavior known as "robot transparency." To achieve such transparency, it is best to minimize the mass of the robot. Force control systems using admittance or impedance closed-loop control concepts is another approach to enhance transparency (Lee et al., 2022; Verdel et al., 2021).

Furthermore, to improve the patient's motivation during the rehabilitation process, strategies are sought to engage in mental activities in order to induce a feeling of "fun" and avoid boredom and frustration. The dopamine release as a result of this bio-cooperative control technique will contribute to improving neuroplasticity and hence more chances of treatment success (Atashzar et al., 2021). A good approach is to engage the patient in fun activities through computer games and virtual reality techniques, which have many applications in medical and surgical training and procedures (Alves et al., 2022; Jafari et al., 2022b). The mental engagement of the user can even be assessed in the control loop as psycho-physiological quantities. One way to do that is by measuring nervous system responses such as heart rate, skin conductance response, skin temperature, joint torques, and breathing rate.

Another aspect of robot control design is introducing a level of autonomy to the system in order to improve human-robot interactions and rehabilitation effectiveness (Garcia-Gonzalez et al., 2022). Semiautonomous machines can remove the physical and mental load from the patient and pass them to the robot. Sharing control between the patient and the robot is the main challenge of such semiautonomous solutions. The machine must provide aid when needed and yet not take control when it is not called for, except for safety issues. In addition, the patient's performance will change, and it is difficult to use such a sharing control system on them in the long run. Another challenge to address toward a better human-robot interface is the type of input to the device by the patient since the robot autonomy and user do not always share the same space to insert control signals. Therefore, it is often problematic to decode the patient's intent correctly for a high-dimensional output space (Garcia-Gonzalez et al., 2022; Udupa et al., 2021). To overcome such challenges, advances in sensing and processing tools have come in handy. Some of these tools are electromyographic data, haptic detectors, strain gauges, and pattern recognition techniques. Robot skins and touch sensing are notable examples of developing technologies in this regard. A list of further control systems of recent rehabilitation robots is reviewed by Zhou et al. (2021).

3.3.4 Evaluation standards

To evaluate the final product of a rehabilitation service provider, a few criteria must be considered, the most significant of which is cost. This is due to the fact that a robot's cost is affected by many other factors, such as accessibility, safety, material, and other design criteria (Li et al., 2021). Moreover, cost affects the type and size of the audience who will buy the service. Some products can only be bought by large rehabilitation and medical centers, whereas others might be affordable by ordinary people. In this regard, third-party providers such as health insurance companies can aid in alleviating the cost to their users. Cost is not the only parameter affecting the market and buyer size. An affordable device with a wide range of needs that has to be maintained at an extremely special location or operated only by the unique knowledge of a trained hand will receive limited customers. Therefore, accessibility is another noteworthy criterion that must be mentioned. A device that can be used simply at home or in any nearby hospital is regarded as easily reached and accessible. Plus, a rehabilitation system or robot with a constant need to repair or calibrate is costly and unavailable, especially when it takes too long to be repaired or replaced. Hence, maintainability becomes another issue, and such systems must be designed in a way that minimizes the need for repair and the frequency and time of calibration. In this regard, since the COVID-19 pandemic and global isolation, new rehabilitation systems have been trying to develop methods to deliver remote assistance, supervision, assessment, and other rehabilitation services due to the need of stroke-recovering patients and the elderly for constant training to improve neural plasticity. Cloud telehealth and digital health technologies are the recent solutions for the distant delivery of such services (Atashzar et al., 2021).

Furthermore, a patient's needs and ability levels may change as the rehabilitation process and training are performed regularly. Thus, the necessity for a new system will be felt if the system is not adaptable. Moreover, the more people with whom the rehabilitation device is compatible and adjustable, the more popular and useful it becomes for a wide range of people, and hence the adaptability factor will be promising. On the other hand, if the device is customized for only one person, the cost of such a design will be significantly affected. A design for a specific group, however, justifies the design and manufacturing cost. Some recent methods of enhancing the adaptability level in rehabilitation robots are assessment automation, machine learning, and the development of user profile models by enhanced feedback systems (Martín et al., 2020).

Another key criterion, especially in the field of neurorehabilitation, is the time during which the patient will see acceptable results and recovery. The shorter the time of rehabilitation, the more favorable is the system. However, if the patient loses positive results after the rehabilitation period, a condition commonly known as patient relapse, the performance of the rehabilitation device will be regarded as unsatisfactory. Therefore, yielding suitable results incorporates both reducing the rehabilitation time period and obviating the need for reutilizing the exercise in the future (Fernández-Vázquez et al., 2021).

In addition to the mentioned criteria, there are environmental and safety concerns that need to be emphasized. A rehabilitation robot must neither harm the environment nor hurt the patient using it (Boubaker, 2020). A device that is reusable or able to be recycled is friendlier to the environment. Moreover, rehabilitation robots or systems, especially wearable ones, should not enforce too heavy loads on the human body, which would otherwise cause other health complications. Therefore, the material must be lightweight, and the structure needs to be safe and reliable. Safety features are imperative and will be described in more detail in the following.

3.3.5 Safety assessments

Safety assessments of rehabilitation robots are possible at two stages. The first one is during the development process, and the second one occurs when the product is in use by the patients. While the latter is evaluated using constant monitoring and reports of adverse events in clinical trials (Bessler et al., 2020), the former faces several difficulties since there are few guidelines and directions regarding the safety assessments of rehabilitation robots in the development phase. When it comes to robots collaborating with humans, there are typical engineering methods of safety design aspects, such as model-based systems using software packages, which ground the assessment on knowing the system, the environment, and the people who will use the system (Saenz et al., 2020). In other words, the standards proposed for such safety assessments vary for different applications and domains. For instance, safety measures for designing an industrial or agricultural system might be centered on developing biosensors to avoid soil and water contamination, whereas a rehabilitation robot must support sensors to measure pressures and watch the patient's weight at times and prevent falling (Al-Awwal et al., 2022; Bessler et al., 2021). Different scopes and unsynchronized domains may trouble roboticists applying the correct standards to their systems. A seemingly suitable approach to deal with such a challenge is proposing safety protocols applicable to multiple domains and applications. The idea is to reduce the risk of a particular hazardous event by suggesting the right corresponding safety protocol, called a safety skill. This method offers engineers a solid conceptual outline for accounting risk mitigation measures as relevant guidelines and protocols for various domains. Identifying hazards is the first step for validating risk-lowering techniques. The most frequent adverse event involves soft tissue injuries, including skin abrasions, skin lesions, irritations, reddening, bruising, and pain (Bessler et al., 2020). The second most frequent adverse events are musculoskeletal ones such as lower back discomfort, muscle pain, joint pain, tibial fracture, and tendinopathy. Adverse events usually happen as a result of too many high forces imposed on the user. Rehabilitation robots are supposed to apply these forces to the patient's body through one or several contact points, such as footplates, cuffs, and harnesses, to accomplish their function of assisting a particular movement. The mechanics of a human–robot interface can be different in ULRERs and LLRERs due to different load-bearing abilities in the upper and lower extremities, and, thus, the robot's design affects the hazard under the study. However, some classifications of hazards are expected to be applicable to many common rehabilitation robots. Dangerous forces at the device–skin interface are categorized into normal forces causing excessive pressure and shear forces resulting in friction. The musculoskeletal system will be involved when the exerted force passes through the skin and soft tissues, in which case the movement of the whole body is expected. However, if the forces are too high, then the musculoskeletal system itself can be seriously injured. Examples of such hazardous forces include misalignment and exceeding the physiological motion range. A safety skill can be assigned to each hazard. For example, regarding a normal or sheering force that exceeds the safety limits, the safety skill is to limit the restraining energy exerted on the user by the rehabilitation robot. For a misalignment hazard, maintaining proper alignment with the help of the robot is a proper safety skill. In addition, limiting the range of robotic arm movements using physical stoppers is the right safety skill to avoid exceeding the normal human motion range (Islam et al., 2020). A more comprehensive list of adverse events, robot parts involving the hazard, potential injuries, and recommended safety skills is available in the review study by Bessler et al. (2021).

It is worth mentioning that in addition to patients and users, safety measures should be employed for bystanders and other people who are in long-term contact with a user of a rehabilitation robot. For instance, an adverse event may occur when a therapist is handling the training course of the patient adjacent to the robot. Furthermore, there are occasions when children or pets jump on the robot's route or put a finger on an unsafe part of the robot when it is moving. Therefore, such hazardous situations must be considered by the robot developers. Force and power-limiting strategies might help in the design. Another way to avoid or minimize hazards in the development phase is through safety by design. Robots must be designed in such a way so as to take in inherent safety as much as possible. Adding a level of redundancy, such as using extra sensors, and compensation mechanisms, like employing passive joints or lightweight material choices, are common methods concerning safety by design (Bessler et al., 2021).

3.4 Conclusions

This study presented the taxonomy, history, and application of rehabilitation robotics and reviewed the latest state-of-the-art research in this field. The prominent examples of assistive and therapeutic robots were reviewed along with the story of their development track in history. The challenges in the way of efficient design of rehabilitation devices were briefly reviewed, and several recent robots were outlined for the upper and lower extremities.

Clinical results and surveys showed that rehabilitation robots play a significant role in restoring the function of injured limbs in patients. In addition, the use of rehabilitation robots is effective in planning the rehabilitation process of patients with mental and physical disorders, and their use is greatly suitable in terms of cost, duration of treatment sessions, required tools, and the availability of therapists. Regarding the control techniques, the master/slave control method is being replaced by systems that are able to keep humans in the loop. The recent advancements in the field of control are focused on creating a fun environment using computer and virtual reality techniques, improving robot transparency, and introducing autonomy and machine learning, as the mentioned techniques are suggested to actively contribute to better therapy results. Affordability, accessibility, compatibility, rehabilitation time, and avoiding patient relapse are a few criteria that need to be considered when evaluating the ultimate rehabilitation system. Another important criterion is safety, the assessment of which is a challenge during the development phase. The literature recommends that more safety protocols should be proposed for several generic applications to avoid a wide range of hazards. The strategy of safety by design and safety measures for bystanders and people in proximity are also encouraged.

Acknowledgments

We are grateful to the Bio-Inspired System Design Laboratory at the Amirkabir University of Technology (Tehran Polytechnic).

Conflict of interest statement

The authors confirm that this book chapter and its corresponding research work involve no conflict of interest.

References

Akbarzadeh, A., Kardan, I., 2021. FUM Center of Advanced Rehabilitation and Robotics Research. Available at: https://www.fumrobotics.ir/projects/fumkneeexoskeleton.

Al-Awwal, N., Masjedi, M., El-Dweik, M., Anderson, S.H., Ansari, J., 2022. Nanoparticle immunofluorescent probes as a method for detection of viable *E. coli* O157: H7. J. Microbiol. Methods 193 (106403).

Almassri, A.M., Wan Hasan, W.Z., Ahmad, S.A., Shafie, S., Wada, C., Horio, K., 2018. Self-calibration algorithm for a pressure sensor with a real-time approach based on an artificial neural network. Sensors 18 (8), 2561.

Alqasemi, R.M., McCaffrey, E.J., Edwards, K.D., Dubey, R.V., 2005. Analysis, evaluation and development of wheelchair-mounted robotic arms. In: 9th International Conference on Rehabilitation Robotics, 2005. ICORR 2005, IEEE, pp. 469–472.

Alves, T., Gonçalves, R.S., Carbone, G., 2022. Serious games strategies with cable-driven robots for rehabilitation tasks. In: International Workshop on Medical and Service Robots, Springer, pp. 3–11.

Asl, H.J., Yamashita, M., Narikiyo, T., Kawanishi, M., 2020. Field-based assist-as-needed control schemes for rehabilitation robots. IEEE/ASME Trans. Mechatron. 25 (4), 2100–2111.

Atashzar, S.F., Carriere, J., Tavakoli, M., 2021. How can intelligent robots and smart mechatronic modules facilitate remote assessment, assistance, and rehabilitation for isolated adults with neuro-musculoskeletal conditions? Front. Robot. AI 8 (610529).

Baritz, M.I., 2020. Analysis of behaviour and movement of the upper limb in the weights handling activities. Procedia Manuf. 46 (850–856).

Bessler, J., Prange-Lasonder, G.B., Schaake, L., Saenz, J.F., Bidard, C., Fassi, I., Valori, M., Lassen, A.B., Buurke, J.H., 2021. Safety assessment of rehabilitation robots: A review identifying safety skills and current knowledge gaps. Front. Robot. AI 8 (602878).

Bessler, J., Prange-Lasonder, G.B., Schulte, R.V., Schaake, L., Prinsen, E.C., Buurke, J.H., 2020. Occurrence and type of adverse events during the use of stationary gait robots—a systematic literature review. Front. Robot. AI, 158.

Boubaker, O., 2020. Medical robotics. Control Theory Biomed. Eng., 153–204.

Colombo, G., Joerg, M., Schreier, R., Dietz, V., 2000. Treadmill training of paraplegic patients using a robotic orthosis. J. Rehabil. Res. Dev. 37 (6), 693–700.

Corker, K., Lyman, J.H., Sheredos, S., 1979. A preliminary evaluation of remote medical manipulators. Bull. Prosth. Res. 10 (32), 107–134.

Das, S.K., 2019. Adaptive Physical Human-Robot Interaction (PHRI) With a Robotic Nursing Assistant.

DiCicco, M., Lucas, L., Matsuoka, Y., 2004. Comparison of control strategies for an EMG controlled orthotic exoskeleton for the hand. In: IEEE International Conference on Robotics and Automation, 2004. Proceedings. ICRA'04. 2004, IEEE, pp. 1622–1627.

Eguren, D., Contreras-Vidal, J.L., 2020. Navigating the FDA medical device regulatory pathways for pediatric lower limb exoskeleton devices. IEEE Syst. J. 15 (2), 2361–2368.

Fernández-Vázquez, D., Cano-de-la-Cuerda, R., Gor-García-Fogeda, M.D., Molina-Rueda, F., 2021. Wearable robotic gait training in persons with multiple sclerosis: a satisfaction study. Sensors 21 (14), 4940.

Gallenberger, D., Bhattacharjee, T., Kim, Y., Srinivasa, S.S., 2019. Transfer depends on acquisition: analyzing manipulation strategies for robotic feeding. In: 2019 14th ACM/IEEE International Conference on Human-Robot Interaction (HRI), IEEE, pp. 267–276.

Garcia-Gonzalez, A., Fuentes-Aguilar, R.Q., Salgado, I., Chairez, I., 2022. A review on the application of autonomous and intelligent robotic devices in medical rehabilitation. J. Braz. Soc. Mech. Sci. Eng. 44 (9), 1–16.

Gates, B., 2007. A robot in every home. Sci. Am. 296 (1), 58–65.

Hesse, S., Uhlenbrock, D., 2000. A mechanized gait trainer for restoration of gait. J. Rehabil. Res. Dev. 37 (6), 701–708.

Hong, E., Gorman, P.H., Forrest, G.F., Asselin, P.K., Knezevic, S., Scott, W., Wojciehowski, S.B., Kornfeld, S., Spungen, A.M., 2020. Mobility skills with exoskeletal-assisted walking in persons with SCI: results from a three center randomized clinical trial. Front. Robot. AI 7 (93).

Islam, M.R., Brahmi, B., Ahmed, T., Assad-Uz-Zaman, M., Rahman, M.H., 2020. Exoskeletons in upper limb rehabilitation: a review to find key challenges to improve functionality. Control Theory Biomed. Eng., 235–265.

Jafari, B., Katoozian, H.R., Tahani, M., Ashjaee, N., 2022a. A comparative study of bone remodeling around hydroxyapatite-coated and novel radial functionally graded dental implants using finite element simulation. Med. Eng. Phys. 102 (103775).

Jafari, B., Shams, V., Esfandiari, M., Sadeghnejad, S., 2022b. Nonlinear contact modeling and haptic characterization of the ovine cervical intervertebral disc. In: 2022 9th IEEE RAS/EMBS International Conference for Biomedical Robotics and Biomechatronics (BioRob), IEEE, pp. 1–6.

Kamali, K., Akbari, A.A., Akbarzadeh, A., 2016. Trajectory generation and control of a knee exoskeleton based on dynamic movement primitives for sit-to-stand assistance. Adv. Robot. 30 (13), 846–860.

Kardan, I., Akbarzadeh, A., 2017. Robust output feedback assistive control of a compliantly actuated knee exoskeleton. Robot. Auton. Syst. 98 (15–29).

Keller, U., Schölch, S., Albisser, U., Rudhe, C., Curt, A., Riener, R., Klamroth-Marganska, V., 2015. Robot-assisted arm assessments in spinal cord injured patients: a consideration of concept study. PLoS One 10 (5), e0126948.

Khalili, D., Zomlefer, M., 1988. An intelligent robotic system for rehabilitation of joints and estimation of body segment parameters. IEEE Trans. Biomed. Eng. 35 (2), 138–146.

Krebs, H.I., Dipietro, L., Levy-Tzedek, S., Fasoli, S.E., Rykman-Berland, A., Zipse, J., Fawcett, J.A., Stein, J., Poizner, H., Lo, A.C., 2008. A paradigm shift for rehabilitation robotics. IEEE Eng. Med. Biol. Mag. 27 (4), 61–70.

Krebs, H.I., Hogan, N., Aisen, M.L., Volpe, B.T., 1998. Robot-aided neurorehabilitation. IEEE Trans. Rehab. Eng. 6 (1), 75–87.

Kwee, H.H., 1998. Integrated control of MANUS manipulator and wheelchair enhanced by environmental docking. Robotica 16 (5), 491–498.

Kyrarini, M., Lygerakis, F., Rajavenkatanarayanan, A., Sevastopoulos, C., Nambiappan, H.R., Chaitanya, K.K., Babu, A.R., Mathew, J., Makedon, F., 2021. A survey of robots in healthcare. Technologies 9 (1), 8.

Lee, K.H., Baek, S.G., Lee, H.J., Lee, S.H., Koo, J.C., 2022. Real-time adaptive impedance compensator using simultaneous perturbation stochastic approximation for enhanced physical human-robot interaction transparency. Robot. Auton. Syst. 147 (103916).

Li, L., Fu, Q., Tyson, S., Preston, N., Weightman, A., 2021. A scoping review of design requirements for a home-based upper limb rehabilitation robot for stroke. Top. Stroke Rehabil., 1–15.

Li, W., Lemaire, E.D., Baddour, N., 2020. Design and evaluation of a modularized ankle-foot orthosis with quick release mechanism. In: 2020 42nd Annual International Conference of the IEEE Engineering in Medicine & Biology Society (EMBC), IEEE, pp. 4831–4834.

Li, X., Pan, Y., Chen, G., Yu, H., 2017. Multi-modal control scheme for rehabilitation robotic exoskeletons. Int. J. Robot. Res. 36 (5–7), 759–777.

Loh, E., 2018. Medicine and the rise of the robots: a qualitative review of recent advances of artificial intelligence in health. BMJ Leader 2 (2), 59–63.

Lum, P.S., Van Der Loos, M., Shor, P., Burgar, C.G., 1999. A robotic system for upper-limb exercises to promote recovery of motor function following stroke. In: Proceedings Sixth Int. Conf. on Rehab. Robotics, pp. 235–239.

Martín, A., Pulido, J.C., González, J.C., García-Olaya, Á., Suárez, C., 2020. A framework for user adaptation and profiling for social robotics in rehabilitation. Sensors 20 (17), 4792.

Mason, C.P., Peizer, E., 1978. Medical manipulator for quadriplegic. In: Proc Int'l Conf. on Telemanipulators for the Physically Handicapped.

Mirmohammad, Y., Khorsandi, S., Shahsavari, M.N., Yazdankhoo, B., Sadeghnejad, S., 2021. Ball path prediction for humanoid robots: combination of k-NN regression and autoregression methods. In: Robot World Cup. Springer, pp. 3–14.

Mišeikis, J., Caroni, P., Duchamp, P., Gasser, A., Marko, R., Mišeikienė, N., Zwilling, F., De Castelbajac, C., Eicher, L., Früh, M., 2020. Lio-a personal robot assistant for human-robot interaction and care applications. IEEE Robot. Automat. Lett. 5 (4), 5339–5346.

Naro, A., Pignolo, L., Calabrò, R.S., 2022. Brain network organization following post-stroke neurorehabilitation. Int. J. Neural Syst. 32 (04), 2250009.

Rehmat, N., Zuo, J., Meng, W., Liu, Q., Xie, S.Q., Liang, H., 2018. Upper limb rehabilitation using robotic exoskeleton systems: a systematic review. Int. J. Intell. Robot. Applicat. 2 (3), 283–295.

Reinkensmeyer, D.J., Kahn, L.E., Averbuch, M., McKenna-Cole, A., Schmit, B.D., Rymer, W.Z., 2014. Understanding and treating arm movement impairment after chronic brain injury: progress with the ARM guide. J. Rehabil. Res. Dev. 37 (6), 653–662.

Rivas, J.G., Toribio-Vázquez, C., Taratkin, M., Marenco, J.L., Grossmann, R., 2021. Autonomous robots: a new reality in healthcare? A project by European Association of Urology-Young Academic Urologist group. Curr. Opin. Urol. 31 (2), 155–159.

Rose, L., Bazzocchi, M.C., de Souza, C., Vaughan-Graham, J., Patterson, K., Nejat, G., 2020. A framework for mapping and controlling exoskeleton gait patterns in both simulation and real-world. In: Frontiers in Biomedical Devices. American Society of Mechanical Engineers. V001T09A001.

Sadeghnejad, S., Elyasi, N., Farahmand, F., Vossughi, G., Sadr Hosseini, S.M., 2020. Hyperelastic modeling of sino-nasal tissue for haptic neurosurgery simulation. Scientia Iranica 27 (3), 1266–1276.

Saenz, J., Behrens, R., Schulenburg, E., Petersen, H., Gibaru, O., Neto, P., Elkmann, N., 2020. Methods for considering safety in design of robotics applications featuring human-robot collaboration. Int. J. Adv. Manuf. Technol. 107 (5), 2313–2331.

Senanayake, C., Senanayake, S.A., 2009. Emerging robotics devices for therapeutic rehabilitation of the lower extremity. In: 2009 IEEE/ASME International Conference on Advanced Intelligent Mechatronics, IEEE, pp. 1142–1147.

Shao, Y., Zhang, W., Xu, K., Ding, X., 2019. Design of a novel compact adaptive ankle exoskeleton for walking assistance. In: IFToMM World Congress on Mechanism and Machine Science, Springer, pp. 2159–2168.

Shi, D., Zhang, W., Zhang, W., Ding, X., 2019. Force field control for the three-dimensional gait adaptation using a lower limb rehabilitation robot. In: IFToMM World Congress on Mechanism and Machine Science, Springer, pp. 1919–1928.

Simpson, R.C., Poirot, D., Baxter, F., 2002. The Hephaestus smart wheelchair system. IEEE Trans. Neural Syst. Rehab. Eng. 10 (2), 118–122.

Singh, G., Singla, A., Virk, G.S., 2016. Modeling and simulation of a passive lower-body mechanism for rehabilitation. In: Conference on mechanical engineering and technology (COMET-2016), IIT (BHU), Varanasi, India.

Sowmiya, S., Ramachandran, M., Chinnasamy, S., Prasanth, V., Sriram, S., 2022. A study on humanoid robots and its psychological evaluation. Des. Model. Fabricat. Adv. Robots 1 (1), 48–54.

Topping, M., 2001. Handy 1, a robotic aid to independence for severely disabled people. In: 7th Int'l Conf. on Rehabilitation Robotics, pp. 142–147.

Udupa, S., Kamat, V.R., Menassa, C.C., 2021. Shared autonomy in assistive mobile robots: a review. In: Disability and Rehabilitation: Assistive Technology, pp. 1–22.

Verdel, D., Bastide, S., Vignais, N., Bruneau, O., Berret, B., 2021. An identification-based method improving the transparency of a robotic upper limb exoskeleton. Robotica 39 (9), 1711–1728.

Wang, W., Siau, K., 2019. Artificial intelligence, machine learning, automation, robotics, future of work and future of humanity: a review and research agenda. J. Database Manage. (JDM) 30 (1), 61–79.

Wu, J., Gao, J., Song, R., Li, R., Li, Y., Jiang, L., 2016. The design and control of a 3DOF lower limb rehabilitation robot. Mechatronics 33 (13–22).

Yousif, J., 2020. Humanoid robot as assistant tutor for autistic children. Int. J. Comput. Appl. Sci. 8 (2).

Zandieh, M., Kazemi, A., Ahmadi, M., 2021. A comprehensive insight into the application of machine learning approaches in predicting the separation efficiency of hydrocyclon. Desalin. Water Treat. 236 (123–143).

Zeelenberg, A., 1986. Domestic use of a training robot-manipulator by children with muscular dystrophy. Interact. Robot. Aids-One Opt. Independ. Living: Int. Perspect. Vol. Monogr. 37 (29–33).

Zhang, F., Cully, A., Demiris, Y., 2019. Probabilistic real-time user posture tracking for personalized robot-assisted dressing. IEEE Trans. Robot. 35 (4), 873–888.

Zhang, L., Huang, Q., Cai, K., Wang, Z., Wang, W., Liu, J., 2020. A wearable soft knee exoskeleton using vacuum-actuated rotary actuator. IEEE Access 8 (61311–61326).

Zhang, T., Tran, M., Huang, H.H., 2017. NREL-Exo: a 4-DoFs wearable hip exoskeleton for walking and balance assistance in locomotion. In: 2017 IEEE/RSJ International Conference on Intelligent Robots and Systems (IROS), IEEE, pp. 508–513.

Zhou, J., Yang, S., Xue, Q., 2021. Lower limb rehabilitation exoskeleton robot: a review. Adv. Mech. Eng. 13 (4). 16878140211011862.

CHAPTER 4

Gait devices for stroke rehabilitation: State-of-the-art, challenges, and open issues

Thiago Sá de Paiva[a], Rogério Sales Gonçalves[a], Giuseppe Carbone[b], and Marco Ceccarelli[c]

[a]School of Mechanical Engineering, Federal University of Uberlândia, Uberlândia, Brazil
[b]Department of Mechanical, Energy, and Management Engineering, University of Calabria, Rende, Italy
[c]Department of Industrial Engineering, University of Rome Tor Vergata, Rome, Italy

4.1 Introduction

The World Health Organization defines stroke as: "rapidly developing clinical signs of focal (or global) disturbance of cerebral function, lasting more than 24 hours or leading to death, with no apparent cause other than that of vascular origin" (Coupland et al., 2017). Current research shows that stroke is a leading cause of death and disability around the world (Krishnamurthi et al., 2020), and the improvement of quality of life and rehabilitation outcomes is a topic of interest for many researchers around the world.

Stroke can have several different and devastating effects on an individual, resulting in death or disability and severely hindering independence for stroke survivors. Robot-assisted gait training (RAGT) is being extensively investigated as a rehabilitation resource for improving mobility and quality of life outcomes following stroke. Assessments of gait rehabilitation after stroke concluded that electromechanically assisted gait training combined with physiotherapy provides better rehabilitation outcomes than gait training using only conventional physiotherapy, although the role of the type of device is still not clear (Mehrholz et al., 2020).

There are many different technologies currently being used in lower limb rehabilitation following stroke, such as the use of a specialized treadmill for gait exercises, body weight support, use of serious games, stationary rehabilitation approach (platform devices), nonstationary rehabilitation approach (exoskeletons), active or powered devices, passive devices, use of control techniques, biofeedback, and so on.

This introduction is followed by an overview of devices useful for gait rehabilitation after stroke, including the ones already in clinical trials and the latest research proposals and advances. This chapter ends with a discussion of the open challenges and directions in the rehabilitation of human gait.

Medical and Healthcare Robotics
https://doi.org/10.1016/B978-0-443-18460-4.00003-2

4.2 Gait devices for stroke rehabilitation

Stroke can result in neurological disorders associated with impaired locomotion. Research studies have shown that repetitive execution of impaired movement, supported by any external help, can improve the motor function of the affected lower limbs. Research indicates that these improvements are based on neuroplasticity and result in compensation for the loss of lesioned brain (Krishnamurthi et al., 2020; Mehrholz et al., 2020; Gonçalves et al., 2019; Calabrò et al., 2016).

There are a variety of devices for lower extremity rehabilitation including many treadmill-based devices and robotic systems. These devices were designed to replicate body weight supported treadmill therapy (BWSTT), a method hypothesized to be the gold standard in gait neurorehabilitation. The BWSTT involves two or three therapists driving the patient's legs while supporting the patient's weight (Gonçalves and Krebs, 2017).

A randomized clinical trial demonstrated that the BWSTT did not lead to results superior to those from traditional therapy, and highlighted that the goal of rehabilitation robotics cannot be to simply automate current rehabilitation practices (Gonçalves and Krebs, 2017). This way, a scientific basis is needed for the development of effective robotic therapy and this chapter presents the state of the art of gait devices for stroke rehabilitation.

Rehabilitation gait devices are used in different approaches and with different features. They can be classified according to their approach to gait training: stationary or overground walking systems, wearable exoskeleton devices, and a single degree of freedom (DOF) type of devices. Stationary devices can be further divided into two main categories: exoskeleton devices and end-effector devices.

4.2.1 Stationary exoskeleton gait training devices

Stationary exoskeleton gait training devices provide automated gait training with reduced clinician effort and are composed of a treadmill and an exoskeleton. Different from traditional overground walking, this type of device maintains a patient stationary, and the treadmill provides a moving surface for walking. Lower limb movement is assisted using an exoskeleton in synchrony with the treadmill, and depending on the type of exoskeleton it is possible to control hip, knee, and ankle joints. Other features include body weight support systems (BWSSs) and integration with serious games and biofeedback systems (Fig. 4.1).

4.2.2 Stationary end-effector gait training devices

Stationary end-effector devices provide also a fixed environment for gait training. The main differences from stationary exoskeleton devices are the control of just the ankle joint and the fact that they do not require a treadmill, since the role of a moving surface is performed by the end effector (Fig. 4.2). The end effector is a platform that guides

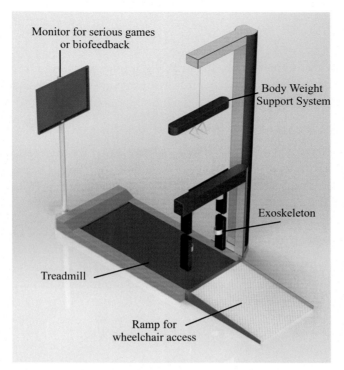

Fig. 4.1 A CAD representation for a stationary exoskeleton gait training device.

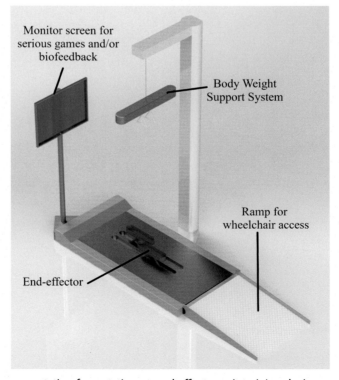

Fig. 4.2 A CAD representation for a stationary end-effector gait training device.

the patient's foot according to a programmable gait trajectory, and it can offer resistance or assistance to the movement. Another feature of an end-effector-type device is its capability of simulating different gait tasks like climbing stairs or walking downhill.

4.2.3 Wearable exoskeleton devices

Wearable exoskeleton devices are exoskeleton-type structures that remain fixed in parallel with a patient's lower limbs. They can be active (motorized) or passive (Gonçalves et al., 2019) (manual structures) (Fig. 4.3), and gait movement is performed overground with or without assistance.

4.3 Robot-assisted gait training

RAGT has shown great advances over the last decades and is based on the motor learning paradigm as a result of task-oriented, repetitive, and intensive motor exercises requiring the subject's attention and effort (Calabrò et al., 2016). There are many devices already in

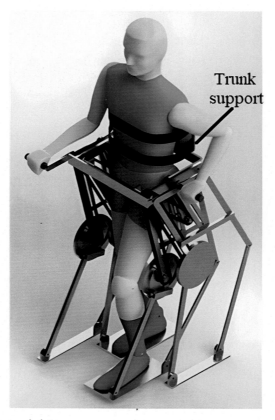

Fig. 4.3 A passive-type exoskeleton.

use around the world, producing data on RAGT rehabilitation outcomes measured by different methodologies. An objective of walking independently is considered as a primary outcome, using the Functional Ambulation Category (Bruni et al., 2018), Barthel Index (Watanabe et al., 2019), Functional Independence Measure (Brown et al., 2015), or Rivermead Mobility Index (Rojek et al., 2020) as performance indicators. Walking speed and walking capacity (Hornby et al., 2020) are used as secondary indicators, exemplifying measures of activity limitations. A brief description of some gait training devices intended for RAGT is provided in this section.

4.3.1 Stationary exoskeleton-type devices
4.3.1.1 Lokomat
The Lokomat is a robotic orthosis or exoskeleton type of gait machine, commercialized by Hocoma (Nam et al., 2017). It was developed to automate treadmill training rehabilitation of lower limbs to increase the duration and improve the quality of treadmill gait training. Introduced first as driven gait orthosis (DGO) (Dietz, 2016), this device set the individual and partially unloads its weight using a suspension system and harness (Fig. 4.4).

It is adjustable to different users, taking into account anatomical positions and size, and utilizes soft pads in all areas of contact between patient and orthosis. For balance control, since not all potential users have trunk stability, a structure for stabilizing the patient in the vertical direction is also provided, using a rotatable parallelogram that prevents tilting and keeping the DGO fixed relative to the treadmill (no backward movement). Only upward and downward movements are allowed since those are the main natural movements during walking. In addition, a gas spring is used with the rotating parallelogram to compensate for the total weight of the orthosis and hold up the parallelogram, to alleviate the patient from the orthosis weight.

Torque capabilities for the movement of joints take into account not only the movement requirements, but also the power that is required to overcome difficulties arising from motor disorders such as spastic muscle hypertonia. Hip and knee joints use custom-designed drives with precision ball screws, where the nut is driven by a toothed belt that in turn is moved by a DC motor. For hip and knee joints, this system can deliver an average torque of 30 and 50 Nm, respectively.

The DGO control architecture comprises three main hardware parts: host personal computer (PC), target PC, and current controller. It utilizes separate control for all four driven joints, using individual position-controller loops implemented in a real-time system that runs on the target PC. Angle measurement is performed by potentiometers and then converted using an analog-to-digital converter to be transferred to the real-time system. Host PC provides a database and a user interface programmed in LabView, which can be used by the therapist to modify the DGO speed. Communication between the

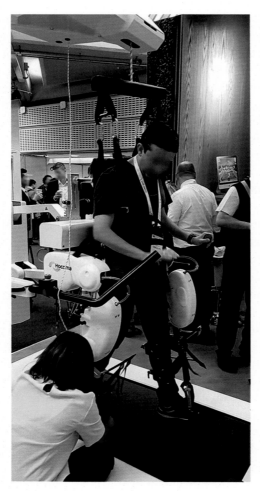

Fig. 4.4 Example of Lokomat device.

host PC and target PC is achieved through the serial bus (RS 232), so planned speed can be read by target PC, and gait pattern adjustments can be made. Force sensors provide interaction torque information between the user and the device, allowing the introduction of various control techniques besides conventional position control. Modifications on treadmill speed are made by the target PC using a second serial port. Further developments were made for the BWSS to achieve constant and precise unloading forces for better therapy outcomes (Lokolift). Also, the use of immersive virtual environments and biofeedback are introduced (Riener, 2016).

4.3.1.2 Robogait

The Robogait by Bama Technology is a robot-assisted gait therapy device that is produced in Turkey and is used in physiotherapy centers for both adult and pediatric patients (Sucuoglu, 2020). It can adjust to the user's body shape and has a hanger system for adjustable dynamic body weight support through a harness. Legs are attached in three different areas to the robotic orthosis and walking speeds are adjustable from 0.2 to 3.2 km/h (Sucuoglu, 2020).

The Robogait also includes a computer for control of settings, and a synchronized treadmill with a speed range of 0.1–5 km/h (manual training) and 0.2–3.2 km/h (with the robotic orthosis). A patient lift system provides the transfer from the wheelchair to the gait device. The device sensors measure the interaction forces between the orthosis and legs constantly, deeming positive if the forces are in the direction of motion and negative if not. This measurement is done by using eight force sensors and four sensor amplifiers at the joints. Serious game integration can be done using this biofeedback system (Erbil et al., 2018).

4.3.1.3 AutoAmbulator

The AutoAmbulator (Encompass Health) is composed of a treadmill, overhead lift, one pair of articulated arms, and two structures that contain parts of the mechanism and computer controls (Fisher et al., 2015). The system has the capability of lifting the user to a standing position so the gait training can begin. The computer controls the gait drive using position, time, and distance to allow smooth, precise, and synchronized motion of the legs and treadmill, at different speeds. The interface is made via a touchscreen display and the system counts with safety interlocks and other features to detect unsafe conditions.

The robot arms, with four DOFs (two for each leg corresponding to the knee and hip), move the user's legs during gait training. Leg attachment is achieved at the mid-level of the thigh and calves. A handlebar is also present to provide balance assistance to the user during gait training.

4.3.1.4 Walkbot_S

The Walkbot_S (P&S Mechanics, Seoul, Republic of Korea) is a treadmill-based, robot-aided rehabilitation system, equipped with two leg orthoses (two DOFs and an additional ankle part), gravity balancing orthosis (GBO), body-weighted support (BWS), and an automatic treadmill. It is adjustable to size, and two joints use an indirect drive actuator mechanism composed of a motor, timing belt, and harmonic drive (Van Tran et al., 2015).

Among other product features, the Walkbot_S can provide adjustable optimal loading, real-time visual feedback for both stiffness and torque, and joint (hip, knee, and ankle) kinematics (Lee et al., 2021). It is also capable of different control strategies, such as trajectory tracking, active-assistive, and other active control strategies linked to the user's performance and real-time visual biofeedback (Park et al., 2018).

4.3.1.5 Gait-Assistance Robot

Described as a robotic arm control system, the Gait-Assistance Robot (GAR), developed jointly by the Yaskawa Electric Corporation and the University of Occupational and Environmental Health from Japan, is composed of four robotic arms intended to control independently thighs and legs, a control panel, power generator, thigh cuffs, and leg apparatuses, foot pressure biofeedback system, and treadmill. Unlike other devices, there is no BWSS, instead, the system permits walking with full body weight on the user's legs (Ochi et al., 2015).

4.3.1.6 Gait Exercise Assist Robot

Changes in motor learning variables' performance and transferability are central to the Gait Exercise Assist Robot (GEAR). They are dependent upon several factors including quantity of exercise (frequency), the difficulty of the exercise, and feedback (Schmidt and Lee, 2013). To avoid limitations to transferability, the idea is to take into account that the final gait and gait exercise must not be too different, so gait machines providing symmetrical gait patterns to hemiplegic poststroke patients might not be ideal. In this way, the design aimed to assist only the affected limb and permits flexible adjustments of the motor learning variables (Hirano et al., 2017).

The GEAR, designed in collaboration between the Fujita University and Toyota Corporation, consists of a knee-ankle-foot robot, treadmill, body weight support, robot weight support device, control panel, and monitor. The knee-ankle-foot orthosis utilizes a motor at the knee joint, and the robot weight support device has the task of eliminating the orthosis's weight on the patient. A pressure sensor, placed at the plantar region of the active orthosis, sends data to the robot, which estimates the phase of the gait cycle based on this information and knee joint angle information to perform flexion or extension appropriately. The orthosis can be adjusted to different leg sizes and knee angles, so it is possible to be used by different patients. Different types of feedback can be provided: image feedback is provided in the form of either a mirror image or foot image, acoustic feedback is a sound of success when the weight of the hemiplegic side surpasses a set target value, or a sound of failure when the knees fail. For the clinician, the control panel can display in real-time weight bearing on the hemiplegic side and the trajectory of the foot's center of pressure. Levels of assistance can be customized for knee extension, swing, body weight support, the timing of knee flexion, and extension in a stepwise fashion (range of levels). GEAR is not indicated for patients with conditions like paraplegia, cerebella ataxia, and muscle weakness (Hirano et al., 2017).

4.3.1.7 LOPES

The LOwer extremity Powered ExoSkeleton (LOPES) device is composed of a two-dimensional (2D) actuated pelvis part with a leg exoskeleton with three actuated rotational joints (two for hips and one for the knee) that are impedance controlled to allow two modes: "patient-in-charge" and "robot-in-charge." A third mode, called "therapist-in-charge," is possible where programming can be performed to achieve a specific therapeutic goal (van Asseldonk and van der Kooij, 2016).

Weight compensation is achieved using an actuated support at the pelvis height (end-effector robot). In this sense, the LOPES robot is a combination of an exoskeleton for leg motion and an end-effector robot for supporting the user's weight in the pelvis. In total, LOPES comprises eight actuated DOFs (three rotational joints per leg—two at the hip and one at the knee, two for horizontal translation of the pelvis). The vertical motion of the pelvis is left free to move without actuation (respecting design limits) and has passive compensation of weight by using a spring mechanism (van Asseldonk and van der Kooij, 2016). Rotational joints are actuated with the use of Bowden-cable-driven series elastic actuators and sideways pelvis translation utilizes a linear actuator of the same kind (Veneman et al., 2016). The device is intended to work with a treadmill.

4.3.2 Stationary end-effector-type devices

4.3.2.1 Gait Trainer GT II

Reha Stim's Gait Trainer GT II, an end-effector type of gait machine, is based on a system with a doubled crank and rocker mechanism that moves two footplates placed on two bars (mechanism's couplers) (Anaya et al., 2018). This way we have for each leg a crank, a rocker, and a coupler moving a footplate. The intention is to simulate the gait phases with a ratio of 40%–60% between the swing and stance phases. The swing phase corresponds to forward movement and the stance phase to low backward movement. This ratio between stance and swing phases is achieved with the use of a planetary gear system strategically attached to the foot bars eccentrically, so the upper half of the revolution (swing phase) has 40% of the gait cycle and the lower half (stance phase) has 60% of the said gait cycle.

The system can work with partial or full support to the patient, being capable to sense either assisted or resisted movement, transmitting the momentum (after analog-to-digital conversion) to a display where it provides biofeedback to both patient and therapist. The center of mass (COM) position is also maintained both vertically and horizontally through the use of the planetary gear and two cranks, one for the vertical oscillations and one for the horizontal oscillations following the literature concerning COM oscillations through the gait cycle (Tesio and Rota, 2019). Thus, the COM oscillates vertically following a sine curve of 2 cm of amplitude and with a frequency of two periods per gait cycle. Horizontally, the COM also oscillates sinusoidally, but with an amplitude of 4 cm and with a frequency of one period per gait cycle. A rope attached to this vertical crank serves as the user's central suspension. A second rope, connected to the horizontal crank was attached to the left of the patient harness, at the pelvic crest level.

4.3.2.2 G-EO System

The G-EO System (Smania et al., 2018) also follows an end-effector principle. The moving mechanism consists of two footplates, connected to two moving sleds (principal sled and relative sled) by a pivoting arm. Sleds in turn are connected to a transmission belt of a linear guide, which is driven by a fixed servomotor placed at the back end of the linear guide. The forward and backward movement of the principal sled controls step length. Step height control utilizes the scissor principle. Under the relative sled, a servomotor in charge of the relative motion between sleds is fixed by a screw axle. A third drive was fixed on the pivoting arm and transfers rotation through a transmission belt to an external axle aligned to the ankle joint to command plantar flexion and dorsiflexion of the feet. Footplate trajectories are completely programmable, given all adjustments that can be made to each parameter of the gait machine.

There are eight programmable DOFs for this machine—three for each limb and two for center of mass (COM) control. This control is performed by an industrial personal computer, which controls gait trajectories and COM vertical and horizontal oscillations, following healthy individual data available in the literature (Bertaux et al., 2022; Prakash et al., 2018). The BWSS consists of an electric lift system using a belt that passes through a three-roll mechanism and attaches to the user's harness. Belt length varies during motion, creating the COM vertical oscillating motion. A different drive moves a rope fixed on its two ends to the user's harness (at hip height), to control the lateral motion of COM.

One of the G-EO system's main features is its capability to simulate stair climbing, both up and down. This feature was developed after a simulation was performed using an active marker system processing of data relative to the markers of interest, and data of trajectories for both feet and COM during stair climbing was obtained. Actual trajectories can be shown by a graphic user interface online, allowing corrections to be made by the therapist if required.

4.3.2.3 Morning Walk

The Morning Walk is a gait trainer designed by Hyundai Heavy Industries and Taeha Mechatronics, both from Korea. It is an end-effector type of device, with a BWSS that uses a type of saddle for support (Kim et al., 2019).

Different from other outstanding end-effector gait trainers, it can simulate different walking experiences like ground walking and stair climbing/descending. In addition, it allows the clinician to adjust various gait parameters like cadence, step height, step length, to-off angle, and initial contact angle. A therapist can also control ankle trajectory with precision. The amount of body weight support and ground reaction forces can be used as biofeedback (Kim et al., 2020).

4.3.2.4 GaitMaster4

With the intent of presenting visual images and creating the impression of walking for rehabilitation activities, the GaitMaster4 is a device developed at the University of Tsukuba, Japan that works using the end-effector principle and that possesses a spherical projection display for user immersion, where images are presented also to the peripheral field of view that works in combination with the footpads of the device (Smania et al., 2018).

The device encompasses a manipulator providing two DOFs for each foot (up and down by ball-screw actuator, back and forth by slider-crank mechanism; Tanaka et al., 2012), an immersive projection display, and allows the use of a weight-bearing device. The overall space required for the system to work is a space of 2 m × 2 m of ground space and a height of 2.5 m. The feet are fixed in such a manner to allow both dorsal flexion and plantar flexion freely. Foot trajectories follow the trajectories of healthy individuals adjusted both spatially and temporally to each individual's needs. Strides can be gradually adjusted for the user to reach target values for speed and other gait parameters (Smania et al., 2018).

4.3.3 Overground exoskeleton type of devices

4.3.3.1 Walking Assist Device with Stride Management Assist (Honda)

Aimed at elderly individuals, the Walking Assist Device with Stride Management System is a walking assist device focused on improving walk ratio, posture, and muscular usage. The intent was to induce independent walking and create a lasting effect in this sense and reduce the stress connected to walking activities (Asbeck et al., 2015; Buesing et al., 2015). Walk ratio, defined as stride/cadence, has a value for optimum energy consumption and this value decreases among the elderly, therefore, hindering optimum walking for them, and causing energy consumption to go up. So, this population would benefit from a device that could successfully induce proper walk ratio by assisting them during walking.

The mechanism that creates this assistance is composed of two small and lightweight actuators (multipolar brushless DC motor combined with a planetary gear mechanism with a reduction ratio of 10 and angular sensors) at the hip joint level, a lumbar frame, and a thigh frame. Control circuits and batteries are placed in the lumbar frame. A front support belt must be fastened around the abdomen basis, along with the fastening of the belt thighs. The user must choose between nine frame sizes, the ones that are most suitable for them (lumbar frames and thigh frames). This design option was made to avoid a necessary weight increase in a frame that was adjustable for different sizes, and to avoid complexity.

A rhythm pattern generator (produced by a pacemaker in the control computer) contained in the computer used for control of the device creates walking trajectories related to specific rhythms. The user's actual rhythm is measured by the angular sensors, which then serve as input to the pacemaker, which enters in synchrony with the user. Measurement of the phase difference between the current pacemaker rhythm and user rhythm is performed and a correction for the next step is made. By using this phase difference, the device can induce cadence (propelling or braking) to maintain an ideal phase difference and therefore control stride and walk ratio. Asymmetries between legs are also detected and stride adjustments are made to correct them.

4.3.3.2 Exowalk (HMH Co. Limited)

Designed for overground walking exercises instead of treadmill-based systems, the Exowalk is an active electromechanical exoskeleton that is mobile, providing a different experience since the robot moves according to the user's directions (Kim et al., 2021). Patient fixation is performed at both paretic and nonparetic limbs, at calf and thigh levels. Similar to an end–effector-type device, walking is performed in a way in which the feet do not touch the ground.

Step length and height can be regulated by the clinician, whereas step length can be modified based on the programming of hip and knee range of movement (Nam et al., 2019). It dispenses the use of walkers or canes but does not provide a conventional harness for body weight support (does not have a pelvic strap). The device has four DOFs (two for each leg), is adjustable to different user sizes, has wireless controls, and provides a selection of speeds.

4.3.3.3 Hybrid Assistive Limb

The Hybrid Assistive Limb (HAL) was created aiming to enhance human capabilities using different research domains: cybernetics, mechatronics, and informatics, integrating others such as physiology, robotics, neuroscience, systems engineering, and others. This new interdisciplinary domain was coined "cybernics" (Watanabe et al., 2021).

The HAL-5 (Type B) weighs approximately 23 kg, uses an AC 100 V battery, can operate continuously for nearly 2 h and 40 min, and can assist/perform the tasks of standing up from a chair, walking, and climbing up and down stairs as activities of daily living.

The exoskeleton device has a hybrid control system that is made of two different types of control systems: "Cybernic Voluntary Control System" and "Cybernic Autonomous Control System." The Cybernic Voluntary Control System supports the user based on voluntary intention detected as bioelectrical signals including myoelectricity. The user's joint torque is estimated from the bioelectrical signals and the device generates an assist torque for multiple joints simultaneously. This system is only suitable for healthy individuals, since people with gait disorders may not even generate some of these bioelectric signals at the start of their gait training. For them, the Cybernic Autonomous Control is the more suitable mode, where other kinds of information such as reaction forces and joint angles can be used to provide adequate physical support. In this case, the user's intention is derived from shifts in the Center Of Gravity position. To avoid collisions of toes and floor, springs are used attached to the ankles. Floor reaction forces sensors and angular sensors are used to provide input to the proportional–derivative (PD) control system, which utilizes walk patterns from healthy individuals as reference patterns for the considered three different phases of walking: swing phase, landing phase, and support phase.

4.3.3.4 EksoGT and EksoNR

The EksoGT and EksoNR are exoskeletons created and marketed by Ekso Bionics that received FDA approvals for their commercialization in 2016 and 2020, respectively, with indications of use for individuals with acquired brain injury (ABI) and spinal cord injury (SCI) (US Food and Drug Administration, 2016a, b, 2020), and a caveat that the device is not intended for sports or stair climbing. The intended use is ambulatory functions performed at rehabilitation institutions supervised by a trained therapist. Described as a powered motorized orthosis, the device is composed of fitted metal braces supporting the legs, feet, and torso. Straps are used to fix the user's body and a solid torso part contains both a computer and a power supply. Hip and knee joints are driven by electric motors, which in turn are powered by the battery already mentioned. Ankles use a passive spring for movement. The devices are used with walking aids such as walkers, crutches, or canes. Ranges of motion described in US Food and Drug Administration (2016a, b, 2020) are hip flexion of 135 degrees, a hip extension of 20 degrees, knee flexion of 130 degrees, knee extension of 0 degrees, ankle flexion of 10 degrees, and ankle extension of 10 degrees. A failsafe feature that occurs in an event of power failure locks knees and frees hips to emulate typical passive leg braces. A predefined reference trajectory is enforced during walking (Lv et al., 2018) and the crutches contain pressure sensors that prompt sit/ stand or walking tasks (Gardner et al., 2017). Other features are posture support, clinician control, smart assistance (adjustments for various impairment degrees), and adaptive gait training (Ekso Bionics, 2020).

4.3.3.5 Anklebot

Developed at MIT, the ankle robot (Anklebot) was intended for ankle rehabilitation following stroke, counteracting the "foot-drop" effect and with the idea that the ankle plays a role of the utmost importance during gait since it contributes to several tasks such as shock absorption, propulsion, and equilibrium. The device is aimed at training the user to surpass the foot-drop issue mentioned. The Anklebot allows a normal range of motion in all three DOFs of the foot (relative to shank): 25 degrees dorsiflexion, 45 degrees plantar flexion, 25 degrees inversion, 20 degrees eversion, and 15 degrees for external or internal rotation (Krebs et al., 2016). It has low mechanical impedance, is low friction, and is backdriveable.

Assistance is provided in two of the three DOFs: dorsi–plantar flexion and inversion-eversion (internal-external rotation passively). The design consists of two linear actuators in parallel, and works as follows: actuation in the same direction creates dorsi–plantar flexion torque, and in different directions creates inversion-eversion. It utilizes brushless DC motors coupled with gear reducers and is equipped with two encoder sensors for position and ankle angle information. Torque is measured by analog current sensors and the control is an impedance control implemented as a PD controller, where the reference or neutral position can be programmable, along with proportional gain (torsional stiffness) and derivative gain (torsional damping).

Besides being used for rehabilitation, the Anklebot can serve as a tracking tool of rehabilitation outcomes since it is capable of measuring different ankle properties—for instance, passive ankle stiffness.

4.3.3.6 Gait Enhancing and Motivating System

Gait Enhancing and Motivating System (GEMS) is a wearable that can be used to hip, knee, or ankle level (GEMS-H for hip, GEMS-K for knee, and GEMS-A for ankle), designed to be lightweight and comfortable to wear, allow motion that does not deviate much from the individual's original gait pattern (Lim et al., 2019).

It utilizes an interaction controller for gait assistance or resistance based on delayed output feedback control, a type of control that is known for stabilizing oscillatory systems when certain conditions are present (Kalani et al., 2021). Samsung GEMS was originally intended for the elderly (rehabilitation) but now has its scope expanded to everyday users as well (Samsung Newsroom, 2019).

4.3.3.7 Bionic Leg

The Bionic Leg is a device produced by AlterG and consists of an independent knee orthosis, created for those cases where lower limb dysfunction is asymmetrical. It utilizes four sensors mounted on a footplate to detect the amount of weight supported by the leg using the Bionic Leg, and then with the assistance of additional sensors (knee extension angle measurement) permits the clinician to adjust the support or resistance applied, area of locomotion, and minimal force for device activation (Guo et al., 2018).

The device makes use of sensors and microprocessor technology, along with its specific software to detect the user's actions like walking or stair climbing to respond accordingly (Authorized Representative Obelis sa Bd Général Wahis, European, 2015).

4.3.3.8 Indego

Indego is a powered exoskeleton device marketed by Parker Hannifin, which is FDA approved with indications for use by individuals with SCI at levels T3 to L5, C7 to

L5 (in rehabilitation facilities), and for individuals with hemiplegia due to stroke. Its use must be supervised by a trained professional and it is not indicated for stair climbing or sports (US Food and Drug Administration, 2018a).

The device is a modular wearable with hip, upper leg, and lower leg segments that connect using a snap–fit concept. The hip houses the controller, battery, and Bluetooth radio, while each upper leg contains two motors, sensors, and controllers (US Food and Drug Administration, 2018a). Control of the device requires postural changes, that is, the user must lean backward or forward to perform the functions of sitting, standing, and walking. Stride length, gait speed, and other control configurations can be adjusted wirelessly by the therapist through the use of a phone or tablet application (Gardner et al., 2017). User feedback is provided in the form of vibratory feedback or LED indicators on the hip module. It also possesses a fall detection and mitigation feature that performs adjustments during the fall event to minimize injury or allow recovery. It also provides a failsafe feature in the event of a power outage, locking knees, and leaving hips free, making the device work similarly to passive leg braces. The device must be used with walking aids or crutches (US Food and Drug Administration, 2018a).

4.3.3.9 ReWalk P6.0

Another exoskeleton, the ReWalk P6.0 marketed by ReWalk Robotics and FDA approved, is recommended for SCI at levels T7 to L5 and T4 to L6 (in rehabilitation facilities) and must be used together with a walker or crutches (US Food and Drug Administration, 2018b).

The ReWalk P6.0 is composed of an exoskeleton, remote control (RC) communicator, a battery charger, and a laptop. The user controls the device using the RC, tilt sensor, and particular body motions, which trigger the gait movements that are executed by gears and motors at the hip and knee joints. The ambulatory functions performed are sitting, standing, and walking, and it can reach a maximum of 2.3 km/h (US Food and Drug Administration, 2018b).

4.3.4 Other devices

4.3.4.1 Robowalk

The Robowalk, an expander part of the treadmill product by h/p/cosmos sports & medical Gmbh (Germany), consists of eight expander cables (four in front and four in the rear) that must be attached to the user via leg cuffs. It can be used as an aid to manual gait therapy or for assisted gait training (still supervised by the therapist) when the system provides traction support. Traction resistance can also be set, and a special cuff can be used to prevent foot-drop problems during walking. Another adjustment that can be made is related to the angle of action of support/resistance cables and the preloading of cables. The system must be used with a harness intended to prevent falls (does not provide

Fig. 4.5 The MIT Skywalker system with saddle BWSS, marker system, and camera to track gait movements.

weight support). Body weight support can be provided with an additional product, airwalk ap (h/p/cosmos sports & medical gmbh, 2014).

4.3.4.2 MIT-Skywalker

Inspired by the concept of passive walkers (Iqbal et al., 2014), the MIT-Skywalker (Fig. 4.5) was developed looking to produce ground leeway for swing and to take advantage of gravity to assist during propulsion, without making the movement too rigid by restraining the limbs and maximizing weight bearing with realistic heel strike (Krebs et al., 2016).

Using body weight support, the gait is achieved when the subject, on top of a hinged split treadmill, starts the swing phase of gait when the treadmill actuates downward, leaving the leg free, and returns to the horizontal position for the heel strike of the support phase (Fig. 4.6). By alternating the downward actuation between both legs, a gait movement is achieved. This actuation is performed by a specially designed cam system, shown in Krebs et al. (2016). The cam system design takes into account different user sizes and actuation speeds for different walking speeds to be able to work properly under a variety of conditions. Also, for this same design, a split is considered between stance (60% of the

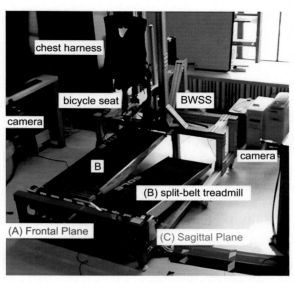

Fig. 4.6 MIT Skywalker system and its main parts—chest harness, bicycle seat (saddle), split-belt treadmill.

total gait cycle) and swing (40% of the total gait cycle) phases, following the known proportion of an average gait cycle.

The BWSS follows another type of design, different from conventional BWSSs that utilize a harness. Drawing from another MIT robot (Elvis-the-Pelvis), the user's weight is supported from below the waist using a seat, and upper body stability is achieved with a more simple chest harness (Krebs et al., 2016). This BWSS was evaluated previously in Gonçalves and Krebs (2017) and was considered advantageous for allowing better balance training, allowing easier don-on and don-off tasks, using a split treadmill with asymmetric speed profiles more safely than overhead support systems, and not requiring major space alterations (differently than many overhead BWSS). Developments and evaluations regarding user comfort were performed in Goncalves et al. (2017), resulting in the selection of a commercial unicycle saddle for the BWSS. Further developments involve a markerless control system for the device using a Microsoft Kinect V2 to track gait movements with accuracy within 120 ms, considered appropriate for therapy use (Gonçalves et al., 2021).

4.3.4.3 ARTHuR and PAM
For gait training, leg motion is not the sole concern. The motion of the pelvis during gait must also be carefully considered if the intent is to allow movement that is as natural as possible. Reinkensmeyer et al. consider both leg and pelvis movements, and two robots working together were the design choice for achieving a more naturalistic gait pattern:

the Ambulation-Assisting Tool for Human Rehabilitation (ARTHuR) and the Pelvic Assist Manipulator (PAM) (Cao et al., 2014).

The ARTHuR, described in Callegaro et al. (2014), is a device with coil forcers driving a two-bar linkage, where its top vertex is attached to the bottom of the foot, similar to an end-effector robot. The movement of the coil forces in a rail defines the apex trajectory. An update (ARTHuR 2.0; Reinkensmeyer et al., 2004) introduced a third coil forcer to make the device with three DOFs: knee position, angle of knee flexion, and angle of knee extension.

The PAM robot, also described in Van der Loos et al. (2016), manipulates pelvic motion by the use of two pneumatic robots (three DOFs each) during gait training using a treadmill by a person with body weight support. The robots attach to the back of an adjustable belt that must be worn by the user. The system has five DOFs, controlling all three translational movements and two rotations (rotation around the vertical axis and obliquity around the sagittal axis).

4.3.4.4 Exoskeleton with electric actuators
In Iancu et al. (2017), a new design of a low-cost lower limb exoskeleton was proposed, with simple construction, easier wearing, and adaptability to various human legs dimensions. The exoskeleton has three DOFs for the hip, knee, and ankle joints. The actuation of the exoskeleton joints is made by one rotational servomotor and two electric linear motors.

4.3.4.5 RECOVER device
Gherman et al. (2019) and Pisla et al. (2021) presented a parallel robot designed for the lower limb rehabilitation of bedridden stroke survivors named RECOVER. The device was developed to achieve a direct correlation between the active joints of the robot and the anatomic joint angles.

The device permits hip flexion/extension, knee flexion/extension, ankle flexion/extension, and inversion/eversion movements. This structure presents important features like control points for the anatomic joints, modularity, good usability, and a compliant mechanism for the limb segments anchor (Fig. 4.7).

4.3.4.6 Cable-driven lower limb rehabilitation robot
Seeking to reproduce the movements performed by the lower limbs, Barbosa et al. (2018) designed a cable-driven rehabilitation robot aimed at individuals who cannot perform lower limb movements and need external assistance (Fig. 4.8). It can use one to six cables, depending on the movement complexity. The device is composed of sets of DC motors, encoders, pulleys, load cells, and cables placed on a fixed platform, situated above a

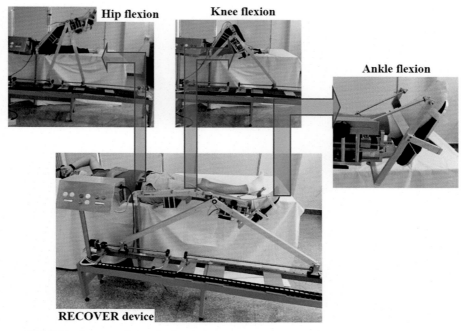

Fig. 4.7 RECOVER device to lower limb rehabilitation.

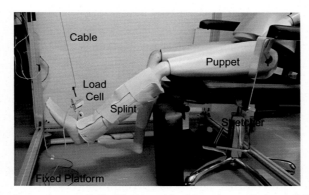

Fig. 4.8 Prototype using a dummy doll with the cable-driven lower limb rehabilitation robot.

stretcher. Control is performed using the approach "teach by showing" in two steps: first, a therapist executes the desired movement, therefore, "teaching" the device, and then the robot runs the predefined movement taught. The device can greatly reduce the therapist's exertion during rehabilitation and also improves movement repeatability, which in situations where the patient cannot execute movements can be of great importance

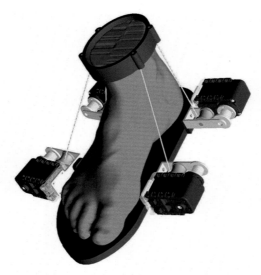

Fig. 4.9 A CABLEankle prototype design.

in improving the range of motion and strengthening the musculature. This device is low cost and easy to use.

In Russo and Ceccarelli (2020), the CABLEankle device is designed for motion ankle exercising in rehabilitation and training. The structure is based on a cable-driven S-4SPS parallel architecture, which enables motion assistance over the large motion range of the human ankle in a walking gait (Fig. 4.9). CABLEankle device is designed to be low cost and easy to use.

The LARM Wire driven EXercising device (LAWEX) is proposed in Lazar et al. (2018) and Pop et al. (2020). The LAWEX is a cable-driven-assisted human upper and lower limbs' exercises device. The advantages of the prototype are: lightweight structure easy to move and store; manufacturing and maintenance cost are low; is easy to set up and operate for clinical or home usage; and it is accessible also by people in wheelchairs and control system with commercial low-cost components. LAWEX has a mobile platform connected to a rigid structure by four cables, with three DOFs, that guide the lower limb during the exercises (Fig. 4.10). Each cable is connected to a 24 V DC motor with encoders, three in the upper part and one below in the fixed platform (Fig. 4.10A). The control system uses a low-cost microcontroller, Arduino Mega. The system can be programmed with an arbitrary trajectory that is defined by a physiotherapist or by starting from the predefined/desired trajectory and applying the mathematical model to control the motors. The authors used the Microsoft Kinect system to obtain a natural limb movement during rehabilitation motions using the LAWEX (Pop et al., 2020).

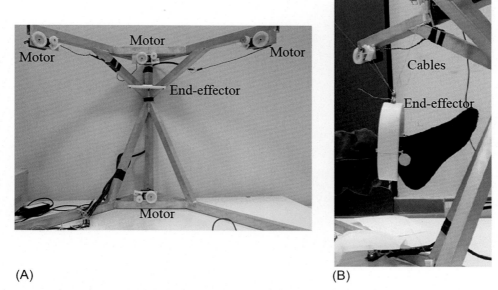

Fig. 4.10 (A) A LAWEX prototype and (B) experimental test with lower limb.

4.3.4.7 Gait analysis

A proper analysis of the kinesiology of human gait is necessary to develop devices for gait rehabilitation. Several methods can be used in gait research including sensors like pressure and force plates, video or optoelectronic-based analysis systems, inertial measurement unit (IMU), instrumented shoes (Gonçalves et al., 2021; Vaida et al., 2020; Rose et al., 2021), to acquire kinematic and kinetic gait data like trajectories, joins limits, ground reactions, and prototypes validation. In Varela et al. (2015), a device for experimental analysis of gait is proposed a cable-driven parallel robot named Cassino Tracking System (CaTraSys) (Fig. 4.11). CaTraSys is a passive device that can be used to determine the pose and exerted force of mechanical systems along large trajectories. This device has the advantages of a low-cost and easy-operation system to acquire, real-time, 2D, and three-dimensional (3D) gait patterns with an accuracy of 2 mm.

4.3.5 Single DOF devices

The UCI gait mechanism, described in Tsuge and McCarthy (2016), is a single DOF mechanism formed by a six-bar linkage that models the lower limb to achieve ankle trajectory, including foot orientation. The device was designed using linkage synthesis theory and optimization techniques (specifically, homotopy-directed optimization as shown

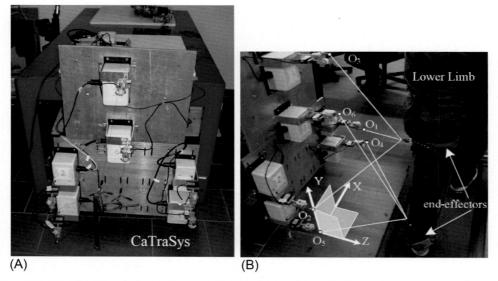

Fig. 4.11 (A) CaTraSys passive device and (B) CaTraSys used to gait characterization.

in Tsuge et al., 2016). It consists of a modifiable six-bar linkage and a cam-driven parallelogram linkage for foot orientation. The apparatus can accommodate small changes but need to be scaled up or down for different sizes (Tsuge and McCarthy, 2016).

Shin et al. (2018) introduced the design of a single DOF gait trainer by using an eight-link Jansen linkage which is further optimized to fit a range of walk patterns when adjusting two of the links.

Another single DOF device for gait training is introduced in Sabaapour et al. (2019), using also an eight-link Jansen mechanism, synthesized to generate ankle trajectory and with a seat-type weight support system. Each user's ankle is attached to one eight-link Jansen mechanism, and the hip must be fixed to the linkage too, to be able to provide the movement relative to the hips. A reciprocal configuration permits the use of a single actuator for moving the linkages. The weight support system (WSS) has two parts: "upper support" using a seat type of structure, and "lower support" placed on a treadmill to maintain walking in one place.

Gonca͵lves and Rodrigues (2022) proposed three four-bar crank-rocker linkages (single DOF each) to generate joint angles manually. They were designed using geometric relations and optimized by a differential evolution algorithm, aiming for suitability for people up to 1.8 m in height. A chain drive mechanism reduces the torque requirements to the upper limbs since the devices were intended to be operated manually (Fig. 4.12).

Table 4.1 summarizes the main characteristics of gait rehabilitation devices described in this chapter.

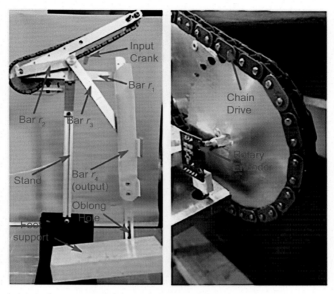

Fig. 4.12 Prototype of knee mechanism and the chain drive mechanism.

Table 4.1 A Summary of characteristics in gait rehabilitation devices.

Device	Active joints/DOFs (1)	Gait surface (2)	Type (3)	Weight support
Lokomat	4 (LH, RH, LK, RK)	T	Exo	Yes/ upper
Robogait	4 (LH, RH, LK, RK)	T	Exo	Yes/ upper
AutoAmbulator	4 (LH, RH, LK, RK)	T	Exo	Yes/ upper
Walkbot_S	4 (LH, RH, LK, RK)	T	Exo	Yes/ upper
GAR	4 (LH, RH, LK, RK)	T	Other (robotic arms)	No
GEAR	2 (LK, RK)	T	Exo	Yes/ upper
LOPES exoskeleton	6 (LH1, LH2, RH1, RH2, RK, LK)	T	Exo	Yes/ lower
Gait Trainer GT II	2 (LF, RF)	FP	*EE*	Yes/ upper

Continued

Table 4.1 A Summary of characteristics in gait rehabilitation devices—cont'd

Device	Active joints/DOFs (1)	Gait surface (2)	Type (3)	Weight support
G-EO System	2 (LF, RF)	FP	*EE*	Yes/upper
Morning Walk	2 (LF, RF)	FP	*EE*	Yes/lower
GaitMaster4	2 (LF, RF)	FP	*EE*	No
Walking Assist Device with Stride Management Assist	2 (LH, RH)	O	Other	No
Exowalk	4 (LH, RH, LK, RK)	FP	Exo	No
HAL	4 (LH, RH, LK, RK)	O	Exo	No
EksoGT and EksoNR	4 (LH, RH, LK, RK)	O	Exo	No
Anklebot	2 per ankle (dorsi-plantar flexion & inversion-eversion) plus 1 passive (internal-external rotation)	O	Other	No
GEMS	2(LH, RH or LK, RK or LF, RF)	O	Exo	No
Bionic Leg	1 (knee)	O	Exo	No
Indego	4 (LH, RH, LK, RK) plus 2 passives (LF, RF)	O	Exo	No
ReWalk P6.0	*4 (LH, RH, LK, RK)*	O	Exo	No
Robowalk	*4 (Left and right thighs; left and right crus)*	O	Other	Yes/upper
MIT-Skywalker	None (passive concept—movement caused by treadmill)	T	Other	Yes/lower
ARTHuR and PAM	1 for ARTHuR (ankle); 5 DOFs (pelvis)	T	Other	Yes/upper
Cable-driven lower limb rehabilitation robot	1 DOF (hip, knee, or ankle—number of cables used varies)	Other	Other	No
Exoskeleton with electric actuators	3 DOFs (hip, knee, and ankle)	O	Exo	No
CABLEankle	4 DOFs	O	Other	No
RECOVER	4 DOFs (hip, knee, and ankle)	Other	Exo	No
LAWEX	3 DOFs (hip, knee or ankle)	Other	Other	No

Note: 1—Active joints: *LH*, left hip; *RH*, right hip; *LK*, left knee; *RK*, right knee; *LF*, left foot; *RF*, right foot. 2—Gait surfaces: *T*, treadmill; *FP*, footplate; *O*, overground. 3—Type: *Exo*, exoskeleton, *EE*, end effector; and other for the remaining types.

4.4 Safety aspects of gait devices

Just as important, if not more important, as the concern with the effectiveness of a device in gait rehabilitation is the assurance of safety for both the user and the therapist conducting the therapy session. There is no good therapy without effectiveness but also there is no good therapy without safety. In this sense, it is important to evaluate risk-to-benefit ratios (when the risks are not too severe or unacceptable), so it makes sense not only to track performance indicators but also adverse events that appear in clinical studies using these gait devices.

A review of adverse events (AEs) in stationary gait devices was performed in Bessler et al. (2020) with a useful classification of these AEs in three main categories: soft tissue-related AEs, musculoskeletal AEs, and physiological AEs. Soft tissue-related AEs can be skin reddening, skin lesions, bruises, and discomfort (related to harnesses attached to the user). Musculoskeletal AEs can be muscle pain, joint pain, or even bone fractures. Physiological AEs can be numerous different events but in the case of gait rehabilitation, a common example in different trials is blood pressure changes. The review showed a higher prevalence of soft tissue-related AEs, followed by musculoskeletal AEs and then physiological AEs. The majority of AEs were of mild severity, followed by moderate and then severe (three different occurrences—one tendinopathy, one open skin lesion, and one tibia fracture—all occurred with Lokomat, an exoskeleton-type stationary gait device).

Another review performed in He et al. (2018) related to overground lower limb exoskeleton devices listed adverse events that occurred during the use of these devices. The adverse events included device malfunction, user error, skin and soft tissue damage, spasticity and hypotension-related events, falling, and bone fracture. From the range of different adverse events that occurred in overground exoskeleton-type devices, it is possible to deduce that the risk profile has some differences (e.g., the risk of falling is present in overground exoskeletons but not in stationary gait devices that use a body weight support) when compared to stationary gait devices, but both have also similarities and common causes (skin lesions and bruises most likely have the same common cause related to the device fixation over user's body and therapy duration that could have a cumulative effect).

Besides safety related to forces, actuation, range of motion, and fixation of a gait device, there are also safety aspects to be taken into account related to balance and stability for these devices. As previously mentioned, some gait devices (typically wearable exoskeletons) can present fall risks during their use and must have some countermeasures to assure the prevention of these risks. For example, in the event of power failure, Ekso exoskeletons lock the knees and let the hips free emulating a typical passive leg brace (US Food and Drug Administration, 2016a, b, 2020) as a failsafe feature. The Indego powered exoskeleton has the capability of detecting forward, backward, and sideways falling and takes actions to minimize the injury risks during the fall or even let the user recover without external assistance (US Food and Drug Administration, 2018a). The ReWalk has a feature called "Graceful Collapse" that is activated when the battery is depleted or in a power failure event. The exoskeleton slowly lowers the user to a seated position or on the ground, while supporting the user's weight (US Food and Drug Administration, 2018b).

In addition, the safety validation process of a rehabilitation robot is a crucial part of the design, given the adverse events that already took place in previous uses of those gait rehabilitation devices. Therefore, the identification of hazards is essential for the design process to be able to act and prevent those hazards, thus creating a safer product. The review from Bessler et al. (2021) listed some of these hazards, such as normal forces (exceeding user safe limits), shear forces (also exceeding user safe limits), misalignments, and movement range overrun (which can be different for each individual), serving as an auxiliary guide to an ever-evolving field seeking for more safety of its users.

4.5 Control techniques used on gait rehabilitation devices

The majority of gait devices, stationary or mobile, exoskeleton-type or end–effector-type, utilize a control system to deliver proper gait therapy and to provide safety during use. Specific pathologies may require specific approaches and a control system that provides those specific approaches can be an important tool in achieving the best results of rehabilitation. Also, a control system is useful if a subject starts gait therapy in a condition where he requires a higher level of assistance, but needs to be challenged progressively to improve his condition over time. Sun et al. (2022) reviewed some control approaches used for lower limb exoskeletons: mode control, gait control, impedance control, metabolic control, force-aware control, vision control, and predictive control.

Using mode control in an exoskeleton, signals that can be prior or posterior are used to obtain the device's actual status aiming to predict torque actuation and motion. It has been applied in situations where the trajectories are predefined according to different stages of the gait cycle, so it is possible to generate optimal control plans applicable to each stage. An example of this approach is shown in Ha et al. (2012), using a finite state machine (FSM) controller coupled with a PD controller.

In gait control, a gait detection algorithm provides real-time information about the subject's present gait, and a module generates the predicted action. The gait detection algorithms can be divided into two approaches: rule-based classifier and pattern recognition classifier. The rule-based classifier approach utilizes a set of possible gait patterns and through discrimination, classifies gait and takes action by switching between predefined gaits. A pattern recognition classifier uses machine learning and statistics to process data from acquired signals and classifies gait accordingly, using various techniques like artificial neural network (ANN) or support vector machine (SVM). This approach can be seen in the REX exoskeleton study conducted in Kwak et al. (2014).

Impedance control is a control technique that adjusts the mechanical impedance of a robot based on force and position control (Akdogan and Aktan, 2020; Cruz et al., 2021). In this modality of control, the controller does not step in unless a deviation between the user's motion and expected motion is detected. Examples can be seen in dos Santos and Siqueira (2019) and Chen et al. (2020).

Metabolic control refers to a control modality that uses sensor data from metabolic variables as feedback, which in turn serves for the adjustment of the control laws used and level of assistance, often to optimize metabolic costs during rehabilitation and improve rehabilitation, as shown in Lee et al. (2020). Force–aware control seeks to optimize a set of indicators (like contact forces between user and device) through the detection of exoskeleton quantities that are representative of that indicator, measured using sensors. An example is the Sarcos exoskeleton (Young and Ferris, 2017), although it is not an exoskeleton used for rehabilitation.

Vision control makes use of cameras and/or infrared sensors to perceive the space in which the exoskeleton is located and to perceive also the present motion state to allow the controller to decide the best approach based on these inputs, as shown in Liu et al. (2021). Predictive control, as the name implies, tries to predict the subject's motion using gait variables like the ground reaction force or, in the case of hemiplegic subjects, using healthy side movement variables. An approach using plantar forces is described in Ding et al. (2020).

Many other control techniques are used in gait devices and the field keeps evolving and proposing new approaches. The control strategies listed in this section provide an overview of the broad and diverse field of research that is the control engineering of these devices.

4.6 Challenges and open issues

There are many devices, both commercially or under research, targeting gait training for a variety of conditions. Stationary exoskeleton devices using treadmill gait training have been extensively tested and can be considered safe and effective when providing gait training with intensity, repeatability, and providing engagement, especially when integrated with serious games (Michaud et al., 2017; Noveletto et al., 2020). A review performed by Mehrholz et al. (2020) showed that RAGT plus physiotherapy produces superior results when compared to conventional gait training for poststroke individuals (but with no clear distinction or role for device type). Also, the first 3 months after a stroke can be particularly beneficial (Mehrholz et al., 2020). Randomized clinical trials (RCTs) showed that this intervention coupled with physiotherapy can provide superior results than the only conventional gait training (Mehrholz et al., 2020) in becoming independent in walking. Some cited limitations of these types of devices are that they are less accommodating of user's gait deviations (Esquenazi and Talaty, 2019) (enforcing a gait pattern) and have shown some Adverse effects (AEs), such as musculoskeletal AEs and blood pressure changes (Bessler et al., 2020). The need for exoskeleton alignment with the patient's joints and several human-robot interfaces presents injuries and discomfort risks.

Stationary end-effector devices can also be considered safe and capable of providing gait training with intensity, repeatability, and engagingly (especially when using serious games). They can generate gait trajectories and reproduce gait patterns well. Besides the task of walking, it is also possible to simulate stair-climbing tasks (Callegaro et al., 2014). They have too superior results when compared to conventional gait training in becoming independent in walking (Mehrholz et al., 2020). However, they do not generate a true swing phase and emulate ground contact, since the footplates are always in contact with the patient's feet. Some risks of AEs are also present, in this case, soft tissue AEs and blood pressure changes (Bessler et al., 2020).

Compared to the stationary type, overground exoskeleton devices can provide more engagement and more realistic contact with the ground, whilst still providing safety, training intensity, and repeatability. However, due to their very nature, they might require more supervision and have autonomy constraints linked to battery duration. They too performed better than conventional gait training regarding walking independence. They can also provide other tasks like sit-to-stand and turning. Some devices might not provide a realistic gait pattern to conserve energy (Sankai, 2010). AEs were also reported in wearable exoskeletons, like fall risks, orthostatic hypertension, skin abrasions, low back pain, and fatigue of the upper extremities (Rodríguez-Fernández et al., 2021).

All devices have their limitations, such as no active or passive ankle joint to prevent drop foot issues and injuries during gait training (AutoAmbulator, LOPES, Walking Assist Device), or no stair-climbing ability in the case of exoskeletons (EksoGT, EksoNR, Indego, ReWalk P6.0), limited battery operation (exoskeletons), no exact reproduction of heel strike ground contact (Gait Trainer GT II, G-EO System, Morning Walk, GaitMaster4, Exowalk), limitations to specific gait rehabilitation conditions (Bionic Leg for unsymmetrical gait and Walking Assist Device for elderly gait retraining) or do not control all joint trajectories to keep them as close to natural gait as possible (knee joint free—Gait Trainer GT II, G-EO System, Morning Walk, GaitMaster4, or ARTHuR). A cited limitation for the safety of users is the lack of knowledge of safe limits for interaction forces (shear forces or pressures), and the ways of measuring them effectively (Bessler et al., 2021), indicating that more research must be done to obtain these answers and increment current device safety. Control challenges are also present, namely stability issues, comfort issues, calibration issues, and energy requirements.

The use of cable-driven design to lower limb rehabilitation makes it inherently safe in human/robot interactions due to the extremely low inertia and the cables can be projected to break at a certain value (Cafolla et al., 2019; Alves et al., 2021, 2022). The risk of injuries due to errors in path planning and possible collision of the robot with a human are significantly reduced.

Despite many advancements, evidence of motor recovery and walking improvement is still considered moderate and the role of the type of device is not yet clear. New directions must be found to improve rehabilitation outcomes like a suggestion from a study to use interlimb coordination mechanisms (Hobbs and Artemiadis, 2020). Although the results do not show great promise on average, there is a portion of patients who experiences slightly better results, enough to be above the minimally clinically important difference (MCID). The minimally clinically important difference is defined as the smallest benefit of value to patients (McGlothlin and Lewis, 2014). A relevant problem consists of segregating which patients respond to which protocol to improve these results.

A positive aspect of robotic rehabilitation devices is that they enabled scientific advances in rehabilitation knowledge. Since they require programming and definition of specific rehabilitation exercises, they enabled clinicians to better limit the scope of treatment with more precision. In addition, many devices allowed the acquisition of exercise data that can be used in studies (present and future). In this sense, these characteristics help improve future hypotheses related to movement training (not only gait) and how to track and predict recovery. Some scientific findings that were achieved due to robotic devices, like the importance of active patient effort (efference) (Hicks, 2020), the tendency of "slack" (Smith et al., 2018) when assistance is provided or the importance of proprioception (Ingemanson et al., 2019) for the effectiveness of motor learning can also be pointed out as positive outcomes of robotic rehabilitation.

There is skepticism around the idea that continuous improvement of devices and fine-tuning are likely to provide meaningful and sizeable improvements in motor recovery. Some argue that plasticity-enhancing interventions are needed. Other criticism is that robotic rehabilitation lacks neuroscience and that the room for improvement is related to better and clearer biological thinking instead of engineering advancements (Krakauer and Reinkensmeyer, 2022).

Some debate also exists on whether devices that offer physical assistance can be more effective than those that do not. In gait rehabilitation devices, this might not even be a choice, since sometimes physical assistance is required or the walking task cannot even be performed. According to Krakauer and Reinkensmeyer (2022), benefits from assistance can come in at least three ways: 1—motivation improvement (Rowe et al., 2017) which can lead to better results, 2—neural strengthening achieved by repeated stimulation (Reinkensmeyer et al., 2012), and 3—promotion of somatosensory stimulation to drive beneficial cortical plasticity (Krakauer and Reinkensmeyer, 2022).

The financial burden is also another challenge for RAGT. Although having moderately better outcomes than conventional therapy, costs for using RAGT can double or even triple the cost of the conventional approach, limiting the adoption of these technologies (Carpino et al., 2018). Lowering this financial burden of gait rehabilitation technology have the potential to improve healthcare outcomes by offering the therapy to more people, which justify the single DOF approaches cited before, and could even provide similar results if they could attain the same level of intensity, repeatability, task-specific, and engagement than RAGT.

Some interesting low-cost solutions are under development in Gonçalves et al. (2019), Barbosa et al. (2018), Gonca,lves and Rodrigues (2022), Iancu et al. (2017), Russo and Ceccarelli (2020), Gherman et al. (2019), Pisla et al. (2021), Vaida et al. (2020), Lazar et al. (2018), Pop et al. (2020), and Cafolla et al. (2019). In common, these have a low number of actuators/DOFs, facilitating the control system design and aiming at the specific rehabilitation of individual joints of the lower limb. These were designed to be easy to use by both patients and healthcare professionals, but still lack further experimental validation with clinical tests.

4.7 Conclusions

Gait devices for stroke rehabilitation provide a promising perspective of improvement if used properly in addition to traditional gait therapy. They can be a powerful ally in gait rehabilitation, for both patients and therapists. The need for this type of device will probably increase in the coming years when population aging is taken into account, stressing healthcare systems around the world, and producing higher demand for rehabilitation. As a result of this growing necessity, robotic rehabilitation must achieve better and safer results. Improvements in gait outcomes are still moderate, and many advances must be made to increment user safety and therapy outcomes. There's progress still to be made in several different areas, from the mechanical construction of a gait device that alleviates stresses in the subject's body and prevents injuries, to advanced control systems that maintain gait stability, prevent falls, and optimize recovery during therapy sessions. More knowledge must be acquired regarding the safe limits of robot-user interaction forces, shear stresses, and misalignments. Also, good methods of measurement of these interactions are paramount to obtaining this knowledge and monitoring these interactions.

The population aging will also create an increased financial burden on healthcare systems, bringing the need for more cost-effective devices for this type of therapy to be broadened and more utilized. Also, understanding which patients should try which type of therapies is of great importance for the future of robotic rehabilitation since that can be a source of increase in recovery outcomes. Therefore, despite the many advancements already achieved, there's still a long way to go to fulfil the promise of robotic rehabilitation increased outcomes.

References

Akdogan, E., Aktan, M.E., 2020. Impedance control applications in therapeutic exercise robots. In: Control Systems Design of Bio-Robotics and Bio-Mechatronics With Advanced Applications, Elsevier, pp. 395–443.

Alves, T., Gonçalves, R.S., Carbone, G., 2021. Quantitative progress evaluation of post-stroke patients using a novel bimanual cable-driven robot. J. Bionic Eng. 18 (6), 1331–1343.

Alves, T., Gonçalves, R.S., Carbone, G., 2022. Serious games strategies with cable-driven robots for bimanual rehabilitation: a randomized controlled trial with post-stroke patients. Front. Robot. AI 9, 739088.

Anaya, F., Thangavel, P., Yu, H., 2018. Hybrid FES-robotic gait rehabilitation technologies: a review on mechanical design, actuation, and control strategies. Int. J. Intell. Robot. Appl. 2 (1), 1–28.

Asbeck, A.T., Schmidt, K., Walsh, C.J., 2015. Soft exosuit for hip assistance. Robot. Auton. Syst. 73, 102–110.

Authorized Representative Obelis sa Bd Général Wahis, European, 2015. Bionic Leg Orthosis User Manual. www.alterg.com.

Barbosa, A.M., Carvalho, J.C.M., Gonçalves, R.S., 2018. Cable-driven lower limb rehabilitation robot. J. Braz. Soc. Mech. Sci. Eng. 40 (5), 1–11.

Bertaux, A., Gueugnon, M., Moissenet, F., Orliac, B., Martz, P., Maillefert, J.-F., Ornetti, P., Laroche, D., 2022. Gait analysis dataset of healthy volunteers and patients before and 6 months after total hip arthroplasty. Sci. Data 9 (1), 399.

Bessler, J., Prange-Lasonder, G.B., Schulte, R.V., Schaake, L., Prinsen, E.C., Buurke, J.H., 2020. Occurrence and type of adverse events during the use of stationary gait robots—a systematic literature review. Front. Robot. AI 7, 158.

Bessler, J., Prange-Lasonder, G.B., Schaake, L., Saenz, J.F., Bidard, C., Fassi, I., Valori, M., Lassen, A.B., Buurke, J.H., 2021. Safety assessment of rehabilitation robots: a review identifying safety skills and current knowledge gaps. Front. Robot. AI 8, 33.

Brown, A.W., Therneau, T.M., Schultz, B.A., Niewczyk, P.M., Granger, C.V., 2015. Measure of functional independence dominates discharge outcome prediction after inpatient rehabilitation for stroke. Stroke 46 (4), 1038–1044.

Bruni, M.F., Melegari, C., De Cola, M.C., Bramanti, A., Bramanti, P., Calabrò, R.S., 2018. What does best evidence tell us about robotic gait rehabilitation in stroke patients: a systematic review and meta-analysis. J. Clin. Neurosci. 48, 11–17. https://doi.org/10.1016/j.jocn.2017.10.048.

Buesing, C., Fisch, G., O'Donnell, M., Shahidi, I., Thomas, L., Mummidisetty, C.K., Williams, K.J., Takahashi, H., Rymer, W.Z., Jayaraman, A., 2015. Effects of a wearable exoskeleton stride management assist system (SMA®) on spatiotemporal gait characteristics in individuals after stroke: a randomized controlled trial. J. Neuroeng. Rehabil. 12 (1), 69.

Cafolla, D., Russo, M., Carbone, G., 2019. CUBE, a cable-driven device for Limb rehabilitation. J. Bionic Eng. 16 (3), 492–502.

Calabrò, R.S., Cacciola, A., Bertè, F., Manuli, A., Leo, A., Bramanti, A., Naro, A., Milardi, D., Bramanti, P., 2016. Robotic gait rehabilitation and substitution devices in neurological disorders: where are we now? Neurol. Sci. 37 (4), 503–514.

Callegaro, A.M., Unluhisarcikli, O., Pietrusinski, M., Mavroidis, C., 2014. Robotic systems for gait rehabilitation. In: Neuro-Robotics: From Brain Machine Interfaces to Rehabilitation Robotics, Springer, Dordrecht, pp. 265–283.

Cao, J., Xie, S.Q., Das, R., Zhu, G.L., 2014. Control strategies for effective robot assisted gait rehabilitation: the state of art and future prospects. Med. Eng. Phys. 36 (12), 1555–1566.

Carpino, G., Pezzola, A., Urbano, M., Guglielmelli, E., 2018. Assessing effectiveness and costs in robot-mediated lower limbs rehabilitation: a meta-analysis and state of the art. J. Healthc. Eng. 2018.

Chen, L., Wang, C., Song, X., Wang, J., Zhang, T., Li, X., 2020. Dynamic trajectory adjustment of lower limb exoskeleton in swing phase based on impedance control strategy. Proc. Inst. Mech. Eng. Part I J. Syst. Control Eng. 234 (10), 1120–1132.

Coupland, A.P., Thapar, A., Qureshi, M.I., Jenkins, H., Davies, A.H., 2017. The definition of stroke. J. R. Soc. Med. 110 (1), 9–12.

Cruz, A.B., Radke, M., Haninger, K., Krüger, J., 2021. How can the programming of impedance control be simplified? Proc. CIRP 97, 266–271.

Dietz, V., 2016. Clinical aspects for the application of robotics in locomotor neurorehabilitation. In: Neurorehabilitation Technology, Springer International Publishing, Cham, pp. 209–222.

Ding, M., Nagashima, M., Cho, S.-G., Takamatsu, J., Ogasawara, T., 2020. Control of walking assist exoskeleton with time-delay based on the prediction of plantar force. IEEE Access 8, 138642–138651.

dos Santos, W.M., Siqueira, A.A.G., 2019. Optimal impedance via model predictive control for robot-aided rehabilitation. Control Eng. Pract. 93, 104177.

Bionics, Ekso, 2020. Eksonr–ekso bionics. https://eksobionics.com/eksonr/.

Erbil, D., Tugba, G., Murat, T.H., Melike, A., Merve, A., Cagla, K., Mehmetali, Ç.C., Akay, Ö., Nigar, D., 2018. Effects of robot-assisted gait training in chronic stroke patients treated by botulinum toxin-a: a pivotal study. Physiother. Res. Int. 23 (3), e1718. https://doi.org/10.1002/pri.1718.

Esquenazi, A., Talaty, M., 2019. Robotics for lower limb rehabilitation. Phys. Med. Rehabil. Clin. 30 (2), 385–397.

Fisher, S., Lucas, L., Thrasher, T., 2015. Robot-assisted gait training for patients with hemiparesis due to stroke. Top Stroke Rehabil. 18 (3), 269–276. https://doi.org/10.1310/tsr1803-269.

Gardner, A.D., Potgieter, J., Noble, F.K., 2017. A review of commercially available exoskeletons' capabilities. In: 2017 24th International Conference on Mechatronics and Machine Vision in Practice, M2VIP 2017, vol. 2017-December. Institute of Electrical and Electronics Engineers Inc, pp. 1–5.

Gherman, B., Birlescu, I., Plitea, N., Carbone, G., Tarnita, D., Pisla, D., 2019. On the singularity-free workspace of a parallel robot for lower-limb rehabilitation. Proc. Rom. Acad. Ser. A Rom. Acad. 20 (4), 383–391.

Gonçalves, R.S., Krebs, H.I., 2017. MIT-Skywalker: considerations on the design of a body weight support system. J. Neuroeng. Rehabil. 14 (1), 1–11.

Gonça,lves, R.S., Rodrigues, L.A.O., 2022. Development of nonmotorized mechanisms for lower limb rehabilitation. Robotica 40 (1), 102–119.

Goncalves, R.S., Hamilton, T., Daher, A.R., Hirai, H., Krebs, H.I., 2017. MIT-Skywalker: evaluating comfort of bicycle/saddle seat. In: IEEE International Conference on Rehabilitation Robotics, August, IEEE Computer Society, pp. 516–520.

Gonçalves, R.S., Soares, G., Carvalho, J.C., 2019. Conceptual design of a rehabilitation device based on cam-follower and crank-rocker mechanisms hand actioned. J. Braz. Soc. Mech. Sci. Eng. 41 (7), 1–12.

Gonçalves, R.S., Salim, V.V., Krebs, H.I., 2021. Development of a markerless control system for the MIT-Skywalker. Int. J. Mech. Control 22 (1), 143–152.

Guo, X.-Y., Li, W.-B., Gao, Q.-H., Mukherjee, B., Dey, S.K., Pradhan, B.B., 2018. The recent trends and inspections about powered exoskeletons. IOP Conf. Ser. Mater. Sci. Eng. 377 (1), 012222.

Ha, K.H., Quintero, H.A., Farris, R.J., Goldfarb, M., 2012. Enhancing stance phase propulsion during level walking by combining FES with a powered exoskeleton for persons with paraplegia. In: Conference Proceedings: …Annual International Conference of the IEEE Engineering in Medicine and Biology Society. IEEE Engineering in Medicine and Biology Society, vol. 2012. NIH Public Access, p. 344.

He, Y., Eguren, D., Luu, T.P., Contreras-Vidal, J.L., 2018. Risk and adverse events related to lower-limb exoskeletons. In: 2017 International Symposium on Wearable Robotics and Rehabilitation, WeRob 2017, Institute of Electrical and Electronics Engineers Inc, pp. 1–2.

Hicks, A.L., 2020. Locomotor training in people with spinal cord injury: is this exercise? Spinal Cord 59 (1), 9–16.

Hirano, S., Saitoh, E., Tanabe, S., Tanikawa, H., Sasaki, S., Kato, D., Kagaya, H., Itoh, N., Konosu, H., 2017. The features of Gait Exercise Assist Robot: precise assist control and enriched feedback. NeuroRehabilitation 41 (1), 77–84.

Hobbs, B., Artemiadis, P., 2020. A review of robot-assisted lower-limb stroke therapy: unexplored paths and future directions in gait rehabilitation. Front. Neurorobot. 14, 19.

Hornby, T.G., Reisman, D.S., Ward, I.G., Scheets, P.L., Miller, A., Haddad, D., Fox, E.J., Fritz, N.E., Hawkins, K., Henderson, C.E., Hendron, K.L., Holleran, C.L., Lynskey, J.E., Walter, A., 2020. Clinical practice guideline to improve locomotor function following chronic stroke, incomplete spinal cord injury, and brain injury. J. Neurol. Phys. Ther. 44 (1), 49–100.

h/p/cosmos sports & medical gmbh, 2014. Original Directions for Use h/p/cosmos® Robowalk® Expander. h/p/cosmos sports & medical gmbh.

Iancu, C.A., Ceccarelli, M., Lovasz, E.C., 2017. Design and lab tests of a scaled leg exoskeleton with electric actuators. In: Proceedings of the 26th International Conference on Robotics in Alpe-Adria-Danube Region, RAAD 2017, July, vol. 49. Springer Netherlands, Torino, pp. 719–726.

Ingemanson, M.L., Rowe, J.R., Chan, V., Wolbrecht, E.T., Reinkensmeyer, D.J., Cramer, S.C., 2019. Somatosensory system integrity explains differences in treatment response after stroke. Neurology 92 (10), e1098–e1108.

Iqbal, S., Zang, X., Zhu, Y., Zhao, J., 2014. Bifurcations and chaos in passive dynamic walking: a review. Robot. Auton. Syst. 62 (6), 889–909.

Kalani, H., Tahamipour-Z, S.M., Kardan, I., Akbarzadeh, A., 2021. Application of DQN learning for delayed output feedback control of a gait-assist hip exoskeleton. In: 2021 9th RSI International Conference on Robotics and Mechatronics (ICRoM), November, IEEE, pp. 341–345.

Kim, J.Y., Kim, D.Y., Chun, M.H., Kim, S.W., Jeon, H.R., Hwang, C.H., Choi, J.K., Bae, S., 2019. Effects of robot-(Morning Walk®) assisted gait training for patients after stroke: a randomized controlled trial. Clin. Rehabil. 33 (3), 516–523.

Kim, D.Y., Jung, C., Kim, Y., Kwon, S., Chun, M.H., Kim, J., Kim, S.H., 2020. Morning walk®-assisted gait training improves walking ability and balance in patients with ataxia: a randomized controlled trial. Brain Neurorehabil. 13 (3). https://doi.org/10.12786/bn.2020.13.e23.

Kim, H.K., Seong, S., Park, J., Kim, J., Park, J., Park, W., 2021. Subjective evaluation of the effect of exoskeleton robots for rehabilitation training. IEEE Access 9, 130554–130561.

Krakauer, J.W., Reinkensmeyer, D.J., 2022. Epilogue: robots for neurorehabilitation—the debate. In: Reinkensmeyer, D.J., Marchal-Crespo, L., Dietz, V. (Eds.), Neurorehabilitation Technology. Springer, Cham, pp. 757–764.

Krebs, H.I., Michmizos, K., Susko, T., Lee, H., Roy, A., Hogan, N., 2016. Beyond human or robot administered treadmill training. In: Neurorehabilitation Technology, Springer International Publishing, Cham, pp. 409–433.

Krishnamurthi, R.V., Ikeda, T., Feigin, V.L., 2020. Global, regional and country-specific burden of ischaemic stroke, intracerebral haemorrhage and subarachnoid haemorrhage: a systematic analysis of the global burden of disease study 2017. Neuroepidemiology 54 (2), 171–179. https://doi.org/10.1159/000506396.

Kwak, N.-S., Muller, K.-R., Lee, S.-W., 2014. Toward exoskeleton control based on steady state visual evoked potentials. In: 2014 International Winter Workshop on Brain-Computer Interface (BCI), February, IEEE, pp. 1–2.

Lazar, V.A., Pisla, D., Vaida, C., Cafolla, D., Ceccarelli, M., Carbone, G., León, J.F.R., 2018. Experimental characterization of assisted human arm exercises. In: 2018 IEEE International Conference on Automation, Quality and Testing, Robotics (AQTR), July, Institute of Electrical and Electronics Engineers Inc, pp. 1–6.

Lee, D., Kwak, E.C., McLain, B.J., Kang, I., Young, A.J., 2020. Effects of assistance during early stance phase using a robotic knee orthosis on energetics, muscle activity, and joint mechanics during incline and decline walking. IEEE Trans. Neural Syst. Rehabil. Eng. 28 (4), 914–923.

Lee, H.Y., Park, J.H., Kim, T.W., 2021. Comparisons between Locomat and Walkbot robotic gait training regarding balance and lower extremity function among non-ambulatory chronic acquired brain injury survivors. Medicine 100 (18), e25125.

Lim, B., Jang, J., Lee, J., Choi, B., Lee, Y., Shim, Y., 2019. Delayed output feedback control for gait assistance and resistance using a robotic exoskeleton. IEEE Robot. Autom. Lett. 4 (4), 3521–3528.

Liu, D.-X., Xu, J., Chen, C., Long, X., Tao, D., Wu, X., 2021. Vision-assisted autonomous lower-limb exoskeleton robot. IEEE Trans. Syst. Man Cybern. Syst. 51 (6), 3759–3770.

Lv, G., Zhu, H., Gregg, R.D., 2018. On the design and control of highly backdrivable lower-limb exoskeletons: a discussion of past and ongoing work. IEEE Control Syst. 38 (6), 88–113.

McGlothlin, A.E., Lewis, R.J., 2014. Minimal clinically important difference. JAMA 312 (13), 1342.

Mehrholz, J., Thomas, S., Kugler, J., Pohl, M., Elsner, B., 2020. Electromechanical-assisted training for walking after stroke. Cochrane Database Syst. Rev. 2020 (10). https://doi.org/10.1002/14651858.CD006185.pub5.

Michaud, B., Cherni, Y., Begon, M., Girardin-Vignola, G., Roussel, P., 2017. A serious game for gait rehabilitation with the Lokomat. In: International Conference on Virtual Rehabilitation, ICVR, August,
vol. 2017-June. Institute of Electrical and Electronics Engineers Inc.

Nam, K.Y., Kim, H.J., Kwon, B.S., Park, J.-W., Lee, H.J., Yoo, A., 2017. Robot-assisted gait training
(Lokomat) improves walking function and activity in people with spinal cord injury: a systematic review.
J. Neuroeng. Rehabil. 14 (1), 24.

Nam, Y.G., Lee, J.W., Park, J.W., Lee, H.J., Nam, K.Y., Park, J.H., Yu, C.S., Choi, M.R., Kwon, B.S.,
2019. Effects of electromechanical exoskeleton-assisted gait training on walking ability of stroke patients:
a randomized controlled trial. Arch. Phys. Med. Rehabil. 100 (1), 26–31.

Noveletto, F., Soares, A.V., Eichinger, F.L.F., Domenech, S.C., Da Hounsell, M.S., Filho, P.B., 2020. Biomedical serious game system for lower limb motor rehabilitation of hemiparetic stroke patients. IEEE
Trans. Neural Syst. Rehabil. Eng. 28 (6), 1481–1487.

Ochi, M., Wada, F., Saeki, S., Hachisuka, K., 2015. Gait training in subacute non-ambulatory stroke patients
using a full weight-bearing Gait-Assistance Robot: a prospective, randomized, open, blinded-endpoint
trial. J. Neurol. Sci. 353 (1–2), 130–136.

Park, J.H., Shin, Y.I., You, J.H., Park, M.S., 2018. Comparative effects of robotic-assisted gait training combined with conventional physical therapy on paretic hip joint stiffness and kinematics between subacute
and chronic hemiparetic stroke. NeuroRehabilitation 42 (2), 181–190.

Pisla, D., Nadas, I., Tucan, P., Albert, S., Carbone, G., Antal, T., Banica, A., Gherman, B., 2021. Development of a control system and functional validation of a parallel robot for lower limb rehabilitation.
Actuators 10 (10), 277.

Pop, N., Ulinici, I., Pisla, D., Carbone, G., Nysibalieva, A., 2020. Kinect based user-friendly operation of
LAWEX for upper limb exercising task. In: Proceedings of 2020 IEEE International Conference on
Automation, Quality and Testing, Robotics (AQTR), May, pp. 185–189.

Prakash, C., Kumar, R., Mittal, N., 2018. Recent developments in human gait research: parameters,
approaches, applications, machine learning techniques, datasets and challenges. Artif. Intell. Rev.
49 (1), 1–40.

Reinkensmeyer, D., Aoyagi, D., Emken, J., Galvez, J., Ichinose, W., Kerdanyan, G., Nessler, J.,
Maneekobkunwong, S., Timoszyk, B., Vallance, K., Weber, R., Wynne, J., de Leon, R.,
Bobrow, J., Harkema, S., Edgerton, V., 2004. Robotic gait training: toward more natural movements
and optimal training algorithms. In: 26th Annual International Conference of the IEEE Engineering in
Medicine and Biology Society, Institute of Electrical and Electronics Engineers (IEEE), pp. 4818–4821.

Reinkensmeyer, D.J., Guigon, E., Maier, M.A., 2012. A computational model of use-dependent motor
recovery following a stroke: optimizing corticospinal activations via reinforcement learning can explain
residual capacity and other strength recovery dynamics. Neural Netw. 29–30, 60–69.

Riener, R., 2016. Technology of the robotic gait orthosis Lokomat. In: Neurorehabilitation Technology,
Springer International Publishing, Cham, pp. 395–407.

Rodríguez-Fernández, A., Lobo-Prat, J., Font-Llagunes, J.M., 2021. Systematic review on wearable lower-
limb exoskeletons for gait training in neuromuscular impairments. J. Neuroeng. Rehabil. 18 (1), 1–21.

Rojek, A., Mika, A., Oleksy, Ł., Stolarczyk, A., Kielnar, R., 2020. Effects of exoskeleton gait training on
balance, load distribution, and functional status in stroke: a randomized controlled trial. Front. Neurol.
10, 1344.

Rose, L., Bazzocchi, M.C.F., Nejat, G., 2021. A model-free deep reinforcement learning approach for control of exoskeleton gait patterns. Robotica 40, 2189–2214.

Rowe, J.B., Chan, V., Ingemanson, M.L., Cramer, S.C., Wolbrecht, E.T., Reinkensmeyer, D.J., 2017.
Robotic assistance for training finger movement using a Hebbian model: a randomized controlled trial.
Neurorehabil. Neural Repair 31 (8), 769–780.

Russo, M., Ceccarelli, M., 2020. Analysis of a wearable robotic system for Ankle rehabilitation. Machines 8
(3). https://www.mdpi.com/2075-1702/8/3/48/htm.

Sabaapour, M.R., Lee, H., Afzal, M.R., Eizad, A., Yoon, J., 2019. Development of a novel gait rehabilitation device with hip interaction and a single DOF mechanism. In: Proceedings—IEEE International
Conference on Robotics and Automation, May, vol. 2019. Institute of Electrical and Electronics Engineers Inc, pp. 1492–1498.

Newsroom, Samsung, 2019. Get a glimpse of the next-generation innovations on display at Samsung's technology showcase. https://news.samsung.com/global/get-a-glimpse-of-the-next-generation-innovations-on-display-at-samsungs-technology-showcase.

Sankai, Y., 2010. HAL: hybrid assistive limb based on cybernics. In: Kaneko, M., Nakamura, Y. (Eds.), Robotics Research. Springer Tracts in Advanced Robotics, vol. 66. Springer, Berlin, Heidelberg, pp. 25–34. https://doi.org/10.1007/978-3-642-14743-2.

Schmidt, R.A., Lee, T.D., 2013. Motor Learning and Performance From Principles to Application, fifth ed. Human Kinetics, pp. 1–482.

Shin, S.Y., Deshpande, A.D., Sulzer, J., 2018. Design of a single degree-of-freedom, adaptable electrome-chanical gait trainer for people with neurological injury. J. Mech. Robot. 10 (4), 1–7.

Smania, N., Geroin, C., Valè, N., Gandolfi, M., 2018. The end-effector device for gait rehabilitation. In: Advanced Technologies for the Rehabilitation of Gait and Balance Disorders, vol. 19. Springer International Publishing, pp. 267–283.

Smith, B.W., Rowe, J.B., Reinkensmeyer, D.J., 2018. Real-time slacking as a default mode of grip force control: implications for force minimization and personal grip force variation. J. Neurophysiol. 120, 2107–2120.

Sucuoglu, H., 2020. Effects of robot-assisted gait training alongside conventional therapy on the development of walking in children with cerebral palsy. J. Pediatr. Rehabil. Med. 13 (2), 127–135.

Sun, Y., Tang, Y., Zheng, J., Dong, D., Chen, X., Bai, L., 2022. From sensing to control of lower limb exoskeleton: a systematic review. Annu. Rev. Control 53, 83–96.

Tanaka, N., Saitou, H., Takao, T., Iizuka, N., Okuno, J., Yano, H., Tamaoka, A., Yanagi, H., 2012. Effects of gait rehabilitation with a footpad-type locomotion interface in patients with chronic post-stroke hemiparesis: a pilot study. Clin. Rehabil. 26 (8), 686–695. https://doi.org/10.1177/0269215511432356.

Tesio, L., Rota, V., 2019. The motion of body center of mass during walking: a review oriented to clinical applications. Front. Neurol. 10, 999.

Tsuge, B.Y., McCarthy, J.M., 2016. An adjustable single degree-of-freedom system to guide natural walking movement for rehabilitation. J. Med. Dev. 10 (4). https://doi.org/10.1115/1.4033329.

Tsuge, B.Y., Plecnik, M.M., McCarthy, J.M., 2016. Homotopy directed optimization to design a six-bar linkage for a lower limb with a natural ankle trajectory. J. Mech. Robot. 8 (6), 1–7.

US Food and Drug Administration, 2016a. 510(k) Summary for the Ekso and Ekso GT exoskeletons. https://www.accessdata.fda.gov/scripts/cdrh/cfdocs/cfpmn/pmn.cfm?ID=K143690.

US Food and Drug Administration, 2016b. 510(k) Summary for the Ekso and Ekso GT exoskeletons. https://www.accessdata.fda.gov/scripts/cdrh/cfdocs/cfpmn/pmn.cfm?ID=K161443.

US Food and Drug Administration, 2018a. 510(k) Summary for the Indego Exoskeleton. PSC Publishing Services. https://www.accessdata.fda.gov/scripts/cdrh/cfdocs/cfpmn/pmn.cfm?ID=K173530.

US Food and Drug Administration, 2018b. 510(k) Summary for the ReWalk Exoskeleton. PSC Publishing Services. https://www.accessdata.fda.gov/scripts/cdrh/cfdocs/cfPMN/pmn.cfm?ID=K200032.

US Food and Drug Administration, 2020. 510(k) Summary for the EksoNR Exoskeleton. PSC Publishing Services. https://www.accessdata.fda.gov/scripts/cdrh/cfdocs/cfpmn/pmn.cfm?ID=K200574.

Vaida, C., Birlescu, I., Pisla, A., Ulinici, I.M., Tarnita, D., Carbone, G., Pisla, D., 2020. Systematic design of a parallel robotic system for lower limb rehabilitation. IEEE Access 8, 34522–34537.

van Asseldonk, E.H.F., van der Kooij, H., 2016. Robot-aided gait training with LOPES. In: Neurorehabilitation Technology, Springer International Publishing, Cham, pp. 461–481.

Van der Loos, H.F.M., Reinkensmeyer, D.J., Guglielmelli, E., 2016. Rehabilitation and health care robotics. In: Springer Handbook of Robotics, Springer Science and Business Media Deutschland GmbH, pp. 1685–1728.

Van Tran, Q., Kim, S., Lee, K., Kang, S., Ryu, J., 2015. Force/torque sensorless impedance control for indirect driven robot-aided gait rehabilitation system. In: IEEE/ASME International Conference on Advanced Intelligent Mechatronics, AIM, August, vol. 2015. Institute of Electrical and Electronics Engineers Inc, pp. 652–657.

Varela, M.J., Ceccarelli, M., Flores, P., 2015. A kinematic characterization of human walking by using CaTraSys. Mech. Mach. Theory 86, 125–139.

Veneman, J.F., Ekkelenkamp, R., Kruidhof, R., Van Der Helm, F.C.T., Van Der Kooij, H., 2016. A series elastic- and Bowden-cable-based actuation system for use as torque actuator in exoskeleton-type robots. Int. J. Robot. Res. 25 (3), 261–281. https://doi.org/10.1177/0278364906063829.

Watanabe, S., Kotani, T., Taito, S., Ota, K., Ishii, K., Ono, M., Katsukawa, H., Kozu, R., Morita, Y., Arakawa, R., Suzuki, S., 2019. Determinants of gait independence after mechanical ventilation in the intensive care unit: a Japanese multicenter retrospective exploratory cohort study. J. Intensive Care 7 (1), 53.

Watanabe, H., Marushima, A., Kadone, H., Shimizu, Y., Kubota, S., Hino, T., Sato, M., Ito, Y., Hayakawa, M., Tsurushima, H., Maruo, K., Hada, Y., Ishikawa, E., Matsumaru, Y., 2021. Efficacy and safety study of wearable cyborg HAL (hybrid assistive limb) in hemiplegic patients with acute stroke (Early Gait Study): protocols for a randomized controlled trial. Front. Neurosci. 15, 785.

Young, A.J., Ferris, D.P., 2017. State of the art and future directions for lower limb robotic exoskeletons. IEEE Trans. Neural Syst. Rehabil. Eng. 25, 171–182.

CHAPTER 5

Robotic devices for upper limb rehabilitation: A review

Kishor Lakshmi Narayanan[a], Tanvir Ahmed[b], Md Mahafuzur Rahaman Khan[b], Tunajjina Kawser[c], Raouf Fareh[d], Inga Wang[b,e], Brahim Brahmi[b,f], and Mohammad Habibur Rahman[b]

[a]Neuro-Rehabilitation Lab, School of Electronics Engineering, Vellore Institute of Technology, Vellore, India
[b]Biorobotics Laboratory, University of Wisconsin-Milwaukee, Milwaukee, WI, United States
[c]Department of Anatomy, Shaheed Tajuddin Ahmad Medical College, Gazipur, Bangladesh
[d]Electrical Engineering Department, University of Sharjah, Sharjah, United Arab Emirates
[e]Rehabilitation Outcomes Research Lab, University of Wisconsin-Milwaukee, Milwaukee, WI, United States
[f]Electrical Engineering Department at the College of Ahuntsic, Montréal, Canada

5.1 Introduction

Upper limb rehabilitation devices are necessary for individuals who suffer from injuries or conditions that affect the use of their arms or hands. These devices can help individuals regain strength, range of motion, and function in their affected limbs. One common condition that requires upper limb rehabilitation is stroke. According to the American Heart Association, around 795,000 strokes occur in the United States annually (Tsao et al., 2022). Similarly, the World Stroke Organization reports that 12.2 million people experience a stroke worldwide each year, with two-thirds of them surviving (Feigin et al., 2022). These figures demonstrate that a significant number of stroke survivors are left with impairments, with 25% experiencing minor impairments and 40% suffering from moderate-to-severe impairments that require special care (American Stroke Association, 2019). Stroke is a leading cause of long-term disability, particularly affecting the upper limbs. Another condition that may require upper limb rehabilitation is spinal cord injury. A spinal cord injury can result in the loss of motor function and sensation in the arms and hands. Other causes of upper limb impairments include sports injuries, traumas, and occupational injuries. Thus, rehabilitation programs that promote functional recovery of individuals with upper limb dysfunction (ULD) are crucial in restoring independence and require dedicated efforts from both the clinician and patient.

To rehabilitate upper limb impairments, task-oriented repetitive movement therapy that typically involves one-on-one interaction with a therapist has been shown to be an effective method (Poli et al., 2013; Demircan et al., 2020). As cases of ULD are increasing, robot-aided therapeutic intervention has the potential to be an effective solution. Studies have found that repetitive robot-assisted rehabilitation programs can decrease upper limb motor impairment significantly (Amirabdollahian et al., 2007; Gandolfi

Medical and Healthcare Robotics
https://doi.org/10.1016/B978-0-443-18460-4.00005-6

123

et al., 2018; Veerbeek et al., 2017; Kim et al., 2017; Lee et al., 2017; Sale et al., 2014; Yoo and Kim, 2015). Robot-aided therapy has advantages over conventional manual therapy, as it can provide therapy for longer periods of time, is more precise, and provides better quantitative feedback (Teasell and Kalra, 2004). There are two types of robot-aided rehabilitation devices: end effector and exoskeleton. An end effector-type rehabilitation robot/device is a kind of rehabilitation device that does not actively support or hold the subject's upper limb but connects with the subject's hand or forearm (see Fig. 5.1A). These types of rehabilitation robots/devices are suitable for multijoint upper limb exercises, such as reaching movements, where the cooperative and simultaneous movements of the shoulder, elbow, and wrist joints are involved.

On the other hand, exoskeletons are wearable robots or powered orthoses designed to be worn on the lateral side of the upper arm (see Fig. 5.1B). Ideally, the kinematic structure (joint articulation, link length, and range of motion) of exoskeleton robots is similar to that of the human upper limb. Thus, exoskeleton robots can mimic the natural range of upper limb motion. These robots are suitable for individual joint movement (such as elbow flexion/extension or forearm pronation/supination) and multijoint movement exercises.

Rehabilitation robots work by providing patients with repetitive, task-oriented movements. These movements are designed to mimic the natural movements of the upper limb and are intended to stimulate the neural pathways that control movement. This type of robot provides resistance or assistance to the patient as they perform movements, helping to improve their strength and coordination. Several types of end effector robots have been developed for upper limb rehabilitation, including MIT-MANUS (Hogan et al., 1992; Krebs et al., 2007, 2016), ARM-Guide (Reinkensmeyer et al., 2000), and GENTLE/s (Coote et al., 2008), each with its own design and capabilities. Studies have found that the use of these devices can lead

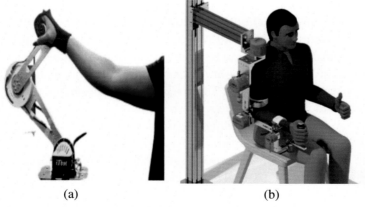

(a)   (b)

Fig. 5.1 Schematic of (A) an end effector-type robot and (B) an exoskeleton-type robot.

to improvements in upper limb function in patients with conditions such as stroke or spinal cord injury. A study by Poli et al. (2013) found that robot-assisted therapy for the upper limb is more effective in improving hand function and motor recovery compared to conventional therapy alone in stroke patients. Another study by Demircan et al. (2020) showed that patients who received robot-assisted therapy had improved range of motion and muscle strength in their affected upper limbs.

In addition to upper limb rehabilitation, end effector-type robots have a wide range of applications. For example, this type of robot can be used to aid in gait rehabilitation of patients who have suffered from neurological disorders such as stroke or spinal cord injury. A study by Mazzoleni et al. (2017) found that the use of a robotic end effector device, specifically the G-EO System (Waldner et al., 2009), improved gait kinematics and muscle activity in stroke patients during gait rehabilitation. Similarly, end effector-type robots can be used for postsurgery rehabilitation, particularly for patients who have undergone orthopedic surgery and need improvement in knee joint kinematics and muscle activity (Kuroda et al., 2021).

Exoskeleton-based robots are designed to provide individual joint movement and are able to map onto human anatomical joints. This allows them to provide therapy that is more specific to the individual needs of the patient. Exoskeleton devices have an advantage over end effector devices as they offer complete control over a patient's individual joint movement and applied torque, more precise guidance of motion, a larger range of motion, and more accurate quantitative feedback. Many exoskeleton prototypes have been developed for upper limb rehabilitation, such as CADEN-7 (Perry et al., 2007), ARMin (Nef et al., 2009), 6-REXOS (Gunasekara et al., 2015), RehabArm (Liu et al., 2016), CAREX-7 (Cui et al., 2017), BLUE SABINO (Perry et al., 2019), u-Rob (Islam et al., 2020a, b), and PWRR (Zhang et al., 2020), among others. However, exoskeleton-based rehabilitation robots are often more expensive and bulkier than end effector-type robots, which can make them less accessible to some patients. Additionally, exoskeleton-based robots may not be as comfortable to wear for long periods of time as end effector-type robots.

In addition to rehabilitation, robotic exoskeletons have been developed for other applications as well. For example, exoskeletons have been designed to augment human strength while handling heavy loads in unstructured environments. Researchers have also developed exoskeletons to be used as body extenders (Marcheschi et al., 2011) and to improve maneuverability (Kazerooni, 2005). Some exoskeletons have been controlled using electromyography (EMG) signals of the agonist muscles to enhance power augmentation (Zhu et al., 2014). Other exoskeletons have been developed as underactuated exoskeletons to augment human power for lifting heavy loads (Walsh et al., 2006). This chapter reviews the design, control, and scope of notable end effector- and exoskeleton-type robots for upper limb rehabilitation.

5.2 End effector-type upper limb rehabilitation robots

This section presents some notable end effector-type robots through case studies.

5.2.1 InMotion2

InMotion2 (see Fig. 5.2) was developed by Interactive Motion Technologies (IMT), a company that specializes in developing robotic devices for rehabilitation and therapy. This device was first developed by MIT's Biomechatronics group, led by Professor Hugh Herr, who is now the founder and CTO of IMT. This device is commercially available and is designed to provide repetitive, task-oriented movements to the upper limb for patients with impairments caused by conditions such as stroke or traumatic injuries.

Design and control: InMotion2 has a 2-degree-of-freedom design and can provide resistance or assistance to the patient as they perform movements. The device is able to mimic the natural movements of the upper limb and is intended to stimulate the neural pathways that control movement. The robot is equipped with sensors that track the patient's movements and provide feedback to the therapist on their progress. It also has adjustable range of motion and resistance settings, allowing the therapist to customize the therapy to the patient's specific needs. The robot also provides real-time data on the patient's performance, which can be used to track progress and adjust the therapy plan.

During the therapy sessions, the patient sits or stands in front of the robot and uses their affected limb to perform exercises such as reaching, grasping, and stepping movements. The robot's robotic arm guides the patient's limb through the desired movements, providing resistance to challenge the patient's muscles and help them regain strength and function.

Fig. 5.2 InMotion2 (Lee et al., 2020).

Clinical studies: Several studies have been conducted to investigate the effectiveness of this device in improving motor function and the overall quality of life of disabled individuals. One study by Lo and Xie (2012) compared the use of the InMotion2 robot to traditional physical therapy in a group of stroke patients. The study consisted of 45 patients who were randomly assigned to either the InMotion2 group or the traditional physical therapy group. Both groups received the same amount of therapy, with the InMotion2 group receiving therapy using the robot for 1 h per day and the traditional group receiving therapy for 1 h per day without the use of the robot. The study found that the InMotion2 group showed significantly greater improvement in upper extremity motor function, as measured by the Fugl–Meyer assessment (FMA), compared to the traditional group. Additionally, the InMotion2 group showed a greater improvement in activities of daily living, as measured by the Barthel Index. The preceding version of InMotion2 is MIT-MANUS, which has been used in several clinical studies and is discussed in the following subsection.

5.2.1.1 MIT-MANUS

MIT-MANUS (see Fig. 5.3) is a robotic exoskeleton developed by researchers at the Massachusetts Institute of Technology (MIT) for upper limb rehabilitation (Hogan et al., 1995). It is a robotic device designed to provide repetitive, task-oriented movements to the upper limb for patients recovering from a stroke or other neurological injuries. The device enables patients to perform reaching movements in a horizontal plane and can assist or resist their movements while tracking arm position and forces applied.

Design and control: The device utilizes a motor to assist or oppose the patient's movement and is equipped with sensors to track hand position and speed (Hogan

Fig. 5.3 MIT-MANUS (Krebs et al., 2004).

et al., 1995). A 6-degree-of-freedom (DOFs) force sensor is located near the handle to measure the forces exerted by the patient along the rail and perpendicular to the intended movement. The device can be adjusted in elevation and yaw, and the range of movements can also be regulated. The patient's affected limb is secured in a splint that connects to the robot through a 6-degree-of-freedom force-torque sensor, allowing for the measurement of interaction forces during reaching tasks and for therapy adjustments. Additionally, the device is fully instrumented, enabling the therapist to infer the patient's limb position from the robot's position. MIT-MANUS is designed to be lightweight and portable, making it easy for patients to use it both at home and in clinical settings.

Clinical studies: Several studies have been conducted using the MIT-MANUS device to evaluate the effectiveness of robot-assisted therapy in upper limb function in patients with acute hemiparetic stroke. The goal of these studies was to determine whether patients who received robot-assisted therapy in addition to conventional therapy showed greater improvements in upper limb function compared to those who received "sham" robot therapy with their conventional therapy (Krebs et al., 1998, 2000; Volpe et al., 1999, 2000, 2001). The studies consisted of acute stroke patients using the MIT-MANUS to reach various targets, with the robot assisting them when they were unable to complete the movements. On average, patients completed 3 sets of 20 repetitions per session, totaling 4–5 h per week over a 7-week period. The control group received an additional hour of therapy per week, during which they used the device for 30 min with their unaffected arm and 30 min with their affected arm, but the motors of the device were not turned on.

Patients were evaluated both before and after the intervention, using measures such as the functional independence measure (FIM), the Fugl-Meyer functional impairment scale, strength measurements using the Medical Research Council Motor Power scale, and the Motor Status Score for the shoulder-elbow and wrist-hand complexes. After studying 96 acute stroke patients at Burke Rehabilitation Hospital, it was found that the group receiving robot-assisted therapy had significantly greater gains in elbow and shoulder motor function and strength compared to the control group. However, no significant differences were observed in Fugl-Meyer scores or FIM scores between the groups.

These studies suggest that robot-assisted therapy with the MIT-MANUS device can lead to functional improvements in stroke patients, but there are some concerns about the design of these studies. For example, the control group received only an additional hour of therapy per week, and 30 min of that hour were spent training the unaffected arm, which raises questions about the true comparison between the two types of interventions. Additionally, there are also concerns about the difference in baseline characteristics between the groups, such as the FIM scores and lesion volume. Despite these limitations, the studies show that patients who received robot-assisted therapy with the MIT-MANUS device had statistically significant gains in function after multiple sessions with the device.

5.2.2 ARM-GUIDE

The Assisted Rehabilitation and Measurement Guide (ARM-GUIDE) is a research device developed (see Fig. 5.4) by a team of researchers at the University of California, Irvine, led by Dr. David J. Reinkensmeyer. The device is a robotic end effector that is designed to provide repetitive, task-oriented movements to the upper limb for patients recovering from a stroke or other neurological injuries.

Design and control: The ARM-GUIDE is a robotic end effector device that enables stroke patients to perform reaching exercises along the rail, which can be adjusted to be either neutral or work against gravity. The device is equipped with a motor that can provide resistance or assistance to the patient during the movements and is also equipped with sensors that monitor hand position and speed (Reinkensmeyer et al., 2014). Additionally, the device has a 6-degree-of-freedom force sensor that measures the forces exerted by the patient along the rail and in other directions. The device can also be adjusted for height and angle, and the extent of the movement can be controlled. It is used as both a diagnostic and a treatment tool for addressing arm impairments in stroke patients, and it helps in providing repetitive, task-oriented movements to the upper limb, which is beneficial to patients recovering from a stroke or other neurological injuries.

Clinical studies: A small study was conducted to compare the effectiveness of long-term arm training between subjects using the ARM-GUIDE and a control group that performed free-reaching movements (Kahn et al., 2001). The study involved a group of chronic stroke patients (more than 1-year poststroke) who were trained for 8 weeks, 3 days per week. The study consisted of six subjects who used the ARM-GUIDE device,

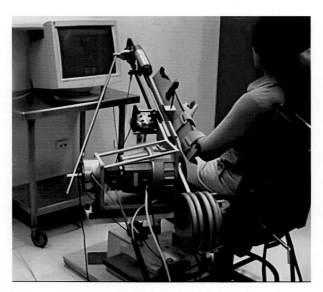

Fig. 5.4 ARM-GUIDE (Qassim and Wan Hasan, 2020).

and four subjects served as controls. The ARM-GUIDE group performed reaching movements with their impaired arm, with the device providing resistance or assistance based on the subjects' movement speed and trajectory. The control group simply reached for targets on a wall at a comfortable speed. Both groups improved in arm function and the ability to perform everyday tasks; however, there were no significant differences in the improvements observed between the two groups. Both groups also demonstrated significant improvements in the active range of reach and reaching speed, as well as decreased passive resistance to movement, but, again, there were no significant differences between the groups. The study suggests that the mode of therapy or the amount of therapy may be more important in restoring arm function than the type of therapy used. However, it is important to note that the sample size of this study was small, and the starting impairment level was slightly higher in the ARM-GUIDE group, which may have influenced the results. Further research is needed to investigate the optimal type of robot-assisted therapy for addressing arm impairment in this patient population.

5.2.3 MIME

The Mirror Image Movement Enabler (MIME) is an upper limb training protocol based on a motion assist robot that was developed (see Fig. 5.5) through a collaborative effort between the Veterans Affairs Medical Center in Palo Alto and Stanford University. The protocol utilizes a modified PUMA 560 industrial robot (manufactured by Stäubli Corporation) to interact with patients in a stable and repeatable manner.

Design and control: The patient's affected limb is placed in a splint, which is connected to the robot through a 6-degree-of-freedom force–torque sensor (Lum et al., 2002). This sensor measures the interaction forces between the patient and the device during reaching tasks, allowing the therapist to adjust the therapy accordingly. The device is also fully instrumented so that the position of the patient's limb can be inferred

Fig. 5.5 MIME (Burgar et al., 2011).

from the robot's position. The idea behind this protocol is to explore the effectiveness of restoring arm function in stroke patients by having them execute movements that mirror one another in both of their upper limbs. The therapy is designed to provide repetitive, task-oriented movements to the patient's affected limb to help them regain arm and hand function.

The device uses four different modes of operation to train patients in reaching and movement tasks. The first mode is passive movement, where the robot moves the patient's arm from a starting position to a target along a predetermined path and the patient is instructed to relax their arm and allow the robot to move it. The second mode is active stabilization, where the patient attempts to move their arm to a target while the robot stabilizes the limb. The third mode is active resistance, where the robot provides resistance as the patient reaches for targets across the work space. The fourth mode is bimanual training, where the patient uses both arms to reach for symmetric targets simultaneously, with one arm connected to the robot and the other to a position-sensing digitizer.

Clinical studies: A study was conducted to evaluate the effects of these robotic modes of therapy in comparison to neurodevelopmental therapy (NDT) on 27 chronic stroke patients (Lum et al., 2002). The patients in the robot group received 24 1-h sessions over a 2-month period, during which they practiced shoulder and elbow movements assisted by the robot. The control group received NDT, during which they practiced functional or self-care tasks. The results showed that the patients who received MIME therapy made statistically higher gains in proximal arm function, strength, and the amount of active reach. However, at the 6-month follow-up, there were no statistical differences in function between the two groups. No changes were found in distal arm function or the ability to perform daily activities. Overall, the study suggests that although MIME therapy may be effective in improving proximal arm function and strength, it may not be more effective than conventional therapy. Another study that specifically examined the effects of the MIME protocol on subjects with stroke found that active reach performance significantly improved in both low- and high-functioning subjects (Lum et al., 2004). The study also found that muscle activation patterns in the elbow and shoulder improved when reaching against gravity, but no changes were seen during movements performed on a tabletop.

5.2.4 Bi-Manu-Track

Bi-Manu-Interact (see Fig. 5.6) was developed at Klinik Berlin/Charite University Hospital in Berlin, Germany. A computerized motor-driven arm trainer called the Bi-Manu-Track allows for bilateral training of two movement patterns, including wrist flexion and extension and forearm pronation and supination. The benefits of this device are its portability, simplicity, low cost, ability to perform bilateral exercises, and it is commercially available at Reha-Stim in Berlin, Germany.

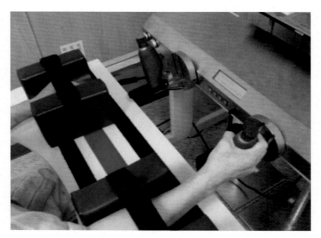

Fig. 5.6 Bi-Manu-Track (Wu et al., 2013).

Design and control: This robot is a 1-degree-of-freedom (DOF) device created for bilateral passive and active exercises, allowing hemiparetic patients to perform two distinct movement cycles bilaterally (Van Delden et al., 2012). The robot handles rotate in either a mirror image or parallel mode, similar to a rocker. There are three different control modes: passive-passive, active-passive (the unaffected limb drives the affected limb), and active-active (the affected limb has to overcome the initial isometric resistance). Individual amplitude, speed, and resistance settings are available. Furthermore, impedance control is used to ensure smooth movement. The handles are connected by an axis that is linked to the motor that controls their position and torque. One handle set has a horizontal axis of rotation for the elbow, whereas the other has a vertical axis of rotation for the wrist. Each handle axis connects to the electric motor. The subjects sat at a table with their elbows bent at an angle of 90° and their forearms in the pronation/supination position. The position of the patient controls the motor drives, and an LCD monitor displays the number of cycles completed to motivate the patient to exercise more (Wu et al., 2013).

Clinical studies: The Bi-Manu-Track was initially evaluated in 12 chronic stroke patients, and 5 of the 12 patients improved (Hesse et al., 2003b). During the first 3 weeks of rehabilitation, patients utilized the Bi-Manu-Track in all three modes for 15 min per day on weekdays. In addition to the exercise program, the patients received a continuing rehabilitation program. Following therapy, the MAS scores demonstrated a significant reduction in muscular tone in the wrist and fingers. However, the scores returned to pretreatment values at the 3-month follow-up. Other scores, such as the Rivermead Motor Assessment, remained stable. An additional randomized controlled trial evaluated the benefits of Bi-Manu-Track training compared to dose-matched active control therapy

(Liao et al., 2012). In all, 2 groups of 10 patients with chronic stroke performed 90–105 min of daily exercises for 4 weeks. The Bi-Manu-Track group performed 300–400 forearm repetitions in both modes (passive-passive and active-passive) in one session as well as 150–200 wrist repetitions in the active-active mode. A software application was integrated into the Bi-Manu-Track training to provide the patients with immediate visual feedback on the actions or forces that they used during the exercises. The Bi-Manu-Track group showed significantly increased FMA scores, upper limb activity ratio, and amount and quality of upper limb use.

5.2.5 Armotion

Armotion (see Fig. 5.7) is a robotic approach for treating severe and moderate upper extremity neuromuscular dysfunction. Armotion, also known as Motore, is a mobile upper limb neuro-orthopedic rehabilitation robot. The robot's motors allow it to actively participate in all the movements required of the patient during treatment. Depending on the needs of each patient, it can assist or impede movement. The robot demands that patients keep their attention on the task at hand at all times and only move if they are provided with some force. This experiment aims to advance the development of Motore, a rehabilitation robot designed to restore upper limb capability.

Design and control: The Armotion device, developed by Reha Technology in Olten, Switzerland, assisted the robot-aided upper limb (UL) training group. This end effector allows for two-dimensional arm movement with visual input, allowing users to conduct activities like reaching and carrying to repair their shoulder and elbow joints (Chang and Kim, 2013). Exercises were done with the shoulder and elbow, which are connected to the device by the wrist. This device helps people move in accordance with their rehabilitation goals. It works in various ways, from guided support for individuals who can move independently with ease to assisted or unassisted movement for those with

Fig. 5.7 A mobile robot for upper limb neuro-orthopedic rehabilitation (Motore) (Paolucci et al., 2021).

severe hemiplegia. The robot has the ability to drive, move, or push against the patient's arm, and treatment parameters, including repetitions, execution speed, and motion resistance, can all be adjusted. Patients can also perform tasks that simulate daily activities (like cleaning dishes or counting money) or train cognitive functions (memory, logical deductive functions). The robot incorporates "impedance control," which changes how the robot moves to match how the patient's upper limb moves. The workouts involved using the elbow, which is linked to the device at the wrist and the shoulder. In all, 10 min of passive mobilization and stretching were followed by robot-assisted exercises for the affected upper limb (35 min). The Armotion software's various exercise categories and the number of repetitions were chosen as follows: (i) "Collect the coins" (45–75 coins/10 min); (ii) "Wash the dishes" (40–60 repetitions/10 min); and (iii) "Burst the balloons" (100–150 balloons/5 min). Elbow flexion-extension and reaching movements were emphasized in all exercises, which were designed to accomplish several objectives in different directions. The robot allows participants to execute exercises through an "assisted-only-as-needed" control strategy. To increase the difficulty, they changed the assisted and unassisted modalities, thus increasing the number of repetitions during the study period (Paolucci et al., 2021).

Clinical studies: A stroke-specific, performance-based impairment evaluation, namely, the Fugl-Meyer assessment (FMA) scale (Fugl-Meyer et al., 1975), is a multiple item Likert-type scale with a total of 226 items. It was intended for use after a stroke. Individuals who have had hemiplegia due to a stroke will be subjected to this test to evaluate their motor function, balance, and sensory and joint functions. The motor domain of the FMA-UL is the section that is used the most frequently and has the most significant value when it comes to evaluating a patient's motor recovery following a stroke, particularly in the upper limb. Each domain contains several tasks, and each task is evaluated using a three-point ordinal scale, with 0 representing the inability to perform, 1 representing partial performance, and 2 representing full performance (Gladstone et al., 2002). The key end measure was the rate of obtaining the minimal clinically meaningful difference in FMA-UE at the T1–T0 change, which was at least 10 points in the subacute phase and 5 points in the chronic phase following treatment (Narayan Arya et al., 2011).

Both clinical and demographic data were obtained during the enrollment procedure. The improvements in the UL-MAS score, which is the sum of evaluations of the shoulder, elbow, and wrist muscles, were the most important consequence that occurred (single-joint score range, 0–4; total score, 0–12; higher scores indicate worse spasticity) (Pennati et al., 2015). The Motricity Index (MI) was utilized in order to quantify the level of muscular weakness present in the paretic upper extremity. It entails assessing a patient's strength according to the patient's capacity to activate a muscle group, to move a limb segment over a range of motion, and to withstand the force of an examiner's pressure. The scale consists of three movements: the pinch grip, elbow flexion, and shoulder abduction. Each of these actions receives a score ranging from 0 to 33, with the maximum potential score of 100 for the upper limb, which is calculated by adding one to the total of the three acts (Bohannon, 1999). The Medical Research Council (MRC) scale assessed changes in UL muscle strength by summing up the scores from the shoulder flexion/abduction/external rotation, elbow flexion/extension, and wrist supination/flexion/extension (single movement score range, 0–5; higher scores indicate muscle strength against full resistance; total score, 0–40; higher scores indicate muscle strength against full resistance) (Gandolfi et al., 2019).

5.3 Exoskeleton-type upper limb rehabilitation robots

This section presents some notable exoskeleton-type robots through case studies.

5.3.1 ArmeoPower

The ArmeoPower (Hocoma, 2022) exoskeleton (see Fig. 5.8) is a medical device used to assist patients with spinal cord injuries or neurological conditions such as stroke in performing upper limb rehabilitation exercises. It was developed by Hocoma AG, a Swiss company that specializes in the development and manufacturing of robotic and sensor-based devices for rehabilitation and therapy.

Design and control: The ArmeoPower exoskeleton is designed to support the patient's arm and hand, and it is controlled by a computer that is operated by a therapist. The device consists of a robotic arm support that is worn by the patient and a computer console that is used by the therapist to control the exoskeleton. The robotic arm support

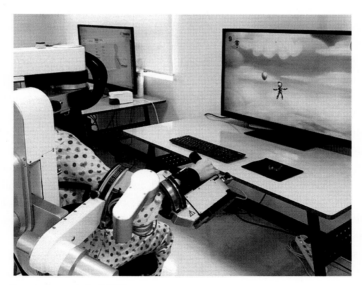

Fig. 5.8 ArmeoPower (Lee et al., 2020).

is made up of a series of joints and actuators that are attached to a lightweight yet sturdy frame. The joints are designed to mimic the natural movements of the human arm, allowing the patient to move their arm in a natural and functional manner. The actuators provide resistance to the patient's movements, helping to strengthen the muscles. The patient's arm is secured in place with straps, and sensors are used to track the patient's movements. The computer console is used by the therapist to control the exoskeleton. The therapist can adjust the level of resistance provided by the actuators as well as the range of motion of the joints. The therapist can also set specific rehabilitation goals for the patient, such as reaching a certain range of motion or a certain level of strength. The computer console also includes a display that shows the patient's progress and allows the therapist to track their progress over time. The ArmeoPower exoskeleton is designed to mimic the natural movements of the human arm; it typically has 7 degrees of freedom, which includes shoulder flexion/extension, shoulder abduction/adduction, internal/external rotation, elbow flexion/extension, forearm pronation/supination, wrist flexion/extension, and radial/ulnar deviation. This allows for a wide range of motion and functional movements during therapy sessions. The therapist can adjust the range of motion of each joint individually, to match the patient's specific needs and goals. Each degree of freedom provides a specific range of motion that corresponds to a specific joint or joints (Rahman et al., 2014a, b) in the human arm, such as the shoulder or elbow. This allows the therapist to target specific areas of the arm and hand for rehabilitation and to progress the therapy as the patient's strength and function improve.

Clinical studies: A study was performed to identify the potential neurophysiological markers that can predict the responsiveness of stroke patients to upper limb robotic treatment (Calabrò et al., 2016a, b). This prospective cohort study took place at the Behavioral and Robotic Neurorehabilitation Laboratory of IRCCS Centro Neurolesi Bonino Pulejo, Messina, Italy. A total of 35 patients who had sustained a first-ever ischemic supratentorial stroke at least 2 months before enrollment and had unilateral hemiplegia were enrolled. All the patients underwent 40 ArmeoPower training sessions that lasted 1 h each (i.e., 5 times a week for 8 weeks). The spasticity and motor function of the upper limb was assessed by means of the Modified Ashworth Scale and FMA, respectively. Additionally, the cortical excitability and plasticity potential of the bilateral primary motor areas in response to the repetitive paired associative stimulation paradigm using transcranial magnetic stimulation and ArmeoPower kinematic parameters were evaluated. The results showed that the patients who showed significant repetitive paired associative stimulation after effects at baseline exhibited an evident increase of cortical plasticity in the affected hemisphere. This was accompanied by a decrease of interhemispheric inhibition and paralleled by clinical improvements and ArmeoPower kinematic improvements.

The data suggested that the use of ArmeoPower may improve upper limb motor function recovery as predicted by the reshaping of cortical and transcallosal plasticity, according to the baseline cortical excitability. The study suggested that the use of the

ArmeoPower exoskeleton can help in the recovery of motor function in stroke patients, with the improvement being predicted by the reshaping of cortical and transcallosal plasticity, according to the baseline cortical excitability.

5.3.2 EXO-UL8

EXO-UL8 (see Fig. 5.9) is a bilateral exoskeleton-type robot developed by Bionics Lab at the University of California, Los Angeles (Shen and Rosen, 2020). This research was conducted by a team of graduate students led by Jacob Rosen. This robot is the latest addition to the EXO-UL series evolved from unilateral 1-DOF EXO-UL1, 3-DOF EXO-UL3, and dual-arm 7-DOF EXO-UL7 (also known as CADEN-7: a cable-driven and back-drivable exoskeleton robot).

Design and control: Each arm of EXO-UL8 has 7 DOFs that cover all seven main DOFs of the human upper limb such as shoulder abduction/adduction, flexion/extension, internal/external rotation, elbow flexion/extension, forearm pronation/supination, wrist flexion/extension, and radial/ulnar deviation. In addition, the researchers added 1 DOF for the hand opening and closing module to this robot. The robot's joints are actuated by DC (direct current) motors (Maxon Motor, Switzerland) and harmonic drives (Harmonic Drive Systems Inc., Japan). The shoulder joint motors are equipped with brakes and encoders to enable freezing functionality during emergency configurations. The researchers chose this gear-motor, nonback-drivable design over its predecessor's cable-driven actuation in this version due to increased torque outputs that allow abnormal movement correction and gravity compensation, more accurate low-level control without unwanted compliance/delay, and, finally, an acceptable torque/volume

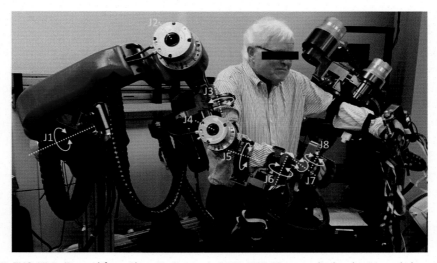

Fig. 5.9 EXO-UL8. *(Reused from Shen, Y., Rosen, J., 2020. EXO-UL upper limb robotic exoskeleton system series: from 1 DOF single-arm to (7 + 1) DOFs dual-arm. In: Wearable Robotics. Elsevier.)*

ratio. EXO-UL8 is equipped with six-axis force-torque sensors (ATI Mini40) at the upper arm, forearm, and handle. The hand opening-closing module has another single-axis force sensor. This robot improved the training modes from solely predefined trajectories to symmetric mirror image movement training and asymmetric bilateral training, leveraging an interactive virtual reality environment. Admittance controllers based on the Kalman filter have been used to interpret the human-applied forces on the robot into estimated joint torques (Shen et al., 2019). These estimated torques are used to generate reference trajectory during the training with the EXO-UL8. The admittance controller is used in conjunction with a PID controller, gravity compensation, and friction compensation components in the control architecture to generate reference torques for the actuators.

Clinical studies: The robot has been tested with one chronic poststroke patient having weakness and spasticity in the right arm in a pilot study (Shen et al., 2018). This study focused on the validation of this robot's functionality in providing single- and dual-arm training, during which a total of three tasks were performed. In the first task, the subject was asked to move his affected arm progressively to harder targets while the robot provided no assistance. In the second task, the subject moved his unaffected (left) arm to reach the target and was then instructed to move his affected arm to reach the same target. The robot provided no assistance in this case. Finally, similar tasks were performed in the second task, but, in this case, the robot assisted the affected arm while reaching for the targets. This study analyzed quantitative data such as task completion time, reachable work space, joint ROM, and display of spatial interaction maps and validated EXO-UL8's functionality in supporting single- and dual-arm movements based on a chronic stroke patient's pHRI.

5.3.3 FLEXO-Arm1

FLEXO-Arm1 (see Fig. 5.10) was collaboratively developed by the Shanghai Engineering Research Center of Assistive Devices and the University of Shanghai for Science and Technology (Xie et al., 2021). The researchers developed this 2-DOF robot for shoulder and elbow rehabilitation, comprising a passive training mode, an active assist mode, and a teaching mode. It also incorporated the use of virtual reality to increase users' interest during their therapy. This robot is unilateral but can be configured to be used with both left and right arms.

Design and control: In FLEXO-Arm1, the shoulder and elbow flexion/extension motions are generated by motors equipped with encoders and torque sensors. This robot has three passive DOFs that allow wrist flexion/extension, elbow joint flexion/extension in the horizontal direction, and shoulder joint adduction/abduction to avoid motion restriction. In this robot, the human exoskeleton coupling dynamic modeling and parameter estimation are carried out using a method of overall parameter identification through collecting torque and motion data of each joint during the motion of the robot. To identify the dynamic parameters, the patients' passive and active training sessions are

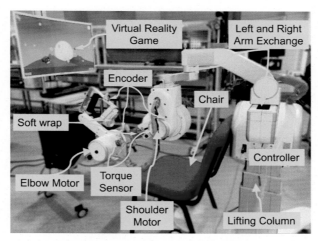

Fig. 5.10 FLEXO-Arm1. *(Source: Xie, Q., Meng, Q., Zeng, Q., Fan, Y., Dai, Y., Yu, H., 2021. Human-exoskeleton coupling dynamics of a multi-mode therapeutic exoskeleton for upper limb rehabilitation training. IEEE Access 9, 61998–62007. https://doi.org/10.1109/ACCESS.2021.3072781.)*

carried out multiple times, and achieved parameter values are used to calibrate the calculated torque for the shoulder and elbow joint motors. The dynamic calibration model has been tested with 11 healthy participants doing passive and active repetitive exercises.

Clinical studies: FLEXO-Arm1 has been used in multicenter, single-blind randomized control trials with a noninferiority design study (Lin et al., 2022). For 3 weeks, 172 stroke survivors received either robot-assisted training or enhanced upper extremity therapy. The Fugl-Meyer assessment for upper extremity subscale (FMA-UE), the Fugl-Meyer assessment for lower extremity subscale (FMA-LE), and the Modified Barthel Index were administered at baseline, midtreatment (1 week after the start of treatment) and posttreatment. Participants in the robot-assisted training group demonstrated significant improvement in the hemiplegia extremity, which was comparable to the enhanced upper extremity therapy group in FMA-UE ($p < 0.05$), while suggesting greater motor recovery of the lower extremity in FMA-LE ($p < 0.05$) compared to the enhanced upper extremity therapy group. The Modified Barthel Index increased significantly within each group, but there was no significant difference between the groups. The study concluded that robot-assisted training is noninferior but not superior to enhanced upper extremity therapy in reducing impairment of the upper extremity in stroke survivors. However, robot-assisted training may be superior in reducing impairment of the lower extremity.

5.3.4 Harmony

Harmony (see Fig. 5.11) is a recent robotic exoskeleton that has been developed for upper limb rehabilitation by Ashish Deshpande and a team of graduate students from the Rehabilitation and Neuromuscular (ReNeu) Robotics Lab, Cockrell School of

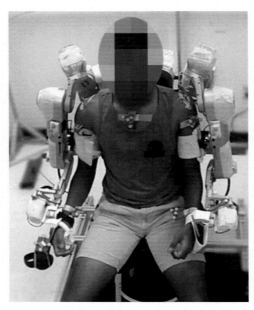

Fig. 5.11 Harmony. *(Source: De Oliveira, A.C., Sulzer, J.S., Deshpande, A.D., 2021. Assessment of upper-extremity joint angles using Harmony exoskeleton. IEEE Trans. Neural Syst. Rehabil. Eng. 29, 916–925. https://doi.org/10.1109/TNSRE.2021.3074101.)*

Engineering at The University of Texas at Austin. The exoskeleton was designed with the intent to enable patients to perform bilateral arm training. Harmony is a lightweight and portable device that is worn on the upper body, and it consists of actuated joints that mimic the human shoulder, elbow, and wrist joints. The exoskeleton is controlled by a computer that receives input from the patient's movements and then provides assistance to the patient to perform the desired task. One of the main advantages of Harmony is that it allows patients to perform bilateral arm training, which is important for patients who have suffered a stroke or other neurological conditions that have affected their ability to use both their arms. Bilateral arm training is known to be more effective than unilateral training in improving the patient's motor function and coordination. Additionally, Harmony offers a high level of adjustability, allowing it to be customized to the specific needs of each patient.

Design and control: The Harmony exoskeleton was designed to accommodate the entire upper body, which sets it apart from other existing technologies that only focus on one arm and limit the possibilities of bilateral training. The exoskeleton connects to patients at 3 different points on each side of the upper body and has 14 axes of motion that allow for a wide range of natural movements. The robot is equipped with a suite of sensors that collected data at a rate of 2000 times per second. These data are then used to provide an instantly personalized robotic interaction by being fed back into the robot's program. With input from physical therapists and doctors, the Cockrell School researchers designed Harmony's shoulder mechanism to facilitate natural, coordinated motions, specifically the scapulohumeral rhythm, which is a critical coordinated rotational motion necessary for upper limb movements and long-term joint stability.

Clinical studies: Harmony has been the subject of several clinical studies to evaluate its effectiveness in upper limb rehabilitation. One of the studies (Kim and Deshpande, 2015), conducted on stroke patients, found that the use of Harmony resulted in significant improvements in upper limb motor function as measured by the FMA and Action Research Arm Test (ARAT) scores. The study also reported that the patients perceived the device to be comfortable and easy to use.

Another study (Kim and Deshpande, 2017) was conducted to evaluate the effectiveness of Harmony in patients with chronic stroke. The results of the study showed that the patients who used Harmony had significant improvements in their upper limb motor function as well as in their ability to perform activities of daily living. The study also reported that the patients found the device easy to use and comfortable to wear.

A more recent study (Hailey et al., 2022) has been conducted to evaluate the effectiveness of Harmony in patients with chronic stroke who had moderate-to-severe upper limb impairment. The results of the study showed that the patients who used Harmony had significant improvements in their upper limb motor function as well as in their ability to perform activities of daily living. Overall, these studies suggest that Harmony is a promising technology for upper limb rehabilitation in stroke patients. The device has been shown to be effective in improving upper limb motor function and in the ability to perform activities of daily living. Additionally, patients found the device comfortable and easy to use. It is important to note that more research is needed to fully evaluate the long-term effectiveness of the device and its potential for other patient populations.

5.3.5 ETS-MARSE and SREx

ETS-MARSE (see Fig. 5.12A), also known as the Motion Assistive Robotic-Exoskeleton for Superior Extremity, is an exoskeleton robot designed to rehabilitate patients with upper limb impairments. Its design is based on the human upper limb anatomy, and it is intended to be worn comfortably by the patient during rehabilitation tasks (Rahman et al., 2015). The second version of ETS-MARSE, known as the Smart Robotic Exoskeleton (see Fig. 5.12B), is designed with an extended range of motion and is 20% lighter than the previous version.

Design and control: The ETS-MARSE/SREx robot system is designed based on the human upper limb anatomy to accurately replicate upper limb joints and movements. It is made to be worn comfortably during rehabilitation tasks. The shoulder section has three joints for vertical and horizontal extension-flexion movements and internal-external rotations. The elbow has one joint for flexion-extension motion, and the wrist has three joints for pronation-supination, ulnar-radial deviation, and flexion–extension motions. The system also has a virtual interface for the subject and therapist to monitor the rehabilitation tasks and offers a virtual reality environment for task-oriented activities in task space, Cartesian space, and free motion.

Force sensor

Fig. 5.12 (A) ETS-MARSE (Rahman et al., 2015). (B) SREx (Ahmed et al., 2022).

The goal of the research on a control strategy for ETS-MARSE/SREx is to design and validate a solution for inverse kinematics and nonlinear control for an upper limb exoskeleton robot to perform various rehabilitation tasks. The idea is that a new control system solution can be achieved by incorporating the entire nonlinear dynamic model of the robot system into its design. However, this presents challenges such as the lack of analytical solutions for nonlinear equations, imperfect kinematic and dynamic models due to difficulties in modeling phenomena such as nonlinear friction and uncertainty caused by visual devices, and the unavailability of feedback signals needed for computing the dynamic parameters of the robot system. The main challenges in this field include high nonlinearity of the robot system, system redundancy, unknown dynamics, and uncertain kinematics. The researchers behind ETS-MARSE/SREx have addressed some of these challenges in a previous work, such as designing an inverse kinematic solution for smooth and human-like motion and controllers that are robust and accurate without sensitivity to uncertain nonlinear dynamics and unexpected disturbances. This gave the control system more flexibility to handle uncertainties and parameter variations in different rehabilitation tasks.

Clinical studies: In one study, an investigation was conducted on the passive and active control strategies to provide physical assistance and rehabilitation by ETS-MARSE, a 7-DOF exoskeleton robot with nonlinear uncertain dynamics and unknown bounded external disturbances caused by the user's physiological characteristics (Brahmi et al., 2018b). A control strategy called an integral backstepping controller with time delay estimation (BITDE) was used, which allowed the exoskeleton robot to achieve the desired performance while working under the mentioned uncertainties and constraints. The time delay estimation (TDE) was employed to estimate the nonlinear uncertain dynamics of the robot and the unknown disturbances. However, the TDE approach has a limitation of the time delay error; thus, a recursive algorithm was used to further reduce its effect. Additionally, integral action was employed to decrease the impact of the unmodeled dynamics. Furthermore, the damped least-squares method was introduced to estimate the desired movement intention of the subject with the objective of providing active rehabilitation. The controller scheme was designed to ensure that the robot system performed passive and active rehabilitation exercises with a high level of tracking accuracy and robustness, despite the unknown dynamics of the exoskeleton robot and the presence of unknown bounded disturbances. The design, stability, and convergence analysis were formulated and proven based on the Lyapunov-Krasovskii functional theory. Experimental results with healthy subjects, using a virtual environment, were conducted to show the feasibility and ease of implementation of the control scheme. Its robustness and flexibility to deal with parameter variations due to unknown external disturbances were also demonstrated. Overall, the results of the research showed that the proposed control strategy BITDE was successful in providing physical assistance and rehabilitation by the exoskeleton robot despite the uncertainties and disturbances.

In another study, a backstepping approach that was integrated with TDE was presented to provide an accurate estimation of the unknown dynamics and to compensate for external bounded disturbances. This control was implemented to perform passive rehabilitation movements with ETS-Motion Assistive Robotic-Exoskeleton for Superior Extremities (Brahmi et al., 2018a). The unknown dynamics and external bounded disturbances had the potential to affect the robotic system in the form of input saturation, time delay errors, friction forces, backlash, and variations in upper limb mass among subjects. The output of the time delay estimator was directly coupled to the control input of the proposed adaptive tracking control through a feed-forward loop. This enabled the control system to ensure highly accurate tracking of the desired trajectory while being robust to uncertainties and unforeseen external forces and flexible with variations in parameters. The proposed strategy did not require accurate knowledge of the dynamic parameters of the exoskeleton robot to achieve the desired performance. The stability of the exoskeleton robot and the convergence of its state errors were established and proved based on the Lyapunov-Krasovskii functional theory. The experimental results and a comparative study were presented to validate the advantages of the proposed strategy. The SREx has been used with five healthy participants executing passive exercises at slower, medium, and faster velocities at isolated, composite joint motions of the shoulder, elbow, forearm, and wrist (Ahmed et al., 2022). These experiments were carried out in three sets; in the first set, the participant remained totally passive, meaning that the participant exerted no force on the robot. During the second set, the participant exerted an intermittent sudden jerk during the motion, and, in the third set, the participant resisted the motion throughout the motion. During this study, the robot was controlled with a conventional PID controller. From statistical analysis, it was found that the simple model-free PID controller was robust in providing therapy to the different subjects at varying velocities and different passive exercise scenarios, including the subject's relaxed state during the exercise, resistance to motion exercise, and sudden involuntary movement during the exercise.

5.3.6 ANYexo 2.0

ANYexo 2.0 (see Fig. 5.13) is one of the latest exoskeleton-type robots for upper limb rehabilitation developed jointly by the Sensory-Motor Systems Lab and the Robotic Systems Lab of ETH Zurich (Zimmermann et al., 2023). The research was led by Marco Hutter and Robert Reiner along with a group of graduate students. It is a serial-type unilateral robot designed to be worn on the right hand of the user. The researchers developed this robot for all stages of rehabilitation and for performing activities of daily living (ADL).

Design and control: This robot is the successor to its previous version, called ANYexo (Zimmermann et al., 2019), which did not include wrist motion modules. ANYexo 2.0 is composed of nine actuators, which provide motion to sternoclavicular protraction/retraction, sternoclavicular, elevation/depression, glenohumeral plane of elevation, glenohumeral elevation, glenohumeral axial rotation, elbow flexion/extension, forearm pronation/supination, wrist flexion/extension, and wrist ulnar/radial bend. Two actuators have been used for the shoulder girdle, three for the glenohumeral joint, one for the elbow joint, and three for the wrist joints. The overall structure of this robot is composed

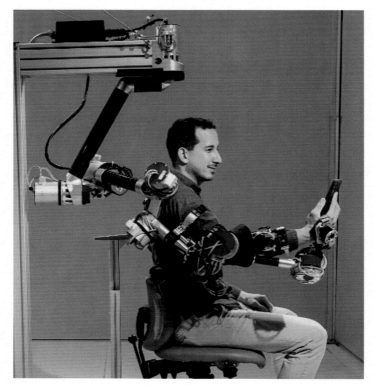

Fig. 5.13 ANYexo 2.0 (Zimmermann et al., 2023).

of carbon fiber-reinforced polymer (CFRP) tubes and machined aluminum parts. This robot uses an SEA actuator, ANYdrive (ANYbotics AG, Switzerland), for generating shoulder and elbow motions and the quasidirect drive Dynadrive Armadillo (Robotic Systems Lab, Switzerland) for wrist motions. All drives are equipped with a torque control bandwidth of up to 60 Hz for small amplitudes. All actuators featured integrated joint bearings, an onboard motor controller, joint encoders, an inertial measurement unit (IMU), and an EtherCAT communication unit. The upper arm, the forearm, and the hand are equipped with 6-DOF force–torque sensors (Rokubi Mini, Bota Systems AG, Switzerland) with an integrated IMU, acquisition boards, and an EtherCAT communication unit. Integrated electronics in all actuators and sensors protect the robot from electromagnetic interference and reduce cabling. The sampling and control rate for the control PC is 800 Hz, whereas the torque, position, and velocity controllers operate at 4 kHz. This robot features superior kinematic synergy with human upper limb movements. ANYexo 2.0 employs active shoulder synergy, coordinate transformation-based trajectory generation, and selective compliance-based position, velocity, and torque controller through hierarchical quadratic programming. The haptic interaction between ANYexo 2.0 and the user is measured using IMU and force-torque sensors, and an RGBD camera is used for state estimation.

Clinical studies: Experiments were carried out with a healthy subject to perform a set of 15 ADLs in the free-space mode. Researchers expect that in a real case scenario with real patients, the free-space controller will be substituted with an assistance controller. The experiments showed that ANYexo 2.0 has adequate range of motion, strength, speed, and haptic-rendering accuracy to provide meaningful occupational (activity-oriented) and physical (joint-oriented) therapy exercises to severely impaired patients. This versatility, coupled with full control over all nine relevant degrees of freedom of the human upper limb, promises to make the device applicable to a wide range of clinic-based therapies (e.g., neurological, orthopedical, and sports injury patients).

5.3.7 CABXLexo-7

CABXLexo-7 (see Fig. 5.14) is a 7-DOF cable-driven upper limb exoskeleton developed by a joint effort of researchers from the Harbin Institute of Technology (HIT) and the Hefei University of Technology (HFUT) in China in 2017 (Xiao et al., 2018).

Design and control: The human upper limb is composed of several joints that each allow for a specific range of motion. This includes the shoulder, which can move through flexion and extension, abduction and adduction, and internal and external rotations. The elbow also has flexion and extension, and the forearm can move through supination and pronation. The wrist has flexion and extension and can also move through radial and ulnar deviations. CABXLexo-7 is designed to replicate these ranges of motion. The exoskeleton has 7 degrees of freedom. The range of motion of the exoskeleton is designed to

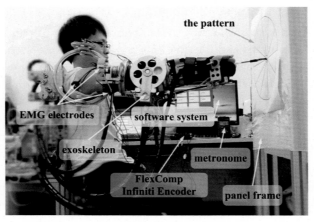

Fig. 5.14 CABXLexo-7 (Islam et al., 2020a, b).

match the range of human motion. The exoskeleton is primarily made of an aluminum alloy and is designed to support a weight of 3.75 kg. The exoskeleton uses motors to power its movements. The shoulder joint requires the most power and therefore uses a 200-W motor. The other joints use 100-W motors. The exoskeleton also uses encoders to accurately track the joint angles. The shoulder joint uses a differential mechanism to achieve flexion and extension and abduction and adduction. The elbow joint uses a different type of differential mechanism, and the wrist and shoulder rotations use a traditional serial mechanism. A new tension device was also designed to ensure that the cables are properly tensioned. The power from the motors is transmitted to the joints using cable-conduit mechanisms.

Clinical studies: The developers of CABXLexo-7 evaluated the effectiveness of the exoskeleton in assisting upper limb movement in healthy subjects (Xiao et al., 2018). In all, 6 participants between the ages of 23 and 45 years, both male and female, were asked to move a pen from the center of a pattern to each of the 12 targets on the pattern and then back to the center while sitting behind a target panel frame. The subjects were asked to follow a metronome's rhythm of 60 beats per minute, which was similar to the reaching speed of poststroke patients. The authors used three different modalities of motion in the experiments: passive, free, and assistive. The subjects repeated the experiments five times under each modality and had a rest of 2 h between experiments to avoid muscle fatigue. The exoskeleton mainly assisted in the movements of shoulder flexion/extension, shoulder abduction/adduction, shoulder internal/external rotation, and elbow flexion/extension. The authors used surface electromyography (sEMG) signals from several muscles of the upper limb as indicators to evaluate the ability of the exoskeleton in assisting movements.

5.3.8 SUEFUL-7

SUEFUL-7 (see Fig. 5.15) is a 7-degree-of-freedom (DOF) upper limb motion assist exoskeleton robot developed by Gopura et al. in 2009 at Dr. Kazuo Kiguchi's Laboratory at Saga University, Japan. The main objective of this robot is to provide motion assistance to individuals who have physical weakness in their daily activities (Gopura et al., 2009).

Design and control: The design of SUEFUL-7 takes into account the moving center of rotation (CR) of the shoulder joint and the deviation of the axes of the wrist. The robot is equipped with a pulley and a cable-driven system to drive the shoulder's vertical and horizontal flexion and extension as well as elbow flexion and extension movements. The actuators are placed on stationary frames for these movements. For the movements of shoulder internal and external rotations, forearm supination and pronation, wrist flexion and extension, and wrist radial and ulnar deviations, the actuators are mounted on the robot itself, either connected directly or *via* a gear drive. The robot also uses a slider–crank mechanism to compensate for the CR of the shoulder joint. The robot weighs approximately 5 kg and is intended to be mounted on a wheelchair, as it is assumed that individuals with physical weakness use wheelchairs.

The SUEFUL-7 rehabilitation robot is controlled using the EMG signals of the user as the primary input information. These signals are used to determine the user's motion intention. The robot also uses subordinate input information such as the forearm force (the force generated between the robot and the user's forearm), the hand force (the force generated between the robot and the user's hand), and the forearm torque (the torque

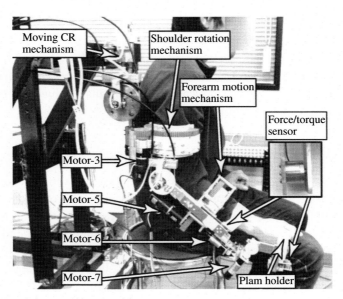

Fig. 5.15 SUEFUL-7 (Islam et al., 2020a, b).

generated between the wrist holder of the robot and the user's forearm) to control the robot's movements. SUEFUL-7 uses a combination of EMG signals, forearm force/ torque, and hand force to control the robot's movements, depending on the user's muscle activation level. When the user's EMG signal level is low, the robot uses forearm force/ torque and hand force to control its movements. When the user's EMG signal level is high, the robot uses the EMG signals to control its movements. When the user's muscle activation level is medium, a combination of EMG signals, forearm force/torque, and hand force is used to control the robot.

Clinical studies: The SUEFUL-7 exoskeleton was tested with only healthy subjects (Gopura et al., 2009). In these experiments, young male subjects performed daily activities involving upper limb movements. The first experiment involved subject A performing cooperative motions of the upper limb to demonstrate the effectiveness of the neuro-fuzzy modifier (a control system that uses EMG signals to interpret and adjust the robot's movements). The goal of the experiment was to show that the activity level of the muscles can be reduced when the neuro-fuzzy modifier is properly trained and the power assist is properly performed. The second experiment involved subjects A and B performing upper limb motions with and without assistance from SUEFUL-7 to demonstrate the effectiveness of the power assist. The EMG levels of the related muscles were measured for both cases. The goal of the experiment was to show that if SUEFUL-7 is properly assisting in the motions, then the EMG levels of the related muscles should be reduced when the robot is assisting with the same motions.

In summary, the experiments were designed to test the effectiveness of the neuro-fuzzy modifier and the power assist of the SUEFUL-7 rehabilitation robot. The first experiment focused on showing that the activity level of the muscles can be reduced when the neuro-fuzzy modifier is properly trained and the power assist is properly performed. The second experiment focused on demonstrating that if SUEFUL-7 is properly assisting the upper limb motions, then the EMG levels of the related muscles should be reduced. The experiments were conducted on young male subjects performing daily activities involving upper limb movements.

5.4 Design specs of some notable rehabilitation robots

The design and control of rehabilitation robots differ based on the therapeutic regime of the upper limb and the types of therapies (mode of operations) provided by those robots. Table 5.1 highlights and compares features (e.g., DOFs, sensors, actuators used, modes of operation, control approach, and therapeutic regime) of some notable end effector-type robots/devices, including those discussed in Section 5.2.

Table 5.2 highlights and compares features (e.g., DOFs, sensors, actuators used, placement of actuators, actuation mechanism, therapeutic regime, control approach) of the exoskeleton-type robots discussed in Section 5.3.

Table 5.1 Notable end effector-type rehabilitation robots.

End effector-type rehabilitation robots	DOFs	Sensors	Actuators	Modes of operation	Control	Therapeutic regime
InMotion2 (Krebs et al., 2007)	2	Six-axis force/torque sensor, DC tachometers	Brushless motors	Passive, active assistive	IMC	S, E
MIT–MANUS (Hesse et al., 2003a)	2	Force sensor, absolute encoder	Brushless motors	Passive	IMC	S, E
ARM-Guide (Reinkensmeyer et al., 2014)	7	Force/torque sensor	DC servo motors	Active assist	PID	S, E, F, W
MIME (Lum et al., 2006)	6	Six-axis force/torque sensor	Brushless DC motors	Active assist Passive assist resist	–	S, E
Bi-Manu–Track (Van Delden et al., 2012)	1	Position and torque sensor	Electric motors	Passive, active, and resist	IMC	F, W
ARMOTION (Motore) (Paolucci et al., 2021; Germanotta et al., 2018)	3	–	DC motors	Passive, active, and resist	IMC	S, E
NeReBot (Rosati et al., 2007)	3	Absolute encoder	Electric motors	Passive and active assist	PID	S, E
DMRbot (Khan et al., 2022)	3	Six-axis force/torque sensor	Brushless DC motors	Passive	POC	E, S

PID = proportional integral derivative; IMC = impedance control; POC = position control; S = shoulder; E = elbow; F = forearm; W = wrist.

Table 5.2 Notable exoskeleton-type rehabilitation robots.

Exoskeleton-type rehabilitation robots	DOFs (Active)	Sensors	Actuators	Actuator placement	Actuation mechanism	Control	Therapeutic regime
ArmeoPower (Hocoma, 2022; Kim and Deshpande, 2015, 2017) ArmeoPower (Hocoma, 2022)	7	Force/torque	Brushless DC motors	Joint	Gear drive	PID	S, E, F, W
EXO-UL8 (Shen and Rosen, 2020; Shen et al., 2018, 2019)	7+1	Force/torque, force	Brushless DC motors	Joint	Gear drive	Admittance, PID, gravity, and friction compensation	S, E, F, W, H
FLEXO-Arm1 (Xie et al., 2021)	2	Joint torque	Electric motor	Joint	Gear drive	Admittance, torque	S, E
Harmony (Kim and Deshpande, 2015, 2017)	7	Joint torque, force/torque	SEA	Joint, base	Gear drive	PID, model-based	S, E, F, W
ETS-MARSE/SREx (Rahman et al., 2009, 2012; Ahmed et al., 2022)	7	Force/torque sensor	Brushless DC motors	Joint	Gear drive	PID, CTC, CC, SMC, SMERL	S, E, F, W
ANYexo 2.0 (Zimmermann et al., 2023)	9	IMU, force/torque	SEA	Joint	Gear drive	Model-based, force	S, E, F, W
CABXLexo-7 (Xiao et al., 2018)	7	EMG	Brushless DC motors	Base	Cable, differential drive	EMG-based	S, E, F, W
SUEFUL-7 (Gopura et al., 2009)	7	Force, EMG	DC servo motors	Joint, remote	Gear drive, cable	Force-based, EMG-based	S, E, F, W

PID = proportional integral derivative; CTC = computed torque control; SMC = sliding mode control; SMERL = sliding mode exponential reaching law; EMG = electromyogram; IMC = impedance control; ADC = admittance control; S = shoulder; E = elbow; F = forearm; W = wrist.

5.5 Research challenges and future directions

A major research challenge facing the development of upper limb rehabilitation robots is the ability to provide personalized therapy to each individual patient. This requires a high degree of adaptability and flexibility in the design of the robot as well as sophisticated algorithms for controlling the robot's movements and adjusting the therapy based on the patient's progress. Another challenge is creating a robot that can provide therapy over a wide range of motion, as the upper limb is capable of a large number of different movements and degrees of freedom. The development of sensors to accurately measure the patient's motion and progress during therapy is also a significant challenge.

One potential future direction for upper limb rehabilitation robots is the integration of machine learning and AI techniques to create more personalized and adaptive therapy. This could involve using data from the patient's therapy sessions to train the robot to better understand their needs and adjust the therapy accordingly. Another potential direction is the development of wearable exoskeleton-type robots that can provide therapy while the patient is performing activities of daily living. Integrating other modalities, such as virtual reality or haptic feedback, to create more immersive and engaging therapy experiences for patients has been researched recently. It is expected that the research and development in the field of soft robotics could lead to the creation of more lightweight, comfortable, and flexible end effectors and exoskeletons for upper limb rehabilitation robots.

However, COVID-19 has exposed the critical need (Li et al., 2020; Qureshi et al., 2021) for home-based rehabilitation: during the pandemic, many patients with acute stroke (who require immediate rehabilitation exercise) avoided or lacked access to hospitals and rehabilitation centers. Indeed, data from 187 stroke rehabilitation centers in 40 countries across 6 continents show that in the 3-month period from March to May 2020 (the COVID-19 era), compared with the immediately preceding 3 months, there was a 19% decline in stroke admissions (Hughes, 2020). In addition, many poststroke patients reported feeling abandoned at home with overwhelmed caregivers (Hughes, 2020). Clearly, many people with stroke-induced disabilities lack access to rehabilitation services, a problem that will exacerbate with age, even as the pandemic abates. Therefore, an alternative to traditional rehabilitation is needed. Home-based/telerehabilitation could help future crises.

5.6 Conclusions

In this chapter, we reviewed the design, control, and studies performed for end effector- and exoskeleton-type robots for upper limb rehabilitation. The rehabilitation programs for upper limb impairments are crucial to restoring independence for stroke survivors and other individuals with ULD. Studies have found that the use of these rehabilitation robots can lead to improvements in upper limb function in patients with conditions such as stroke or spinal cord injury. However, more research is needed to understand the long-term effects of robot-aided therapy and to evaluate the cost-effectiveness of these interventions.

References

Ahmed, T., Islam, M.R., Brahmi, B., Rahman, M.H., 2022. Robustness and tracking performance evaluation of PID motion control of 7 DOF anthropomorphic exoskeleton robot assisted upper limb rehabilitation. Sensors 22, 3747.

American Stroke Association, 2019. https://www.stroke.org/we-can-help/survivors/stroke-recovery/first-steps-to-recovery/rehabilitation-therapy-after-a-stroke/.

Amirabdollahian, F., Harwin, W.S., Loureiro, R.C., 2007. Analysis of the fugl-meyer outcome measures assessing the effectiveness of robot-mediated stroke therapy. In: 2007 IEEE 10th International Conference on Rehabilitation Robotics, IEEE, pp. 729–735.

Bohannon, R.W., 1999. Intertester reliability of hand-held dynamometry: a concise summary of published research. Percept. Mot. Ski. 88 (3), 899–902.

Brahmi, B., Saad, M., Luna, C.O., Archambault, P.S., Rahman, M.H., 2018a. Passive and active rehabilitation control of human upper-limb exoskeleton robot with dynamic uncertainties. Robotica 36, 1757–1779.

Brahmi, B., Saad, M., Ochoa-Luna, C., Rahman, M.H., Brahmi, A., 2018b. Adaptive tracking control of an exoskeleton robot with uncertain dynamics based on estimated time-delay control. IEEE-ASME Trans. Mechatron. 23, 575–585.

Burgar, C.G., Lum, P.S., Scremin, A.M., Garber, S.L., Van der Loos, H.F., Kenney, D., Shor, P., 2011. Robot-assisted upper-limb therapy in acute rehabilitation setting following stroke: department of Veterans Affairs multisite clinical trial. J. Rehabil. Res. Dev. 48 (4), 445–458.

Calabrò, R.S., Cacciola, A., Berté, F., Manuli, A., Leo, A., Bramanti, A., Naro, A., Milardi, D., Bramanti, P., 2016a. Robotic gait rehabilitation and substitution devices in neurological disorders: where are we now? Neurol. Sci. 37, 503–514.

Calabrò, R.S., Russo, M., Naro, A., Milardi, D., Balletta, T., Leo, A., Filoni, S., Bramanti, P., 2016b. Who may benefit from armeo power treatment? A neurophysiological approach to predict neurorehabilitation outcomes. PM R 8, 971–978.

Chang, W.H., Kim, Y.-H., 2013. Robot-assisted therapy in stroke rehabilitation. J. Stroke 15, 174.

Coote, S., Murphy, B., Harwin, W., Stokes, E., 2008. The effect of the GENTLE/s robot-mediated therapy system on arm function after stroke. Clin. Rehabil. 22 (5), 395–405.

Cui, X., Chen, W., Jin, X., Agrawal, S.K., 2017. Design of a 7-DOF cable-driven arm exoskeleton (CAREX-7) and a controller for dexterous motion training or assistance. IEEE/ASME Trans. Mechatron. 22 (1), 161–172.

Demircan, E., Yung, S., Choi, M., Baschshi, J., Nguyen, B., Rodriguez, J., 2020. Operational space analysis of human muscular effort in robot assisted reaching tasks. Robot. Auton. Syst. 125, 103429.

Feigin, V.L., Brainin, M., Norrving, B., Martins, S., Sacco, R.L., Hacke, W., Fisher, M., Pandian, J., Lindsay, P., 2022. World Stroke Organization (WSO): global stroke fact sheet 2022. Int. J. Stroke 17 (1), 18–29.

Fugl-Meyer, A.R., Jääskö, L., Leyman, I., Olsson, S., Steglind, S., 1975. A method for evaluation of physical performance. Scand. J. Rehabil. Med. 7, 13–31.

Gandolfi, M., Formaggio, E., Geroin, C., Storti, S.F., Boscolo Galazzo, I., Bortolami, M., Saltuari, L., Picelli, A., Waldner, A., Manganotti, P., Smania, N., 2018. Quantification of upper limb motor recovery and Eeg power changes after robot-assisted bilateral arm training in chronic stroke patients: a prospective pilot study. Neural Plast., 2018.

Gandolfi, M., Valè, N., Dimitrova, E.K., Mazzoleni, S., Battini, E., Filippetti, M., Picelli, A., Santamato, A., Gravina, M., Saltuari, L., 2019. Effectiveness of robot-assisted upper limb training on spasticity, function and muscle activity in chronic stroke patients treated with botulinum toxin: a randomized single-blinded controlled trial. Front. Neurol. 10, 41.

Germanotta, M., Cruciani, A., Pecchioli, C., Loreti, S., Spedicato, A., Meotti, M., Mosca, R., Speranza, G., Cecchi, F., Giannarelli, G., 2018. Reliability, validity and discriminant ability of the instrumental indices provided by a novel planar robotic device for upper limb rehabilitation. J. Neuroeng. Rehabil. 15, 1–14.

Gladstone, D.J., Danells, C.J., Black, S.E., 2002. The Fugl-Meyer assessment of motor recovery after stroke: a critical review of its measurement properties. Neurorehabil. Neural. Repair. 16 (3), 232–240.

Gopura, R.A.R.C., Kiguchi, K., Yang, L., 2009. Sueful-7: a 7 DOF upper-limb exoskeleton robot with muscle-model-oriented EMG-based control. In: 2009 IEEE/RSJ International Conference on Intelligent Robots and Systems (IROS 2009), 11–15 Oct. 2009, Piscataway, NJ, USA, pp. 1126–1131.

Gunasekara, M., Gopura, R., Jayawardena, S., 2015. 6-Rexos: upper limb exoskeleton robot with improved pHRI. Int. J. Adv. Robot. Syst. 12 (4), 47.

Hailey, R.O., De Oliveira, A.C., Ghonasgi, K., Whitford, B., Lee, R.K., Rose, C.G., Deshpande, A.D., 2022. Impact of gravity compensation on upper extremity movements in harmony exoskeleton. In: 2022 International Conference on Rehabilitation Robotics (ICORR), IEEE, pp. 1–6.

Hesse, S., Schmidt, H., Werner, C., Bardeleben, A., 2003a. Upper and lower extremity robotic devices for rehabilitation and for studying motor control. Curr. Opin. Neurol. 16, 705–710.

Hesse, S., Schulte-Tigges, G., Konrad, M., Bardeleben, A., Werner, C., 2003b. Robot-assisted arm trainer for the passive and active practice of bilateral forearm and wrist movements in hemiparetic subjects. Arch. Phys. Med. Rehabil. 84, 915–920.

Hocoma, 2022. Armeo®Power—Hocoma. Available at: https://www.hocoma.com/us/solutions/armeo-power. (Accessed).

Hogan, N., Krebs, H.I., Charnnarong, J., Srikrishna, P., Sharon, A., 1992. Mit-Manus: a workstation for manual therapy and training. I. In: [1992] Proceedings IEEE International Workshop on Robot and Human Communication, IEEE, pp. 161–165.

Hogan, N., Krebs, H.I., Sharon, A., Charnnarong, J., Massachusetts Institute of Technology, 1995. Interactive Robotic Therapist. U.S. Patent 5,466,213.

Hughes, S., 2020. The Drastic Effects of Covid-19 on Stroke Services. Medscape. Available at: https://www.medscape.com/viewarticle/940804.

Islam, M.R., Rahmani, M., Rahman, M.H., 2020a. A novel exoskeleton with fractional sliding mode control for upper limb rehabilitation. Robotica, 1–22.

Islam, M.R., Brahmi, B., Ahmed, T., Assad-Uz-Zaman, M., Rahman, M.H., 2020b. Exoskeletons in upper limb rehabilitation: a review to find key challenges to improve functionality. In: Control Theory in Biomedical Engineering, pp. 235–265.

Kahn, L.E., Averbuch, M., Rymer, W.Z., Reinkensmeyer, D.J., 2001. Comparison of robot-assisted reaching to free reaching in promoting recovery from chronic stroke. In: Proceedings of the International Conference on Rehabilitation Robotics, pp. 39–44.

Kazerooni, H., 2005. Exoskeletons for human power augmentation. In: 2005 IEEE/RSJ International Conference on Intelligent Robots and Systems, August, pp. 3459–3464.

Khan, M.M.R., Rahman, M.M., De Caro, J.S., Wang, I., Rahman, M., 2022. An end-effector type therapeutic robot for home-based upper limb rehabilitation. Arch. Phys. Med. Rehabil. 103, e146–e147.

Kim, B., Deshpande, A.D., 2015. Controls for the shoulder mechanism of an upper-body exoskeleton for promoting scapulohumeral rhythm. In: 2015 IEEE International Conference on Rehabilitation Robotics (ICORR), IEEE, pp. 538–542.

Kim, B., Deshpande, A.D., 2017. An upper-body rehabilitation exoskeleton Harmony with an anatomical shoulder mechanism: design, modeling, control, and performance evaluation. Int. J. Robot. Res. 36 (4), 414–435.

Kim, G., Lim, S., Kim, H., Lee, B., Seo, S., Cho, K., Lee, W., 2017. Is robot-assisted therapy effective in upper extremity recovery in early stage stroke? A systematic literature review. J. Phys. Ther. Sci. 29 (6), 1108–1112.

Krebs, H.I., Ferraro, M., Buerger, S.P., Newbery, M.J., Makiyama, A., Sandmann, M., Lynch, D., Volpe, B.T., Hogan, N., 2004. Rehabilitation robotics: pilot trial of a spatial extension for MIT-Manus. J. Neuroeng. Rehabil. 1 (1), 1–15.

Krebs, H.I., Hogan, N., Aisen, M.L., Volpe, B.T., 1998. Robot-aided neurorehabilitation. IEEE Trans. Rehabil. Eng. 6 (1), 75–87.

Krebs, H.I., Volpe, B.T., Aisen, M.L., Hogan, N., 2000. Increasing productivity and quality of care: robot-aided neuro-rehabilitation. J. Rehabil. Res. Dev. 37 (6), 639–652.

Krebs, H.I., Volpe, B.T., Williams, D., Celestino, J., Charles, S.K., Lynch, D., Hogan, N., 2007. Robot-aided neurorehabilitation: a robot for wrist rehabilitation. IEEE Trans. Neural Syst. Rehabil. Eng. 15, 327–335.

Krebs, H.I., Edwards, D., Hogan, N., 2016. Forging mens et manus: the Mit experience in upper extremity robotic therapy. In: Reinkensmeyer, D.J., Dietz, V. (Eds.), Neurorehabilitation Technology. Springer International Publishing, Cham, pp. 333–350.

Kuroda, Y., Young, M., Shoman, H., Punnoose, A., Norrish, A.R., Khanduja, V., 2021. Advanced rehabilitation technology in orthopaedics—a narrative review. Int. Orthop. 45 (8), 1933–1940.

Lee, K.W., Kim, S.B., Lee, J.H., Lee, S.J., Kim, J.W., 2017. Effect of robot-assisted game training on upper extremity function in stroke patients. Ann. Rehabil. Med. 41 (4), 539–546.

Lee, S.H., Park, G., Cho, D.Y., Kim, H.Y., Lee, J.Y., Kim, S., Park, S.B., Shin, J.H., 2020. Comparisons between end-effector and exoskeleton rehabilitation robots regarding upper extremity function among chronic stroke patients with moderate-to-severe upper limb impairment. Sci. Rep. 10 (1), 1–8.

Li, Y., Li, M., Wang, M., Zhou, Y., Chang, J., Xian, Y., Wang, D., Mao, L., Jin, H., Hu, B., 2020. Acute cerebrovascular disease following Covid-19: a single center, retrospective, observational study. Stroke Vasc. Neurol. 5 (3).

Liao, W.-W., Wu, C.-Y., Hsieh, Y.-W., Lin, K.-C., Chang, W.-Y., 2012. Effects of robot-assisted upper limb rehabilitation on daily function and real-world arm activity in patients with chronic stroke: a randomized controlled trial. Clin. Rehabil. 26, 111–120.

Lin, Y., Li, Q.-Y., Qu, Q., Ding, L., Chen, Z., Huang, F., Hu, S., Deng, W., Guo, F., Wang, C., 2022. Comparative effectiveness of robot-assisted training versus enhanced upper extremity therapy on upper and lower extremity for stroke survivors: a multicentre randomized controlled trial. J. Rehabil. Med. 54, jrm00314.

Liu, L., Shi, Y.-Y., Xie, L., 2016. A novel multi-DOF exoskeleton robot for upper limb rehabilitation. J. Mech. Med. Biol. 16 (8), 1640023.

Lo, H.S., Xie, S.Q., 2012. Exoskeleton robots for upper-limb rehabilitation: state of the art and future prospects. Med. Eng. Phys. 34 (3), 261–268.

Lum, P., Reinkensmeyer, D., Mahoney, R., Rymer, W.Z., Burgar, C., 2002. Robotic devices for movement therapy after stroke: current status and challenges to clinical acceptance. Top. Stroke Rehabil. 8 (4), 40–53.

Lum, P.S., Burgar, C.G., Shor, P.C., 2004. Evidence for improved muscle activation patterns after retraining of reaching movements with the MIME robotic system in subjects with post-stroke hemiparesis. IEEE Trans. Neural Syst. Rehabil. Eng. 12 (2), 186–194.

Lum, P.S., Burgar, C.G., Van Der Loos, M., Shor, P.C., Majmundar, M., Yap, R., 2006. MIME robotic device for upper-limb neurorehabilitation in subacute stroke subjects: a follow-up study. J. Rehabil. Res. Dev. 43, 631.

Marcheschi, S., Salsedo, F., Fontana, M., Bergamasco, M., 2011. Body extender: whole body exoskeleton for human power augmentation. In: 2011 IEEE International Conference on Robotics and Automation, May, pp. 611–616.

Mazzoleni, S., Focacci, A., Franceschini, M., Waldner, A., Spagnuolo, C., Battini, E., Bonaiuti, D., 2017. Robot-assisted end-effector-based gait training in chronic stroke patients: a multicentric uncontrolled observational retrospective clinical study. NeuroRehabilitation 40 (4), 483–492.

Narayan Arya, K., Verma, R., Garg, R.K., 2011. Estimating the minimal clinically important difference of an upper extremity recovery measure in subacute stroke patients. Top. Stroke Rehabil. 18 (1), 599–610.

Nef, T., Guidali, M., Klamroth-Marganska, V., Riener, R., 2009. Armin—exoskeleton robot for stroke rehabilitation. In: Dossel, O., Schlegel, W.C. (Eds.), World Congress € on Medical Physics and Biomedical Engineering, September 7–12, 2009, Munich, Germany, Springer, Berlin, Heidelberg, pp. 127–130.

Paolucci, T., Agostini, F., Mangone, M., Bernetti, A., Pezzi, L., Liotti, V., Recubini, E., Cantarella, C., Bellomo, R.G., D'Aurizio, C., 2021. Robotic rehabilitation for end-effector device and botulinum toxin in upper limb rehabilitation in chronic post-stroke patients: an integrated rehabilitative approach. Neurol. Sci. 42, 5219–5229.

Pennati, G., Da Re, C., Messineo, I., Bonaiuti, D., 2015. How could robotic training and botolinum toxin be combined in chronic post stroke upper limb spasticity? A pilot study. Eur. J. Phys. Rehabil. Med. 51, 381–387.

Perry, J.C., Rosen, J., Burns, S., 2007. Upper-limb powered exoskeleton design. IEEE/ASME Trans. Mechatron. 12 (4), 408–417.

Perry, J.C., Maura, R., Bitikofer, C.K., Wolbrecht, E.T., 2019. Blue Sabino: development of a bilateral exoskeleton instrument for comprehensive upper-extremity neuromuscular assessment. In: Masia, L., Micera, S., Akay, M., Pons Jose, L. (Eds.), Converging Clinical and Engineering Research on Neurorehabilitation III. Springer International Publishing, Cham, pp. 493–497.

Poli, P., Morone, G., Rosati, G., Masiero, S., 2013. Robotic technologies and rehabilitation: new tools for stroke patients' therapy. BioMed Res. Int.

Qassim, H.W., Wan Hasan, W.Z., 2020. A review on upper limb rehabilitation robots. Applied Sciences 10 (19), 6976.

Qureshi, A.I., Baskett, W.I., Huang, W., Shyu, D., Myers, D., Raju, M., Lobanova, I., Suri, M.F.K., Naqvi, S.H., French, B.R., Siddiq, F., 2021. Acute ischemic stroke and Covid-19: an analysis of 27 676 patients. Stroke 52 (3), 905–912.

Rahman, M.H., Saad, M., Kenne, J.P., Archambault, P.S., 2009. Modeling and control of a 7DOF exoskeleton robot for arm movements. In: 2009 IEEE International Conference on Robotics and Bbiomimetics (ROBIO), pp. 245–250.

Rahman, M.H., Ouimet, T.K., Saad, M., Kenne, J.P., Archambault, P.S., 2012. Development and control of a robotic exoskeleton for shoulder, elbow and forearm movement assistance. Appl. Bionics Biomech. 9 (3), 275–292. https://doi.org/10.3233/ABB-2012-0061.

Rahman, M.H., Ochoa-Luna, C., Rahman, M.J., Saad, M., Archambault, P., 2014a. Force-position control of a robotic exoskeleton to provide upper extremity movement assistance. Int. J. Model. Identif. Control. 21, 390–400.

Rahman, M.H., Rahman, M.J., Cristobal, O.L., Saad, M., Kenne, J.P., Archambault, P.S., 2014b. Development of a whole arm wearable robotic exoskeleton for rehabilitation and to assist upper limb movements. Robotica 33 (1), 19–39.

Rahman, M.H., Rahman, M.J., Cristobal, O., Saad, M., Kenné, J.-P., Archambault, P.S., 2015. Development of a whole arm wearable robotic exoskeleton for rehabilitation and to assist upper limb movements. Robotica 33, 19–39.

Reinkensmeyer, D.J., Kahn, L.E., Averbuch, M., Kenna-Cole, A., 2000. Understanding and treating arm movement im Pairment after chronic brain injury: progress with the arm guide. J. Rehabil. Res. Dev.

Reinkensmeyer, D.J., Kahn, L.E., Averbuch, M., Mckenna-Cole, A., Schmit, B.D., Rymer, W.Z., 2014. Understanding and treating arm movement impairment after chronic brain injury: progress with the ARM guide. J. Rehabil. Res. Dev. 37, 653–662.

Rosati, G., Gallina, P., Masiero, S., 2007. Design, implementation and clinical tests of a wire-based robot for neurorehabilitation. IEEE Trans. Neural Syst. Rehabil. Eng. 15, 560–569.

Sale, P., Franceschini, M., Mazzoleni, S., Palma, E., Agosti, M., Posteraro, F., 2014. Effects of upper limb robot-assisted therapy on motor recovery in subacute stroke patients. J. Neuroeng. Rehabil. 11 (1), 104.

Shen, Y., Rosen, J., 2020. EXO-UL upper limb robotic exoskeleton system series: from 1 DOF single-arm to (7+ 1) DOFs dual-arm. In: Wearable Robotics. Elsevier.

Shen, Y., Ma, J., Dobkin, B., Rosen, J., 2018. Asymmetric dual arm approach for post stroke recovery of motor functions utilizing the EXO-UL8 exoskeleton system: a pilot study. In: 2018 40th Annual International Conference of the IEEE Engineering in Medicine and Biology Society (EMBC), IEEE, pp. 1701–1707.

Shen, Y., Sun, J., Ma, J., Rosen, J., 2019. Admittance control scheme comparison of EXO-UL8: a dual-arm exoskeleton robotic system. In: 2019 IEEE 16th International Conference on Rehabilitation Robotics (ICORR), IEEE, pp. 611–617.

Teasell, R.W., Kalra, L., 2004. What's new in stroke rehabilitation. Stroke 35 (2), 383–385.

Tsao, C.W., Aday, A.W., Almarzooq, Z.I., Alonso, A., Beaton, A.Z., Bittencourt, M.S., Boehme, A.K., Buxton, A.E., Carson, A.P., Commodore-Mensah, Y., Elkind, M.S., 2022. Heart disease and stroke statistics—2022 update: a report from the American Heart Association. Circulation 145 (8), e153–e639.

Van Delden, A., Peper, C.L.E., Kwakkel, G., Beek, P.J., 2012. A systematic review of bilateral upper limb training devices for poststroke rehabilitation. Stroke Res. Treat. 2012.

Veerbeek, J.M., Langbroek-Amersfoort, A.C., Van Wegen, E.E., Meskers, C.G., Kwakkel, G., 2017. Effects of robot-assisted therapy for the upper limb after stroke: a systematic review and meta-analysis. Neurorehabil. Neural Repair 31 (2), 107–121.

Volpe, B.T., Krebs, H.I., Hogan, N., Edelsteinn, L., Diels, C.M., Aisen, M.L., 1999. Robot training enhanced motor outcome in patients with stroke maintained over 3 years. Neurology 53 (8), 1874.

Volpe, B.T., Krebs, H.I., Hogan, N., Edelstein, L., Diels, C., Aisen, M., 2000. A novel approach to stroke rehabilitation: robot-aided sensorimotor stimulation. Neurology 54 (10), 1938–1944.

Volpe, B.T., Krebs, H.I., Hogan, N., 2001. Is robot-aided sensorimotor training in stroke rehabilitation a realistic option? Curr. Opin. Neurol. 14 (6), 745–752.

Waldner, A., Tomelleri, C., Hesse, S., 2009. Transfer of scientific concepts to clinical practice: recent robot-assisted training studies. Funct. Neurol. 24 (4), 173–177.

Walsh, C.J., Paluska, D., Pasch, K., Grand, W., Valiente, A., Herr, H., 2006. Development of a lightweight, underactuated exoskeleton for load-carrying augmentation. In: Proceedings 2006 IEEE International Conference on Robotics and Automation, ICRA, May 2006, pp. 3485–3491.

Wu, C.-Y., Yang, C.-L., Chen, M.-D., Lin, K.-C., Wu, L.-L., 2013. Unilateral versus bilateral robot-assisted rehabilitation on arm-trunk control and functions post stroke: a randomized controlled trial. J. Neuroeng. Rehabil. 10, 1–10.

Xiao, F., Gao, Y., Wang, Y., Zhu, Y., Zhao, J., 2018. Design and evaluation of a 7-DOF cable-driven upper limb exoskeleton. J. Mech. Sci. Technol. 32, 855–864.

Xie, Q., Meng, Q., Zeng, Q., Fan, Y., Dai, Y., Yu, H., 2021. Human-exoskeleton coupling dynamics of a multi-mode therapeutic exoskeleton for upper limb rehabilitation training. IEEE Access 9, 61998–62007.

Yoo, D.H., Kim, S.Y., 2015. Effects of upper limb robot-assisted therapy in the rehabilitation of stroke patients. J. Phys. Ther. Sci. 27 (3), 677–679.

Zhang, L., Li, J., Cui, Y., Dong, M., Fang, B., Zhang, P., 2020. Design and performance analysis of a parallel wrist rehabilitation robot (PWRR). Robot. Auton. Syst. 125, 103390.

Zhu, C., Okada, Y., Yoshioka, M., Yamamoto, T., Yu, H., Yan, Y., Duan, F., 2014. Power augmentation of upper extremity by using agonist electromyography signals only for extended admittance control. IEEJ J. Ind. Appl. 3 (3), 260–269.

Zimmermann, Y., Forino, A., Riener, R., Hutter, M., 2019. ANYexo: a versatile and dynamic upper-limb rehabilitation robot. IEEE Robot. Automat. Lett. 4, 3649–3656.

Zimmermann, Y., Sommerhalder, M., Wolf, P., Riener, R., Hutter, M., 2023. ANYexo 2.0: a fully actuated upper-limb exoskeleton for manipulation and joint-oriented training in all stages of rehabilitation. IEEE Trans. Robot.

Further reading

Bertani, R., Melegari, C., De Cola, M.C., Bramanti, A., Bramanti, P., Calabro, R.S., 2017. Effects of robot-assisted upper limb rehabilitation in stroke patients: a systematic review with meta-analysis. Neurol. Sci. 38, 1561–1569.

Calabrò, R.S., Naro, A., Russo, M., Leo, A., De Luca, R., Balletta, T., Buda, A., La Rosa, G., Bramanti, A., Bramanti, P., 2017. The role of virtual reality in improving motor performance as revealed by EEG: a randomized clinical trial. J. Neuroeng. Rehabil. 14, 1–16.

Editor, I.O., 2021. Who will pay for robotic rehabilitation? The growing need for a cost-effectiveness analysis. Innov. Clin. Neurosci. 17 (10–12), 14–16.

Sae-Media-Group, 2022. Advanced Motor Technology Enables Medical Exoskeleton for Rehabilitation. Available at: https://www.medicaldesignbriefs.com/component/content/article/mdb/pub/features/technology-leaders/46142. (Accessed).

Xiao, F., Gao, Y., Wang, Y., Zhu, Y., Zhao, J., 2017. Design of a wearable cable-driven upper limb exoskeleton based on epicyclic gear train's structure. Technol. Health Care 25 (S1), 3–11.

CHAPTER 6

Failure of total hip arthroplasty (THA): State of the art

Atef Boulila[a], Lanouar Bouzid[b], and Mahfoudh Ayadi[c]
[a]University of Carthage, National Institute of Applied Sciences and Technology, Tunis, Tunisia
[b]University of Tunis El Manar, Tunis Faculty of Medicine, Tunis, Tunisia
[c]University of Carthage, National Engineering School of Bizerte, Menzel Abderrahman, Tunisia

6.1 Introduction

Total hip arthroplasty (THA) is one of the most successful and cost-effective procedures in orthopedics. It offers reliable pain relief and significant improvement in function for patients with hip failure (Crawford et al., 2021; Tokgoz, 2022). THA was introduced in the 1960s as an excellent and reliable treatment option for end-stage hip disease, with satisfactory clinical results after a follow-up of 15–20 years (Pakvis et al., 2011; Sentuerk et al., 2016).

In due course, orthopedic surgeons were confronted with the initial problems of THA failures; other problems were detected regarding material selection. Today, it has become clear that the long-term survival of THA is a multifactorial issue. Factors other than the implant, such as diagnosis (etiology), patient, surgeon, and surgical technique, also play an important role in survival. Despite good results, the revision rates for total hip replacements (THRs) have been steadily increasing in recent years. Fig. 6.1 shows the interactions between the major factors in the failure of THA as depicted and improved by some authors such as Boulila et al. (2008) and Kumar et al. (2014).

The severity of the failure depends on all three factors. Indeed, failure can be an interaction between the patient and the surgeon and technique (FL_{PST}), the patient and the implant (FL_{PI}), or the technique and the implant (FL_{IST}) or an interaction of all the three factors (FL_{IPST}).

As the number of primary arthroplasty implants increases each year, a revision prosthesis will be indicated more often, with the associated complications. Understanding the mechanism and location of THA failure is essential to determine the appropriate treatment (Bäcker et al., 2022).

According to Karachalios et al. (2018), biomechanical studies have shown that critical stress within the thin titanium dioxide layer can lead to microcracks that then develop

Medical and Healthcare Robotics
https://doi.org/10.1016/B978-0-443-18460-4.00012-3

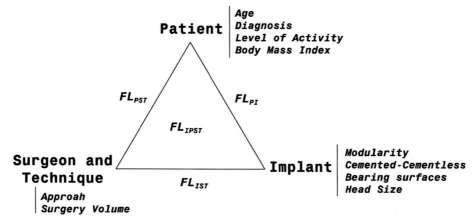

Fig. 6.1 Factors related to the failure modes of THA.

into structural failure. In some cases, a corrosive environment such as oxidative stress can lead to implant failure (Bergmann et al., 2004, 2016).

Kheir et al. (2022), in their case study, showed that a histopathological diagnosis can have a significant impact on the life of patients during the implantation of a primary prosthesis.

The objective of this chapter was to present the state of the art on THA failures. The idea was to gather in one chapter the main failures that can affect the initial total hip prosthesis. This will help the orthopedic surgeon better understand the mechanisms that can lead to THR failure during this post-COVID period.

6.1.1 Natural hip joint

Anatomical planes are used to describe the location of structures in the human anatomy. To study the biomechanics of the hip, three planes are considered: sagittal (SP), frontal (FP), and transverse (TP), as shown in Fig. 6.2. The hip is considered the most important joint in the human body. It represents a patella associating the head of the femur with the acetabulum (Ng et al., 2019). The bony surfaces of the patella are covered with articular cartilage, a smooth tissue that cushions the ends of the bones and allows them to move easily (Fig. 6.3A).

A thin synovial membrane surrounds the hip joint. In a healthy hip, this membrane produces a quantity of fluid that lubricates the cartilage and eliminates any friction during hip movements (Liang, 2018). Bands of tissue called ligaments (the hip capsule) that connect the ball to the socket provide stability to the hip.

The healthy biomechanics of the hip are well-known from the Pauwels balance (Fig. 6.3B), which combines the weight of the patient, the action of the muscles, and

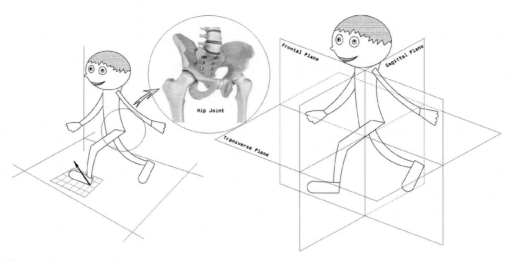

Fig. 6.2 Anatomical planes.

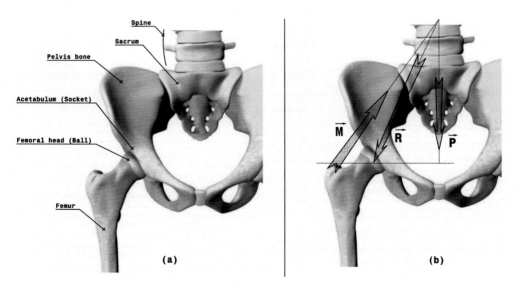

Fig. 6.3 (A) The hip joint in the human body in the frontal plane. (B) The Pauwels balance.

the reaction applied to the femoral head–acetabulum junction as presented by Jacquemaire (1991) and Kummer (1993).

The hip is a synovial spheroid joint that allows movement in all three directions of space (Gold et al., 2022). Its function is to orient the lower limb in all these directions, and it is extremely well-adapted to the standing position. However, its movements remain limited in amplitude.

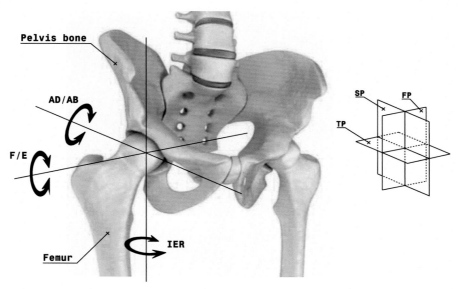

Fig. 6.4 Hip joint mobility.

As mentioned in the literature, the hip joint has three degrees of freedom like the ball–and–socket joint: flexion/extension (F/E), abduction/adduction (AB/AD), and external/internal rotation (EIR) (Jamari et al., 2017; Zaghloul, 2018; Sebti et al., 2020; Ozcadirci et al., 2021) (Fig. 6.4).

Based on biomechanical studies (Boulila, 2010; Boulila et al., 2010), the mechanical load applied to the hip joint $(\vec{R})$ can reach four times the body weight in a single-pod support for a gait cycle.

The gait cycle begins with the initial contact of the heel and ends by definition when the foot regains contact with the ground. Two different phases are defined during the gait: the stance, which corresponds to the entire period when the foot is in contact with the ground, and the oscillating phase when the foot is no longer in contact with the ground and which corresponds to the advancement of the lower limb (Di Gregorio and Vocenas, 2021).

During walking, the load applied on the femoral head can reach seven times the human body weight (Bergmann et al., 2001, 2016) as shown in Fig. 6.5. Instrumented prostheses that allow real-time visualization of the intensity of the applied load have been implanted in patients.

Indeed, as Martelli and Costi (2021) indicate, real-time in vitro reproduction of the 3D, time-varying loading profiles that act on human bones during physical activity could advance bone and implant testing protocols and thereby reduce THA failures.

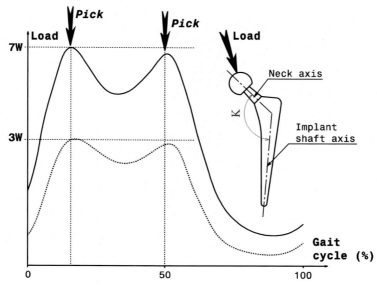

Fig. 6.5 The walking cycle.

Computer-assisted navigation is an effective means used in complex prosthesis revision in which altered anatomy and landmarks are inaccurate references for cup positioning. In these complex cases, this technique reduces the risk of component malposition and joint instability (Rankin et al., 2022).

6.1.2 Total hip replacement

A hip implant is a medical device used to replace a damaged hip joint (Boulila, 2010). As previously mentioned, the hip joint consists of a convex femoral head inserted into a concave acetabulum inside the pelvis (Fig. 6.6). The articular cartilage within a synovial joint capsule cushions the load applied to the femoral head during walking.

The study of indications for revision THP requires an examination of the causes of failure. The lack of stable fixation of an implant leads to the micromobility of the prosthetic components, resulting in loosening. This may be due to aging of the cement, a defect in the initial fixation, or poor implant design (Khanna et al., 2017).

6.1.3 Biomaterials in THA

As depicted by several authors (Boulila et al., 2018; Bouzidi et al., 2022), hip joint prostheses are made with metals, ceramics, and plastic materials. The materials that are most commonly used are titanium alloys, stainless steel, special high-strength alloys, composite materials, ceramics, and UHMWPE.

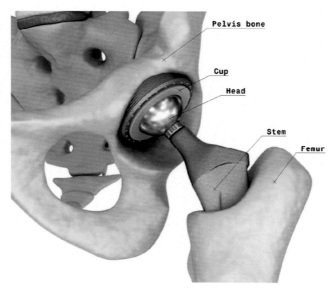

Fig. 6.6 Total hip arthroplasty.

The materials used need to satisfy important requirements; they must have high strength in order to cope with static and dynamic loads and high resistance to mechanical and chemical wear. Biocompatibility and fatigue resistance are also necessary. Usually, stems and necks are composed of metals, whereas the femoral head is made of metal or ceramic materials. For the acetabulum, metals, polymers, and ceramic materials are used.

Biomaterials commonly used in orthopedic surgery, mainly in THA, are divided into four families: metallic materials (stainless steel, titanium alloys, cobalt–chromium alloys, etc.), plastics (polyethylene, acrylic cement (PMMA), etc.), ceramics (alumina, zirconia), and composites. PMMA is used to fix the stem and cup of THA on the femur and the acetabulum. Table 6.1 summarizes all biomaterials used in THA and their corresponding mechanical properties, meeting the criteria of biocompatibility, biofunctionality, and strength (Boulila, 2010; Jeffers and Walter, 2012; Moghadasi et al., 2022).

In this paragraph, we briefly describe the biomaterials used in THA.

6.1.3.1 Stainless steel

Stainless steels are iron–carbon–based alloys that contain Cr, Ni, Mo, Mn, and C. Austenitic alloys (316 series) are generally used for THA. Its resistance to oxidation combined with the relative ease of machining and forming make stainless steel a potential candidate

Table 6.1 Materials used in total hip arthroplasty and the associated main properties.

Prosthetic materials		Cup	Head	Stem	Density (g/cm³)	Tensile strength (MPa)	Elastic modulus (GPa)	Poisson's ratio	Elongation (%)
Polyethylene	SD UHMWPE	✓			0.941–0.965	30–40	0.45–1.3	0.46	130–500
	XL UHMWPE	✓							
	XLPE UHMWPE	✓							
AISI 316L		✓	✓	✓	7.9	515–620	200	–	–
CoCrMo		✓	✓	✓	7.8	600–1795	200–230	–	–
Alumina (Al$_2$O$_3$)		✓	✓		3980	–	380	0.23	–
Zirconia (ZrO$_2$)		✓	✓		6040	–	210	0.30	–
Titanium (Ti-6Al-4V)			✓	✓	–	895–930	110	–	6–10

for material selection. For this reason, this material continues to be used in hip replacements (Priyana Soemardi et al., 2019; Jamari et al., 2022).

6.1.3.2 Cobalt–chromium alloys

Cobalt–Chromium alloys (CoCr alloys), originally used in dentistry, are now one of the main materials used for hip replacements. Their favorable strength, corrosion, and wear characteristics make them one of the leading choices as an implant material (Das and Chakraborti, 2018). They are mainly used as a material for the cemented femoral stem because their elastic moduli are higher than those of titanium alloys and for the joint head because of their wear resistance.

6.1.3.3 Titanium alloys

Titanium and its alloys are popular metallic biomaterials used in THA. Titanium ($\alpha + \beta$) alloys, such as titanium alloy Ti-6Al-4V, have been the most widely used alloys for the cemented acetabulum stem and THA components because of their comparatively low density, high mechanical strength, excellent corrosion resistance, and biocompatibility with the bone (Milovanović et al., 2020). However, titanium alloys are not used for femoral head fabrication because of their poor wear resistance.

As depicted by Hu and Yoon (2018), in the last two decades, vanadium-free titanium alloys, such as ($\alpha + \beta$) titanium-6Al-7Nb alloy, with improved biocompatibility have been developed by incorporating biocompatible elements such as niobium. Much research has been devoted to the development of bulk metallic materials with a lower Young's modulus, among which β-titanium alloys have attracted significant attention.

6.1.3.4 Other materials

Hydroxyapatite, other metals, and porous coatings are used to ensure effective osseointegration between the bone and the implant body and reduce the risk of loosening. As depicted by Castagnini et al. (2019) and Moghadasi et al. (2022), titanium, some of its alloys, and tantalum are the most suitable porous metallic materials for orthopedic applications compared to other conventional metals.

6.1.3.5 Ultrahigh-molecular-weight polyethylene (UHMWPE)

The use of polyethylene as a cup of THA dates back to 1962 when it was introduced in the Charnley hip prosthesis. Conventional PE is sterilized by gamma irradiation in air. This process allows molecular cross-linking but can also produce free radicals that are oxidized in the presence of air. So, they turned to highly cross-linked UHMWPE (XLPE) to reduce PE wear while maintaining its mechanical properties and eliminating the oxidation process (Hu and Yoon, 2018; Bistolfi et al., 2021; Pietrzak, 2021).

A second generation of XLPE was developed by adding antioxidants such as vitamin E to prevent free radical oxidation in order to increase wear resistance (Dumbleton et al., 2006).

Kyomoto et al. (2011) marked a major advance in the tribological aspect of XLPE by treating the seal surface. This treatment consists of coating the surface with a chemically thin layer (100–200 nm) of poly(2-methacryloyloxyethyl phosphorylcholine) (PMPC) to improve the abrasion resistance.

6.1.3.6 Ceramics

Alumina and zirconia have been used as a bearing surface in THA since 1970 (Hannouche et al., 2005; Giuseppe et al., 2022). These ceramics have biocompatibility, high wear resistance, and chemical durability.

Although alumina shows better wear characteristics than those of the metal-polyethylene pair, the incidence of fracture is high. This has led to improved manufacturing processes by decreasing grain size and porosity and annealing the process to increase strength (Jeffers and Walter, 2012).

Zirconia femoral heads were introduced in THA in 1985. The reason for the switch from alumina to zirconia as a component of the femoral head was due to the high incidence of fractures in alumina heads and also the higher fracture and flexural strength than alumina (Arena et al., 2019; Bouville, 2020).

Despite the long clinical history of these two ceramics in THA, they have drawbacks. Attempts to overcome the weaknesses of these materials by combining the hardness of alumina with the toughness of zirconia led to the development of zirconia-hardened alumina (ZTA), which was first marketed by CeramTec as BIOLOX Delta around 2000 (Tateiwa et al., 2020).

6.1.3.7 Other materials

In a recent study (Hu and Yoon, 2018), other materials have been used in THA, allowing to improve the resistance of the implants and increasing their life span:
- Silicon nitride, a nonoxide ceramic material with high strength and toughness, has been used as bearings and turbine blades for more than 50 years.
- The new zirconium alloy Zr-2.5Nb is commercialized as Oxinium.
- Diamond-like carbon (DLC).
- Titanium nitride (TiN).

6.2 Failures related to total hip arthroplasty

The current trend in THA is to develop new artificial hip joints with high wear resistance and mechanical reliability that can last at least 25–30 years for active patients, young or old (Khanna et al., 2017). As depicted by several authors such as Boulila et al. (2008), the failure of a THA necessitates that the patient undergoes revision surgery, despite the fact that prosthesis materials and design have made enormous progress.

Burke et al. (2019) present the failure mechanism of a THA. They confirmed that infection is a common mode of early and late primary arthroplasty failure compared with aseptic loosening and prosthesis instability.

The reasons for the failure of such an implant are: (a) mismatch of the elastic modulus between the implant and the host bone; (b) poor mechanical properties of the implant; (c) generation of corrosion products; (d) dissolution of the material; (e) wear debris; (f) bacterial infection, etc. Therefore, these issues must be considered when designing a biocompatible material for orthopedic application.

A THA failure limits durability and life span and is considered an important aspect that has motivated research on the development of new implants.

In the following, we present the different failures encountered at orthopedic centers. These failures can be attributed to a multitude of factors and depend on the health conditions of the patient.

6.2.1 Wear of polyethylene

Polyethylene wear is the biggest clinical problem in THR, and too many patients are still subjected to the problems associated with prosthesis failure (McGee et al., 2000; Boulila et al., 2008; Bracco et al., 2017).

Material removal in biomechanically loaded joints is mainly due to wear of the interacting surfaces. Wear of an artificial hip joint can be defined as a process characterized by a mechanical action related to friction, resulting in a progressive loss of material from the surfaces in contact. Wear can occur during relative movement between lubricated or nonlubricated solids. It is often accompanied by the inseparable phenomena of tribocorrosion.

Particulate wear debris has been shown to be associated with a locally aggressive biological response that can lead to periprosthetic bone loss and aseptic loosening of implants (Apostu et al., 2018).

Wear initiates an inflammatory reaction, ultimately leading to osteolysis (Khanna et al., 2017). It is defined as a cumulative phenomenon of surface deterioration in which the material is removed from the body in the form of small particles, primarily by mechanical processes. The wear mechanism is the transfer of energy with the removal or displacement of material (Fig. 6.7). Wear changes from adhesive to abrasive during hip joint movements.

In prosthetic tribology, the factors that influence the wear resistance or the wear phenomenon are the characteristics of the materials in contact, their surface roughness, the lubrication method, the type and speed of movement, and the generation and removal of solid particles from the contact area.

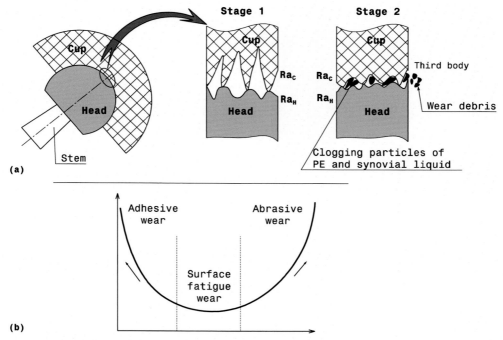

Fig. 6.7 (A) Illustration of the wear phenomenon. (B) Wear types.

Depending on the nature and the type of material in contact, the wear particles coming from abrasion, fatigue, corrosion, or delamination will follow a tribological circuit and will constitute a new active surface whose maintenance depends on the contact geometry, its rheological behavior, and the mechanical environment studied (Boulila et al., 2010).

Multiple factors that contribute to polyethylene wear and wear debris formation have been identified in the literature (Pace et al., 2013; Shim et al., 2022). These factors include articular surface compliance, polyethylene thickness, and femoral head diameter, method of polyethylene fixation, nature of the polyethylene, its sterilization technique, manufacturing method, and surgical implantation technique.

When the inner surface is worn, the head center C_C and the cup center C_H are different, as presented in Fig. 6.8. Then appears the linear wear defined by the Charnley and Cupic model and the Dorr model, as depicted by Boulila et al. (2010).

This radiological method measures the linear wear p of the prosthetic cup based on the position of the center of the THP head (C_H) and that of the cup (C_C). Although convenient, this method depends on the quality of the X-ray images and their orientations.

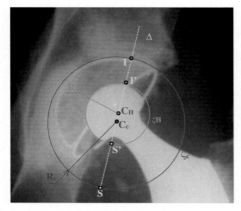

Caption:
Δ : Straight line passing through C_C and C_H
I, I', S, S' belong to Δ
With d(I, I') < d(S, S')
d(I, I')= d_{min}
d(S, S')= d_{Max}
p=R_C-(R_H+d_{min})   (Charnley & Cupic)
p= (d_{Max}-d_{min})/2   (Dorr)

Fig. 6.8 Geometric construction of X-ray images of linear wear.

Table 6.2 Critical weight loss of UHMWPE cups.

Paired material	ID × OD (mm)	Critical weight loss and wear rate (mg/10^6 cycles)		
		At 15.10^6 cycles	At 20.10^6 cycles	At 25.10^6 cycles
316L/UHMWPE	$\varnothing$22 × 50	41.26	40.83	37.69
	$\varnothing$28 × 50	47.76	42.27	38.46
CoCrMo/UHMWPE	$\varnothing$22 × 50	28.93	26.25	24.35
	$\varnothing$22 × 56	26.77	24.37	22.65

However, recent studies have shown a significant difference between the linear wear of the X-ray images method and the volumetric wear of hip simulators (Hua et al., 2019; Bhalekar et al., 2020).

As an indication, Table 6.2 summarizes the different types of wear rates for different paired materials tested on a Tunisian Hip Simulator TN-3DOF using the standard on wear of total hip prostheses (ISO 14242) (Boulila, 2010). It should be noted that one million cycles on the simulator correspond to 1 year of walking by a patient.

This standard ISO 14242 mainly consists of four parts and gives general comments and some technical requirements on the wear of total hip prostheses (Torabnia et al., 2021).

In fact, the osteolysis threshold was defined by a wear rate equal to $80 \, mg/10^6$ cycles. The wear phenomenon was assessment using other types of polyethylene. Wear on the contact surfaces between the head and the THP cup leads to the formation and accumulation of wear debris that can trigger an osteolysis reaction around the implant. The prosthesis becomes unstable and requires revision surgery (Elke and Rieker, 2018).

Thus, owing to in vitro tests on simulators, the surgeon can immediately estimate the rate of osteolysis and therefore the life span of the prosthesis before its revision.

Table 6.3 Correspondence between the wear phenomenon and the appearance of worn surfaces.

Wear type	Worn cup surface
Adhesive	- Transfers, generally from the softer body to the harder material - Plastic deformations - Metallurgical transformations
Two-body abrasive	- Textured surface appearance - Stripes parallel to the direction of travel
Three-body abrasive	- Random surface appearance - Indentations - Plastic deformations
Corrosive	- Colored surface films - Nonuniform appearance (islands)
Delamination—fatigue (ductile materials)	- Fractures parallel to the surface - Pitting, spalling - Plastic deformations

Then, the correspondence between the wear phenomenon and the appearance of worn surfaces can be viewed in Table 6.3.

6.2.2 Loosening

Loosening of components over time is a problem with both cemented and noncemented fixation components and usually results in increased pain during arthroplasty for the patient.

Aseptic loosening of a total hip prosthesis can occur due to inadequate initial fixation, loss of mechanical fixation over time, or loss of biological fixation caused by particle-induced osteolysis around the implant.

Some authors have described many causes of mechanical loosening of cemented prostheses that are distributed over two loosening modes related to the femoral and cup components, as presented in Fig. 6.9:
- Loosening between the cement and the bone
- Loosening between the implant and the cement

This loosening is evidenced by a change in the fixation cement sheath, micromobility between the implant and the cement, or periprosthetic femur fractures (prosthetic stem fractures).

The primary causes of particulate accumulation vary: wear of the implant interface, micromovement in response to corrosion, oxidative reactions, and minor pathogenic contamination (Weerakkody, 2014). The initial response to particulate debris may include a subtle inflammatory response, which becomes more pronounced as osteolysis

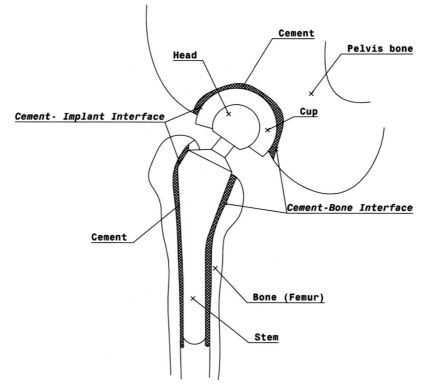

Fig. 6.9 (A) Mode 1 loosening. (B) Mode 2 loosening.

progresses. This inflammatory environment elicits a cellular response characterized by elevated levels of secreted factors such as TNF, RANKL, IL-6, IL-1, and IL-11.

Highly porous-coated titanium acetabulum components have a high coefficient of friction and ultraporous surfaces to enhance bone ingrowth and osseointegration in THA. There have been concerns with the development of early radiolucent lines and aseptic loosening of highly porous acetabular components. It is unclear whether these concerns relate to a specific implant or the entire class. The aim of this study is to compare the revision rates for aseptic loosening of highly porous acetabular combinations in primary THA using data from a large joint replacement registry (Hoskins et al., 2022).

Currently, we have various techniques by which we can reduce the rate of aseptic loosening. Nevertheless, further randomized clinical trials are needed to expand the recommendations for aseptic loosening prevention (Apostu et al., 2022).

Fig. 6.10 shows how prosthetic micromovements cause focal osteolysis. In effect, as depicted by Mjöberg (2018), these micromovements pump high-pressure joint fluid from the space between the femoral stem and the cement through a defect in the cement

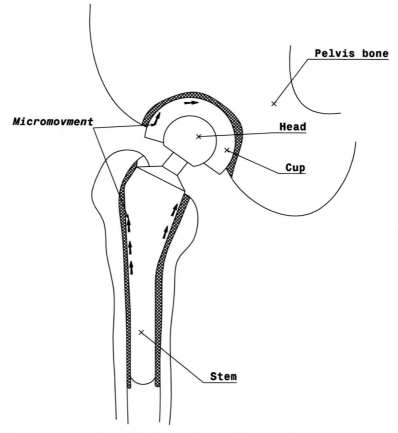

Fig. 6.10 Micromovements observed at the implant and bone interface.

mantle. The pressure can devitalize the adjacent bone tissue, which is resorbed, causing focal osteolysis.

6.2.3 Installation defects

The selection of a THP should take into account the patient's age, health status, body shape, and underlying bone disease. Using the patient's X-rays, it is then possible to determine the size of the prosthesis, landmarks, and placement tips that may be appropriate for the hip.

However, despite the careful choice of the prosthesis, the technique of fitting the prosthesis, and the prescribed postoperative rehabilitation steps, problems of failure of this artificial joint are encountered.

A poor fit of the prosthesis, both in the femoral and acetabular parts, leads to a variation at the center of rotation of the hip prosthesis. This can lead to a variation on the

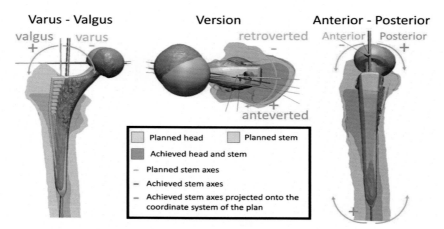

Fig. 6.11 Illustration of installation defects (Belzunce et al., 2020).

contact surfaces between the head and the acetabular cup, a poor distribution of stresses, and thus excessive wear of this artificial joint. In the short or long term, the THA may fail and a revision of the prosthesis is considered with all the problems that this entails.

As depicted by some authors (Du et al., 2020), defects in the cup prosthesis components may be:
- Anteversion/retroversion of the cup from the horizontal
- Abnormalities of the obliquity with respect to the frontal plane (abnormal standing up or horizontalization of the cup)
- Anomalies of the offset (lateralization and medialization of the cup)

Defects in the femoral part of the prosthesis may be:
- Anteversion/retroversion of the neck
- Valgisation/variation of the neck
- Anomalies of the femoral offset

Fig. 6.11 illustrates different defect situations in the placement of the femoral part of THRs (Belzunce et al., 2020).

6.2.4 Implant fracture

An implant fracture is one of the most common failures of THA. Indeed, under excessive physiological loads, the femoral or acetabular component may be subjected to fracture. Clinical cases of femoral stem fractures after primary and revision hip arthroplasty have been observed by radiology (Köksal, 2020).

However, ceramic joint surfaces are increasingly used for THRs, although there are still concerns about the brittleness of ceramic and its potential for mechanical failure (Howard et al., 2017). Fracture of prosthetic components can also cause prosthesis failure,

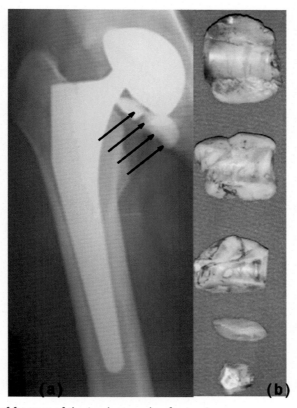

Fig. 6.12 Examples of fracture of the implant made of ceramic.

especially for alumina or zirconia ceramic heads or cups (Traina et al., 2012, 2013). Examples of fractures of ceramic prosthetic components are shown in Fig. 6.12.

6.2.5 Dislocation of THA

Dislocation is a common complication of total arthroplasty that occurs when the head moves relative to the cup, as shown in Fig. 6.13. The majority of dislocations are due to surgical and patient-related factors and occur in the early postoperative period (Lu et al., 2019).

Dislocation of the THA cup can also occur due to advanced wear, geometric shape defect, or malposition. These different factors can lead to a redistribution of residual stresses favoring a lever arm mechanism. Early dislocations are generally due to defects in the placement of the prosthesis and failure of the patient to follow instructions. They are considerably reduced by the use of large heads with a diameter greater than 32 mm, which increase stability but at the cost of an increase in the wear coefficient of the polyethylene

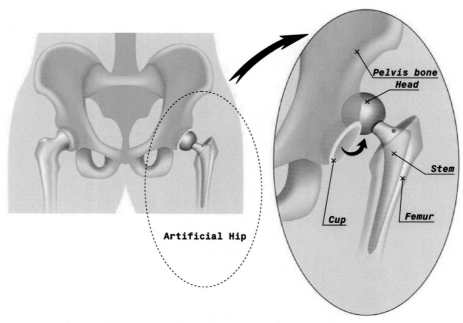

Fig. 6.13 The dislocation phenomenon.

(Boulila, 2010). At an advanced stage, these dislocations will be induced by the patient's muscular weakness and the intraprosthetic instability by a strong wear of the polyethylene.

Dislocation of the THA cup can also occur, usually due to advanced wear and muscle weakness of the patient (geometric shape defect or poor positioning). These different factors can lead to a redistribution of residual stresses favoring a lever arm mechanism.

Experimental and numerical studies have analyzed the effect of THR head size on dislocation and stress distribution in THA (Lu et al., 2019; Markel et al., 2022). These studies show that approximately one-third of patients suffer from dislocations requiring revision prosthesis implantation. The increase in the head of the THP induces severe tribological conditions and accelerated wear of the UHMWPE. For metal/metal or ceramic/ceramic pairs, dislocations are reduced by choosing a clearance between the head and the acetabulum of about 24 μm.

A dual mobility prosthesis was designed to increase the degrees of mobility and stability. The Bousquet system is a dual mobility head–polyethylene–metal cup prosthesis (Boyer et al., 2015). The polyethylene insert that holds the femoral head moves within the noncemented metal cup, thus increasing both mobility and stability.

The choice of this type of cup has greatly reduced the risk of dislocation. The term "intraprosthetic dislocation (IPD)" was used by Bousquet to define the specific complication of dissociation between the femoral head and the dual mobility insert (Boyer et al., 2015).

Intraprosthetic dislocation of dual mobility cups results from wear of the polyethylene, which leads to the inability of the insert to retain the prosthetic head. Wear is favored by direct phenomena (direct contact between the neck and the insert, which can be early if there is a small difference between the diameters of the head and the neck) or indirect phenomena (factors limiting the mobility of the polyethylene of the metal cup).

6.3 Future directions

Digitization has rapidly changed the biomedical field, which has always relied on technology. THR surgery is undergoing radical changes owing to robotics, virtual reality, artificial intelligence, and big data, which are transforming the way surgeons perform implants and engineers design and develop prostheses.

In the past, the design of new biomedical implants and devices often required multiple prototypes and trials. As data have become available, it can speed up the process of designing and manufacturing these even more sophisticated devices. There is no doubt that a major transformation of the biomedical industry is underway.

In this section, we present the major trends and innovations in this area in terms of surgical techniques, prosthetic materials, manufacturing and sterilization processes, and design of these devices:

- In terms of surgical techniques

Image-guided intervention (IGI), developed a few years ago, evolved from stereotactic neurosurgery techniques. Much of the visualization in image-guided procedures is achieved by creating a virtual image of the surgical or therapeutic environment, based on preoperative data, and displaying it on a workstation remote from the patient (Desai et al., 2018b).

The assistance of robotics has made it possible to accurately reconstruct the positioning of the prosthetic components (Zhang et al., 2022). Other authors such as Desai et al. (2018a) have opted for the development of minimally invasive surgical techniques. They have relied on robotic devices.

As described by Boubaker (2020), control theory is fundamental to understanding feedback pathways in physiological systems (neurological systems, etc.) and is a concept for the construction of artificial organs (prosthetics, etc.). The work highlights the importance of control and feedback theory in our lives and explains how this theory is central to future medical developments.

Integrating artificial intelligence into the development cycle of THRs and their implantation into the patient could reduce the causes of THA failure.

Walking aids have been developed and designed for the recovery or rehabilitation of human walking using new materials and intelligent materials such as shape memory alloys (Carbone et al., 2018).

• In terms of prosthetic materials

The current focus is on increasing the life span of implants through their resistance to wear and corrosion. Hard coatings may have the potential to reduce wear and corrosion. DLC coatings have properties that could make them viable for implants (Love et al., 2013; Nagashima et al., 2015; Makarov et al., 2019).

Other authors have focused on customized prostheses that optimally fit the morphology of the joint using magnetic resonance imaging (MRI) (Jacquet et al., 2020). Indeed, as mentioned by some authors such as Corona-Castuera et al. (2021), the use of additive manufacturing techniques and the realization of self-supporting cellular structures in the biomedical field have opened up significant opportunities to control the mechanical properties of THPs, resulting in specific benefits for patients.

The development of biocomposite materials has also improved resistance to wear, loosening, and other forms of structural failures. The use of hydroxyapatite, for example, has promoted biological anchoring of the prosthesis to the bone and preserved the patient's bone capital.

Optimization of manufacturing processes has allowed engineers to develop ceramic matrix composites (CMCs). One example is the alumina matrix composite, which combines good ceramic characteristics with improved mechanical strength to create a potentially more flexible alternative to traditional alumina for hip replacements (Bougherara et al., 2007; Solarino et al., 2021).

da Costa et al. (2019) developed prosthesis models with a biopolymeric matrix, namely, polyurethane (PU), reinforced with glass fibers and calcium carbonate. The reinforcement was constructed as a core inserted into the THP.

Some prosthesis manufacturers have developed highly cross-linked polyethylene (HXLPE) to increase the degree of cross-linking of the material by gamma ray or electron beam processes. This remelting process is designed to reduce the amount of oxidation associated with cross-linking. As a result, the wear resistance of the polyethylene has increased significantly compared to standard polyethylene (Mori and Tsukamoto, 2018; Roedel et al., 2021) Other authors, such as Li et al. (2021), have incorporated vitamin E into HXLPE, again with the aim of improving wear resistance. HXLPE incorporated with vitamin E (VEPE) has been used in THA with significant results.

• In terms of prosthetic manufacturing processes

Improvement of the manufacturing processes has been efficiently achieved with new techniques such as additive manufacturing and incremental forming (Boulila et al., 2018).

6.4 Conclusions

This chapter summarizes the state of the art on failures observed in THA. The idea is to help the orthopedic surgeon to better understand the mechanisms that can lead to THR failure during the post-COVID period, since that there was no significant hindsight at the Tunisian orthopedic centers.

These examples have deep social and economic connections to the healthcare sector and are representative of the efforts developed by the scientific community (Sequeira, 2019).

Wear of the polyethylene associated with the cup is the primary cause of failure in THA. Indeed, this wear accelerates the loosening phenomenon by the formation of micromovements at the implant-bone or implant-cement interface.

The use of new generations of polyethylene has considerably reduced the wear phenomenon. The use of hydroxyapatite-coated prostheses has considerably improved the tightness of the THP stem in the femoral canal.

The use of rotating prostheses has reduced the phenomenon of prosthesis dislocation. The use of other grades of ceramic composites has increased its toughness and reduced its fragility.

References

Apostu, D., et al., 2018. Current methods of preventing aseptic loosening and improving osseointegration of titanium implants in cementless total hip arthroplasty: a review. J. Int. Med. Res. 46 (6), 2104–2119. https://doi.org/10.1177/0300060517732697.

Apostu, D., et al., 2022. How to prevent aseptic loosening in cementless arthroplasty: a review. Appl. Sci. 12 (3), 1571. https://doi.org/10.3390/app12031571.

Arena, A., et al., 2019. Nanostructured zirconia-based ceramics and composites in dentistry: a state-of-the-art review. Nanomaterials 9 (10), 1393. https://doi.org/10.3390/nano9101393.

Bäcker, H.C., et al., 2022. Mechanical failure of total hip arthroplasties and associated risk factors. Arch. Orthop. Trauma Surg. https://doi.org/10.1007/s00402-022-04353-0.

Belzunce, M.A., et al., 2020. Uncemented femoral stem orientation and position in total hip arthroplasty: a CT study. J. Orthop. Res. 38 (7), 1486–1496. https://doi.org/10.1002/jor.24627.

Bergmann, G., et al., 2001. Hip contact forces and gait patterns from routine activities. J. Biomech. 34 (7), 859–871. https://doi.org/10.1016/S0021-9290(01)00040-9.

Bergmann, G., Graichen, F., Rohlmann, A., 2004. Hip joint contact forces during stumbling. Langenbeck's Arch. Surg. 389 (1), 53–59. https://doi.org/10.1007/s00423-003-0434-y.

Bergmann, G., et al., 2016. Standardized loads acting in hip implants. PLoS One 11 (5), e0155612. https://doi.org/10.1371/journal.pone.0155612.

Bhalekar, R.M., Smith, S.L., Joyce, T.J., 2020. Hip simulator testing of the taper-trunnion junction and bearing surfaces of contemporary metal-on-cross-linked-polyethylene hip prostheses. J. Biomed. Mater. Res. B Appl. Biomater. 108 (1), 156–166. https://doi.org/10.1002/jbm.b.34374.

Bistolfi, A., et al., 2021. Ultra-high molecular weight polyethylene (UHMWPE) for hip and knee arthroplasty: the present and the future. J. Orthop. 25, 98–106. https://doi.org/10.1016/j.jor.2021.04.004.

Boubaker, O., 2020. Control Theory in Biomedical Engineering: Applications in Physiology and Medical Robotics. Academic Press, London.

Bougherara, H., et al., 2007. Design of a biomimetic polymer-composite hip prosthesis. J. Biomed. Mater. Res. A 82A (1), 27–40. https://doi.org/10.1002/jbm.a.31146.

Boulila, A., 2010. Optimisation et aide à la conception d'implants orthopédiques—Etude de l'usure des Prothèses Totales de Hanche. de la Méditerranée (Aix-Marseille II) & Tunis El Manar.

Boulila, A., et al., 2008. Contribution à l'étude de l'usure des prothèses totales de hanche. Approche expérimentale sur simulateur. Mécan. Ind. 9 (4), 323–333. https://doi.org/10.1051/meca:2008037.

Boulila, A., et al., 2010. Comportement mécanique des prothèses totales de hanche au pic de chargement. Mécan. Ind. 11 (1), 25–36. https://doi.org/10.1051/meca/2010013.

Boulila, A., et al., 2018. Contribution to a biomedical component production using incremental sheet forming. Int. J. Adv. Manuf. Technol. 95 (5–8), 2821–2833. https://doi.org/10.1007/s00170-017-1397-4.

Bouville, F., 2020. Strong and tough nacre-like aluminas: process-structure-performance relationships and position within the nacre-inspired composite landscape. J. Mater. Res. 35 (8), 1076–1094. https://doi.org/10.1557/jmr.2019.418.

Bouzidi, S., Ayadi, M., Boulila, A., 2022. Feasibility study of the SPIF process applied to perforated sheet metals. Arab. J. Sci. Eng. 47 (7), 9225–9252. https://doi.org/10.1007/s13369-022-06570-6.

Boyer, B., et al., 2015. La luxation intra-prothétique—fossile de l'historique de la double mobilité ? Réponse des explants. Rev. Chirurgie Orthop. Traumatol. 101 (7), S186–S187. https://doi.org/10.1016/j.rcot.2015.09.118.

Bracco, P., et al., 2017. Ultra-high molecular weight polyethylene: influence of the chemical, physical and mechanical properties on the wear behavior. A review. Materials 10 (7), 791. https://doi.org/10.3390/ma10070791.

Burke, N.G., et al., 2019. Total hip replacement—the cause of failure in patients under 50 years old? Irish J. Med. Sci. (1971-) 188 (3), 879–883. https://doi.org/10.1007/s11845-018-01956-8.

Carbone, G., Ceccarelli, M., Pisla, D., 2018. New Trends in Medical and Service Robotics. Springer Berlin Heidelberg, New York, NY.

Castagnini, F., et al., 2019. Highly porous titanium cups versus hydroxyapatite-coated sockets: midterm results in metachronous bilateral total hip arthroplasty. Med. Princ. Pract. 28 (6), 559–565. https://doi.org/10.1159/000500876.

Corona-Castuera, J., et al., 2021. Design and fabrication of a customized partial hip prosthesis employing CT-scan data and lattice porous structures. ACS Omega 6 (10), 6902–6913. https://doi.org/10.1021/acsomega.0c06144.

Crawford, D.A., et al., 2021. Does activity level after primary total hip arthroplasty affect aseptic survival? Arthropl. Today 11, 68–72. https://doi.org/10.1016/j.artd.2021.07.005.

da Costa, R.R.C., et al., 2019. Design of a polymeric composite material femoral stem for hip joint implant. Polímeros 29 (4), e2019057. https://doi.org/10.1590/0104-1428.02119.

Das, S.S., Chakraborti, P., 2018. development of biomaterial for total hip joint replacement. IOP Conf. Ser.: Mater. Sci. Eng. 377, 012177. https://doi.org/10.1088/1757-899X/377/1/012177.

Desai, J.P., Patel, R.V., et al., 2018a. The Encyclopedia of Medical Robotics: Volume 1: Minimally Invasive Surgical Robotics. World Scientific, https://doi.org/10.1142/10770-vol1.

Desai, J.P., Patel, R., et al., 2018b. The Encyclopedia of Medical Robotics: Volume 3: Image-guided Surgical Procedures and Interventions. World Scientific, https://doi.org/10.1142/10770-vol3.

Di Gregorio, R., Vocenas, L., 2021. Identification of gait-cycle phases for prosthesis control. Biomimetics 6 (2), 22. https://doi.org/10.3390/biomimetics6020022.

Du, Y., et al., 2020. Acetabular bone defect in total hip arthroplasty for Crowe II or III developmental dysplasia of the hip: a finite element study. Biomed. Res. Int. 2020, 1–12. https://doi.org/10.1155/2020/4809013.

Dumbleton, J.H., et al., 2006. The basis for a second-generation highly cross-linked UHMWPE. Clin. Orthop. Relat. Res. 453, 265–271. https://doi.org/10.1097/01.blo.0000238856.61862.7d.

Elke, R., Rieker, C.B., 2018. Estimating the osteolysis-free life of a total hip prosthesis depending on the linear wear rate and head size. Proc. Inst. Mech. Eng. H J. Eng. Med. 232 (8), 753–758. https://doi.org/10.1177/0954411918784982.

Giuseppe, M., et al., 2022. Ceramic-on-ceramic versus ceramic-on-polyethylene in total hip arthroplasty: a comparative study at a minimum of 13 years follow-up. BMC Musculoskelet. Disord. 22 (S2), 1062. https://doi.org/10.1186/s12891-021-04950-x.

Gold, M., Munjal, A., Varacallo, M., 2022. Anatomy, bony pelvis and lower limb, hip joint. In: StatPearls. StatPearls Publishing, Treasure Island (FL). http://www.ncbi.nlm.nih.gov/books/NBK470555/. (Accessed 27 October 2022).

Hannouche, D., et al., 2005. Ceramics in total hip replacement. Clin. Orthop. Relat. Res. 430, 62–71. https://doi.org/10.1097/01.blo.0000149996.91974.83.

Hoskins, W., et al., 2022. Revision for aseptic loosening of highly porous acetabular components in primary total hip arthroplasty: an analysis of 20,993 total hip replacements. J. Arthroplast. 37 (2), 312–315. https://doi.org/10.1016/j.arth.2021.10.011.

Howard, D.P., et al., 2017. Ceramic-on-ceramic bearing fractures in total hip arthroplasty: an analysis of data from the National Joint Registry. Bone Joint J. 99-B (8), 1012–1019. https://doi.org/10.1302/0301-620X.99B8.BJJ-2017-0019.R1.

Hu, C.Y., Yoon, T.-R., 2018. Recent updates for biomaterials used in total hip arthroplasty. Biomater. Res. 22 (1), 33. https://doi.org/10.1186/s40824-018-0144-8.

Hua, Z., et al., 2019. Wear test apparatus for friction and wear evaluation hip prostheses. Front. Mech. Eng. 5, 12. https://doi.org/10.3389/fmech.2019.00012.

Jacquemaire, B., 1991. Rappel des principales notions de biomécanique. Orthop. Traumatol. 1 (3), 145–150. https://doi.org/10.1007/BF01791537.

Jacquet, C., et al., 2020. Long-term results of custom-made femoral stems. Orthopade 49 (5), 408–416. https://doi.org/10.1007/s00132-020-03901-z.

Jamari, J., et al., 2017. Range of motion simulation of hip joint movement during salat activity. J. Arthroplast. 32 (9), 2898–2904. https://doi.org/10.1016/j.arth.2017.03.056.

Jamari, J., et al., 2022. Computational contact pressure prediction of CoCrMo, SS 316L and Ti6Al4V femoral head against UHMWPE acetabular cup under gait cycle. J. Funct. Biomater. 13 (2), 64. https://doi.org/10.3390/jfb13020064.

Jeffers, J.R.T., Walter, W.L., 2012. Ceramic-on-ceramic bearings in hip arthroplasty: state of the art and the future. J. Bone Joint Surg. Br. Vol. 94-B (6), 735–745. https://doi.org/10.1302/0301-620X.94B6.28801.

Karachalios, T., Komnos, G., Koutalos, A., 2018. Total hip arthroplasty: survival and modes of failure. EFORT Open Rev. 3 (5), 232–239. https://doi.org/10.1302/2058-5241.3.170068.

Khanna, R., et al., 2017. Progress in wear resistant materials for total hip arthroplasty. Coatings 7 (7), 99. https://doi.org/10.3390/coatings7070099.

Kheir, M.M., Bauer, T.W., Westrich, G.H., 2022. Diagnosis of prostate adenocarcinoma on routine pathology after a primary total hip arthroplasty. Arthropl. Today 15, 19–23. https://doi.org/10.1016/j.artd.2022.02.025.

Köksal, A., 2020. Femoral stem fractures after primary and revision hip replacements: a single-center experience. Joint Dis. Rel. Surg. 31 (3), 557–563. https://doi.org/10.5606/ehc.2020.76162.

Kumar, N., Arora, N.C., Datta, B., 2014. Bearing surfaces in hip replacement—evolution and likely future. Med. J. Armed Forces India 70 (4), 371–376. https://doi.org/10.1016/j.mjafi.2014.04.015.

Kummer, B., 1993. Is the Pauwels' theory of hip biomechanics still valid? A critical analysis, based on modern methods. Ann. Anat. Anatomisch. Anzeiger 175 (3), 203–210. https://doi.org/10.1016/S0940-9602(11)80002-6.

Kyomoto, M., et al., 2011. Cartilage-mimicking, high-density brush structure improves wear resistance of crosslinked polyethylene: a pilot study. Clin. Orthop. Relat. Res. 469 (8), 2327–2336. https://doi.org/10.1007/s11999-010-1718-5.

Li, Z., et al., 2021. Vitamin E highly cross-linked polyethylene reduces mid-term wear in primary total hip replacement: a meta-analysis and systematic review of randomized clinical trials using radiostereometric analysis. EFORT Open Rev. 6 (9), 759–770. https://doi.org/10.1302/2058-5241.6.200072.

Liang, W., 2018. Mechanical design and control strategy for hip joint power assisting. J. Healthcare Eng. 2018, 1–7. https://doi.org/10.1155/2018/9712926.

Love, C.A., et al., 2013. Diamond like carbon coatings for potential application in biological implants—a review. Tribol. Int. 63, 141–150. https://doi.org/10.1016/j.triboint.2012.09.006.

Lu, Y., Xiao, H., Xue, F., 2019. Causes of and treatment options for dislocation following total hip arthroplasty (Review). Exp. Therap. Med. https://doi.org/10.3892/etm.2019.7733.

Makarov, V., et al., 2019. Diamond-like carbon coatings in endoprosthetics (literature review). Orthop. Traumatol. Prosthetics 0 (2), 102–111. https://doi.org/10.15674/0030-598720192102-111.

Markel, J.F., et al., 2022. Causes of early hip revision vary by age and sex: analysis of data from a statewide quality registry. J. Arthroplast. https://doi.org/10.1016/j.arth.2022.03.014. S0883540322002844.

Martelli, S., Costi, J.J., 2021. Real-time replication of three-dimensional and time-varying physiological loading cycles for bone and implant testing: a novel protocol demonstrated for the proximal human femur while walking. J. Mech. Behav. Biomed. Mater. 124, 104817. https://doi.org/10.1016/j.jmbbm.2021.104817.

McGee, M.A., et al., 2000. Implant retrieval studies of the wear and loosening of prosthetic joints: a review. Wear 241 (2), 158–165. https://doi.org/10.1016/S0043-1648(00)00370-7.

Milovanović, A., et al., 2020. Design aspects of hip implant made of Ti-6Al-4V extra low interstitials alloy. Procedia Struct. Integr. 26, 299–305. https://doi.org/10.1016/j.prostr.2020.06.038.

Mjöberg, B., 2018. Does particle disease really exist? Acta Orthop. 89 (1), 130–132. https://doi.org/10.1080/17453674.2017.1373491.

Moghadasi, K., et al., 2022. A review on biomedical implant materials and the effect of friction stir based techniques on their mechanical and tribological properties. J. Mater. Res. Technol. 17, 1054–1121. https://doi.org/10.1016/j.jmrt.2022.01.050.

Mori, T., Tsukamoto, M., 2018. Highly cross-linked polyethylene in total hip arthroplasty, present and future. Ann. Joint 3, 67. https://doi.org/10.21037/aoj.2018.07.05.

Nagashima, S., Moon, M.-W., Lee, K.-R., 2015. Diamond-like carbon coatings for joint arthroplasty. In: Sonntag, R., Kretzer, J.P. (Eds.), Materials for Total Joint Arthroplasty. Imperial College Press, pp. 395–412, https://doi.org/10.1142/9781783267170_0013.

Ng, K.C.G., Jeffers, J.R.T., Beaulé, P.E., 2019. Hip joint capsular anatomy, mechanics, and surgical management. J. Bone Joint Surg. 101 (23), 2141–2151. https://doi.org/10.2106/JBJS.19.00346.

Ozcadirci, A., Caglar, O., Coskun, G., 2021. Range of motion and muscle strength deficits of patients with total hip arthroplasty after surgery. Baltic J. Health Phys. Activ. 13 (2), 67–77. https://doi.org/10.29359/BJHPA.13.2.07.

Pace, T.B., et al., 2013. Comparison of conventional polyethylene wear and signs of cup failure in two similar total hip designs. Adv. Orthop. 2013, 1–7. https://doi.org/10.1155/2013/710621.

Pakvis, D., et al., 2011. Is there evidence for a superior method of socket fixation in hip arthroplasty? A systematic review. Int. Orthop. 35 (8), 1109–1118. https://doi.org/10.1007/s00264-011-1234-6.

Pietrzak, W.S., 2021. Ultra-high molecular weight polyethylene for total hip acetabular liners: a brief review of current status. J. Investig. Surg. 34 (3), 321–323. https://doi.org/10.1080/08941939.2019.1624898.

Priyana Soemardi, T., et al., 2019. Development of total hip joint replacement prostheses made by local material: an introduction. In: Setyobudi, R.H., et al. (Eds.), E3S Web of Conferences. vol. 130, p. 01032, https://doi.org/10.1051/e3sconf/201913001032.

Rankin, K.A., et al., 2022. Computer-assisted navigation for complex revision of unstable total hip replacement in a patient with post-traumatic arthritis. Arthropl. Today 15, 153–158. https://doi.org/10.1016/j.artd.2022.03.015.

Roedel, G.G., et al., 2021. Total hip arthroplasty using highly cross-linked polyethylene in patients aged 50 years and younger: minimum 15-year follow-up. Bone Joint J. 103-B (7 Supple B), 78–83. https://doi.org/10.1302/0301-620X.103B7.BJJ-2020-2443.R1.

Sebti, R., Boulila, A., Hamza, S., 2020. Ergonomics risk assessment among maintenance operators in a Tunisian railway company: a case study. Hum. Factors Ergon. Manuf. Serv. Ind. 30 (2), 124–139. https://doi.org/10.1002/hfm.20828.

Sentuerk, U., von Roth, P., Perka, C., 2016. Ceramic on ceramic arthroplasty of the hip: new materials confirm appropriate use in young patients. Bone Joint J. 98-B (1_Supple_A), 14–17. https://doi.org/10.1302/0301-620X.98B1.36347.

Sequeira, J.S. (Ed.), 2019. Robotics in Healthcare: Field Examples and Challenges. Advances in Experimental Medicine and Biology, vol. 1170. Springer, Cham, Switzerland.

Shim, B.-J., Park, S.-J., Park, C.H., 2022. The wear rate and survivorship in total hip arthroplasty using a third-generation ceramic head on a conventional polyethylene liner: a minimum of 15-year follow-up. Hip Pelvis 34 (2), 115. https://doi.org/10.5371/hp.2022.34.2.115.

Solarino, G., et al., 2021. Outcomes of ceramic composite in total hip replacement bearings: a single-center series. J. Compos. Sci. 5 (12), 320. https://doi.org/10.3390/jcs5120320.

Tateiwa, T., et al., 2020. Burst strength of BIOLOX®delta femoral heads and its dependence on low-temperature environmental degradation. Materials 13 (2), 350. https://doi.org/10.3390/ma13020350.

Tokgoz, E., 2022. TOTAL HIP ARTHROPLASTY Medical and Biomedical Engineering and Science Concepts. Springer, S.l.

Torabnia, S., Mihcin, S., Lazoglu, I., 2021. Parametric analysis for the design of hip joint replacement simulators. In: 2021 IEEE International Symposium on Medical Measurements and Applications (MeMeA). IEEE, Lausanne, Switzerland, pp. 1–6, https://doi.org/10.1109/MeMeA52024.2021.9478689.

Traina, F., et al., 2012. Risk factors for ceramic liner fracture after total hip arthroplasty. HIP Int. 22 (6), 607–614. https://doi.org/10.5301/HIP.2012.10339.

Traina, F., et al., 2013. Fracture of ceramic bearing surfaces following total hip replacement: a systematic review. Biomed. Res. Int. 2013, 1–8. https://doi.org/10.1155/2013/157247.

Weerakkody, Y., 2014. Aseptic Loosening of Hip Joint Replacements. Radiopaedia.org, https://doi.org/10.53347/rID-30533.

Zaghloul, A., 2018. Hip joint: embryology, anatomy and biomechanics. Biomed. J. Scientific Tech. Res. 12 (3). https://doi.org/10.26717/BJSTR.2018.12.002267.

Zhang, S., et al., 2022. Revision total hip arthroplasty with severe acetabular defect: a preliminary exploration and attempt of robotic-assisted technology. Orthop. Surg. 14 (8), 1912–1917. https://doi.org/10.1111/os.13368.

CHAPTER 7

Artificial intelligence in robot-assisted surgery: Applications to surgical skills assessment and transfer

Abed Soleymani, Xingyu Li, and Mahdi Tavakoli
Electrical and Computer Engineering Department, University of Alberta, Edmonton, AB, Canada

7.1 Introduction

Robot-assisted surgery (RAS) is becoming more popular in modern clinical practice. A surgeon must acquire a variety of skills to conduct RAS safely and effectively since inadequate preparation may negatively affect clinical outcomes (Birkmeyer et al., 2013). To help surgical trainees, accurate and reliable methods of assessment and transfer of surgical skills should be available with informative and instructive feedback.

As a convention, RAS skills assessments are conducted through outcome-based analyses, specially designed checklists, and specific scores (Ahmidi et al., 2017). For instance, Martin et al. (1997) created objective structured assessment of technical skill (OSATS), which incorporated operation-specific checklists for pass/fail judgments of the trainees. Another conventional method for identifying levels of robotic surgery expertise is the global evaluative assessment of robotic skills (GEARS) proposed by Goh et al. (2012). Variability in the human's interpretation of similar events makes such evaluation methods expensive, time demanding, less efficient, and less reliable. In addition, such observational methods neglect small but potentially important changes in the trainee's skills, preventing them from providing insights and targeted feedback into the surgical outcomes.

Autonomous skills evaluation approaches, however, have the potential to resolve all of the above-mentioned limitations (Funke et al., 2019). Surgical robotics technologies are making surgical procedure data more accessible, allowing artificial intelligence (AI) (e.g., machine learning [ML] and deep learning [DL] models) to be incorporated in a variety of RAS skills evaluation and transfer tasks. Recent advances in AI have opened the way for using highly complex surgical recordings to extract meaningful features and build a computerized model of users based on their performance during operations and use the model to classify users' level of expertise. The AI model can identify skills-associated features and then transfer them to the trainee's trajectory to better reflect skillful behavior. A robotic surgery platform may utilize the enhanced trajectory as a reference to generate a virtual fixture that serves as a skillful guide for the user's hand toward a better executive trajectory.

Medical and Healthcare Robotics
https://doi.org/10.1016/B978-0-443-18460-4.00014-7

This chapter, in the beginning, will discuss categories of automated methods that extract salient skills-related features from surgical recordings to use them to rate the user performance. In addition, the benefits and drawbacks of the proposed methods, as well as device regularity and patient safety issues will be discussed. We will also introduce a variety of haptic cue-based skills transfer methods to enhance the skillful behavior of less-experienced users using surgical robot platforms. Finally, in Section 7.5, we provide several insights and promising research areas related to surgical skills assessment and transfer.

7.2 Surgical skills assessment

Inductive learning-based models and domain knowledge-based models are main AI categories used in autonomous RAS skills assessment (Muralidhar et al., 2018). Inductive learning-based models use data-driven approaches with minimal field knowledge to avoid user bias in the learning process. Because the structure of such methods and even their hyperparameters are mostly determined by the input data (which is usually big data), they have the advantage of reduced training effort (see Fig. 7.1).

Conversely, domain knowledge-based models do not rely on statistical models (e.g., DL or ML models) to discover known features from the system dynamics or human experiences. Surgical robot platforms are very complex physical systems, which cannot be accurately modeled using limited training data. Such limitations lead us to model uncertainties and unmodeled dynamics. Using field knowledge as a prior decreases uncertainty and makes it easier to solve modeling problems with fewer training data points (von Rueden et al., 2019). As a result of incorporating field knowledge in the training stage, the model is often more transparent (i.e., can be clearly understood and explained in human terms), which in turn increases the reliability of the final solution in safety-critical applications such as robotic surgery (von Rueden et al., 2019).

7.2.1 Inductive learning-based models

Traditional machine learning methods were used to build the first autonomous surgical skills assessment systems. As surgical trials are composed of a sequence of several

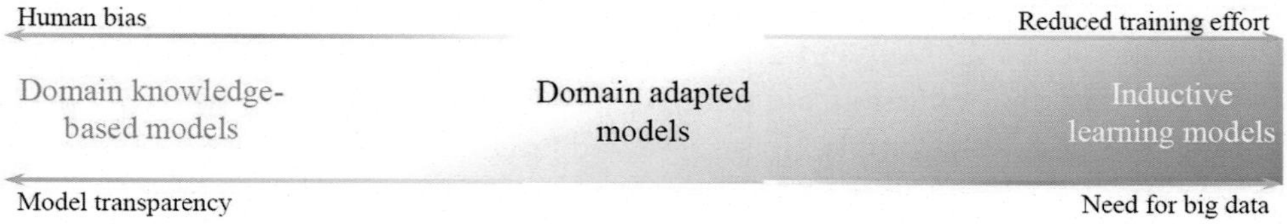

Fig. 7.1 Different AI model training paradigms according to the strength of prior assumptions about the data, model structure, and task dynamics. As we incorporate more domain knowledge and human bias into the training procedure, we achieve better model transparency over fewer training data samples with the cost of having extra feature engineering effort.

predefined subtasks, Rosen et al. proposed the Markov structure of a given surgical task to reveal the user's skills level (Rosen et al., 2001, 2002; MacKenzie et al., 2001). Various methods were later used to extend basic hidden Markov models (HMMs) by training a unique HMM for each skill level (Reiley and Hager, 2009; Tao et al., 2012). Specifically, these studies train separate HMMs for each user and assess their distance from an ideal HMM trained over the data of an expert user. A user's performance is measured by the distance between his or her model and the expert's model. Besides HMMs limited recognition rate and challenge for determining the true number of hidden states, they need manual annotations over trajectories, which are time consuming. Furthermore, HMMs map a given trajectory to static descriptor space which makes it possible to lose important time-related information within the trajectory. Skills-related temporal features will be discussed in more depth in the following sections.

In recent years, deep learning models have become increasingly popular for RAS surgical skill evaluation applications. In some approaches, kinematic data (the translational and/or rotational trajectories of the robot end effector) are fed into the convolutional neural networks (CNNs) and used to learn desired patterns for skill assessment on surgical training platforms (Jian et al., 2020). Wang and Fey (2018) utilized a deep CNN to highlight the skills levels of individual users using the motion kinematics data of a given surgical operation. Moreover, Fawaz et al. (2019) developed a CNN architecture for identifying the surgical skills level of the user via latent pattern extraction of kinematics data of surgical trainees performing basic robotic surgery tasks. Nguyen et al. (2019) developed a classifier network via CNN and long short-term memory (LSTM) models with inertial measurement units (IMU) sensors to highlight user's skills level in a given surgical training data.

There is ongoing extensive research on using easy-to-capture video data as the input for AI models, which provide rich contextual details compared to previously mentioned kinematic data. For instance, Kim et al. (2019) proposed a temporal CNN to evaluate the intraoperative skills level of capsulorhexis video trials. Funke et al. (2019) proposed a DL model using a pretrained three-dimensional (3D) CNN as a temporal segmentation network on the sequence of video frames and optical flow fields for technical surgical skills evaluation tasks. Liu et al. (2019) incorporated a supervised regression loss for video input as well as an unsupervised rank loss to train a DL model for RAS skills assessment.

Recent techniques identify the relative variation in skills between pairs of surgeries by creating a pairwise ranking problem (Jian et al., 2020). For instance, Doughty et al. (2018) incorporated an approach for predicting skills ranking based on video data sets. Using a novel loss function, they utilized both spatial and temporal features (i.e., visual features within each video frame and time-related features along the sequence of consecutive frames) to assess and rate skills. Doughty et al. (2019) introduced a novel model for long videos that determine relative skills level by learnable time-related attention modules. Li et al. (2019) introduced a spatial attention-based approach for skills assessment of video data. Authors introduced a new recurrent neural network (RNN) that incorporates high-level progress information of an ongoing task in addition to the stacked attention states from past frames.

Inductive methods discussed so far yield a global performance measure of the user for the entire task (i.e., expert, intermediate, or novice labels as illustrated in Fig. 7.2). To provide a more tailored and informative feedback to the users about their surgical performance and skills level (e.g., their mistakes and parts of the task they need to improve their skills), one effective and common approach is to decompose user's movements into blocks called surgemes (van Amsterdam et al., 2021) and apply state-of-the-art RAS skills evaluation approaches at the subtask level of the operation. By using this approach, instead of having a global performance metric, a high-resolution surgical workflow will be analyzed, which returns more elaborate feedback about the performance of the different parts of the entire surgical task. Many publications have attempted to perform autonomous analysis of surgical activities in a fine-grained manner (Lea et al., 2016; Menegozzo et al., 2019; DiPietro et al., 2016, 2019; Itzkovich et al., 2019; van Amsterdam et al., 2020) and reinforcement learning (RL) (Liu and Jiang, 2018). These approaches in addition to having the same limitations caused by their black-box nature also suffer from the over-segmentation problem (i.e., producing a large number of false action boundaries) and a poor prediction accuracy that prevents them from making accurate predictions, particularly for unpredicted events (e.g., sudden failures and restarts) (van Amsterdam et al., 2021). The over-segmentation problem may arise since high-capacity inductive models (capacity refers to the model's ability to accommodate variations in input data, which are largely dependent on how many parameters it can learn)

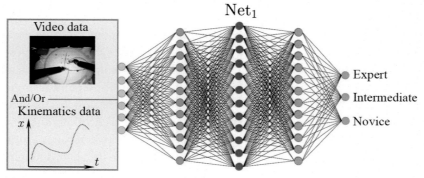

Fig. 7.2 Inductive learning-based AI models for surgical skills assessment applications. In this paradigm, raw surgical data which can be in the form of video and/or kinematics data (i.e., translational and rotational trajectories of the end effector of the robotic platform) will be fed into the input of the model (mostly deep model). The model's depth, hyperparameters, and the structure of the network will be adjusted according to the complexity and expressiveness of the input data.

mainly focus on the local variations of the data (i.e., small and unimportant details), instead of the global structure of the input trajectory. Furthermore, these methods heavily rely on hand-crafted gesture annotations to evaluate segmentation accuracy, which can take a long time and be subject to human bias. In addition, these approaches segment trajectories without providing any meaningful interpretation about the user's dexterity level and behavior at the subtask level.

Inspired by the above-mentioned limitations, Soleymani et al. (2022b) proposed an intuitive, explainable, and unsupervised ML-based approach for approximate decomposition of structured surgical trajectories such in retrospective studies. The introduced dual-sparse dictionary learning algorithm decomposes each trajectory into dictionary atoms that capture the main variations of the data, which are one general trend and several seasonal patterns. The proposed floating atoms concept is further utilized to accommodate temporal structures within trajectories and preserves information within the trajectory while mapping the data to embedding space. By reconstructing each trajectory according to the generated atoms of the training set (mainly expert trajectories as a benchmark), a vector of codes will be generated that is representative of the data in the low-dimensional embedding space. The code vector conveys important information about the skills level of the user and his/her abnormal behaviors within the task. The proposed approach does not need manual annotation and medicates oversegmentation problem since it captures main variations within the input trajectory and neglects local contents. On the other hand, segmentation borders in the proposed method are not as accurate as other related work with delicate annotated training data sets.

7.2.2 Domain knowledge-based models

While the end-to-end learning approaches presented in Section 7.2.1 have shown acceptable classification accuracies, they are black–box models with opaque decision-making procedures that are incomprehensible even for human experts. As a consequence, it is hard to provide meaningful feedback to the user's surgical performance or intuition about the contributing factors to the surgical outcome (here, intuitiveness and explainability mean how much the function or decision of a model are intuitive and explainable from the perspective of human logic, respectively). Moreover, DL models with large capacity require big data to prevent the final model to become overfitted. Since in the field of robotic surgery reliable, clean, and large data sets are very scarce, DL models have a tendency to overfit. This damages the model's generalization and results in models with poor performance in unpredictable situations (e.g., aborting and restarting a task).

In addition, for safety-critical applications such as robotic surgery, the human user must understand whether the model is developed based on meaningful features or irrelevant clues and biases in the training set. Therefore, it is crucial to enhance the explainability and interpretability of the model to meet the ethical requirements of skills assessment methods for robotic surgery (Molnar, 2020). AI models that incorporate domain knowledge not only enhance interpretability and explainability, but they also improve learning performance especially when training data are not large (Islam et al., 2021). Utilizing domain knowledge as a prior in data-scarce surgical tasks not only reduces uncertainty about the success of the operation but also makes the model easy-to-learn and more generalizable with smaller training data sets (von Rueden et al., 2019). The final skills assessment model that integrates extracted manually engineered features, as shown in Fig. 7.3, provides further clarity to the explanation and interpretation of the calculations since the effect of each extracted feature to the final generated outcome is more transparent.

The domain knowledge-based approach to skills classification incorporates meaningful features as evaluation metrics including execution time (Judkins et al., 2009; Liang et al., 2018), motion jerk (Liang et al., 2018), total path length (Judkins et al., 2009), etc., and run comparative statistical analysis on a single metric between different participants. Due to the statistical variations between and within participating users, there are significant overlaps between the extracted metrics. As a result, there is no reliable statistical difference among participants in terms of surgical skills level. It is primarily because some domain knowledge-based features including but not limited to motion jerk are very noisy for surgical tasks. Other features such as the total path length or task execution time are not informative enough to indicate the true level of skill of the user as a single factor.

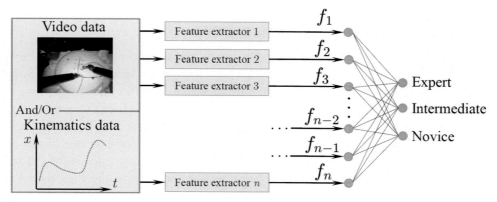

Fig. 7.3 Domain knowledge-based AI models for surgical skills assessment applications. In this paradigm, the surgical raw data which can be in the format of video and/or kinematics data will be fed into the several manually engineered feature extraction blocks to extract clinically meaningful features f_i that reflect the skills level of a surgeon (e.g., total path length, smoothness, fluidity of the motion, etc.). The number of features (i.e., the value of n) and the structure of feature extraction functions totally depends on the task and the amount of available knowledge in that particular field.

As a result, the methods presented in the aforementioned papers remain case-specific and ungeneralizable for new tasks as they ignore time-related patterns and do not create a universal model for a variety of RAS skills assessment.

Furthermore, domain knowledge-based approaches do not consider detailed events within a given task of an operation; they mostly take into account general metrics over the whole process. Such domain knowledge-based studies, for example, ignore critical time-related features such as trajectory nonsmoothness (i.e., the presence of random movements including hand tremors or uncontrolled rapid motions) as an important factor to the skills assessment of a given trajectory. The smoothness of a trajectory appears to be one of the key features that can be extracted and utilized for skills assessment purposes. Smoothness evaluation is challenging because nonsmoothness is a temporal characteristic that occurs frequently within all trajectories. The result is that the nonstructured pattern of smoothness often gets indetectable by other dominant time-domain characteristics including general trend and seasonal patterns. In addition, there is no general and accurate domain knowledge-based approach for searching, detecting, and quantifying smoothness across the entire time series.

Various domain knowledge-based features can be concatenated to create a rich high-dimensional feature space to discover a more expressive and performant representation of RAS trajectories for the sake of skills assessment. Ensembling all of the above-mentioned clinically meaningful features returns an informative long feature vector that meaningfully highlights the skills level of the participant and shows subtle but important information within the surgical trajectory. As depicted in Fig. 7.3, all of these concatenated metrics can be fed into a downstream ML classifier (e.g., support vector machine [SVM] model; Cortes and Vapnik, 1995, which is much simpler than the sophisticated classifier model introduced in Fig. 7.2. These features can also be fed into dimensionality reduction models (e.g., t-distributed stochastic neighbor embedding [tSNE]; Van der Maaten and Hinton, 2008) to visualize the high-dimensional features in a two-dimensional (2D) or 3D map to let the user investigate the internal mechanism of the proposed model.

7.2.3 Domain-adapted models

In the aforementioned papers, authors trained a network using an end-to-end learning paradigm or pure field knowledge-based approaches based on a popular dataset (e.g., JIGSAWS data set; Gao et al., 2014). A good approach is to combine inductive and knowledge-based models to develop a domain-adapted model that jointly incorporates both manually engineered metrics and data-driven end-to-end models for more efficient skills assessment purposes (see Fig. 7.4). In this way, thanks to informative features generated from domain knowledge, the end-to-end model does not need to be very complicated compared to inductive learning approaches (i.e., Net_2 in Fig. 7.4 is lighter than

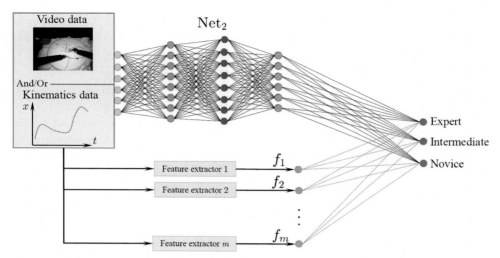

Fig. 7.4 Domain-adapted AI models for surgical skills assessment applications. In this paradigm, the surgical raw data which can be in the format of video and/or kinematics data will be simultaneously fed into a stack of manually engineered feature extraction blocks and inductive learning-based model Net_2 to extract clinically meaningful and data-driven features that reflect the skills level of a surgeon. The advantages of such modeling approach are the reduced number of manually made features (i.e., $m < n$) compared to domain knowledge-based model presented in Section 7.2.2 which means less feature engineering effort and light-weighted structure of Net_2 relative to presented Net_1 in Section 7.2.1 which yields less training and implementation complications and low chance of overfitting.

Net_1 in Fig. 7.3). Moreover, since a substantial number of informative features are extracted by end-to-end model, there is no need to expend too much time and effort for engineering informative domain knowledge-based features (i.e., the number of f_i in Fig. 7.4 in less than that of Fig. 7.3, or in other words $m < n$).

For example, Soleymani et al. extracted spatiotemporal features in the sequence of RAS video data by incorporating a pretrained ResNet50 model (He et al., 2016). Since the feature extraction network is not trained on the specific data set of surgical trials, it can extract the key features related to the skills level and surgical behaviors of the users. This approach at the crucial and prone-to-bias stage of feature extraction makes the model robust as well as generalizable to unseen test data. Moreover, fast Fourier transforms (FFTs) were used in the proposed method as an extra feature extraction layer to decompose the entire representation learning procedure into two phases: spatial-feature learning and temporal-feature extraction. This model uses FFT to represent a commonly accepted piece of domain knowledge that experts have dominant low–frequency components and negligible high–frequency activities (i.e., they have smooth movements). Compared with experts and intermediate users, novice users have smaller low–frequency coefficients (i.e., they show more hand tremors and unwanted random actions). The intermediate group falls somewhere in between expert and novice behavior. Extracted and manually manipulated features were fed into a downstream CNN to extract and learn features in an inductive learning paradigm to classify the skills level of each input trajectory. Due to their method's two-stage learning, they achieved a less complex model than other arts with sophisticated 3D CNNs or complicated CNN+RNN models.

The core limitation of the mentioned approaches in this chapter is that the trajectory of each hand is treated as an independent set of data with no consideration given to possible collaboration between two hands (i.e., interaction between different data channels such as roll rotation of the right hand and yaw rotation of the left hand in knot tying task). However, when a surgeon performs sophisticated bimanual tasks (i.e., collaborative tasks requiring delicate coordination of both hands) what defines him/her as an expert is not just the independent performance of each hand but how he/she manages to synchronize the motions and rotations of both hands (Vedula et al., 2016). The two hands pulling apart the two ends of the suture to tie a knot is a good example of hands collaboration in surgical operations. It seems considering and measuring the collaboration quality between two hands is a promising area in the field of surgical skills assessment for future work.

Another work which utilizes both data-driven and field knowledge-based models is Soleymani et al. (2022a). In the proposed approach, a data-driven learning process extracts smoothness features from the input data, and clinically approved features such as fluidity and economy of motion are used as well-known domain knowledge-based features to detect the true skill level of the given executive trajectory. The fluidity of movement quantifies how quickly and accurately a RAS task trajectory is executed in transnational or rotational space. Following is a method for calculating this metric which incorporates the time derivative of the input trajectory

$$f_{\text{fluid}} = \left(\frac{1}{T} \int_{t=0}^{T} |\dot{u}(t)| dt \right)^{-1}, \text{ or } f_{\text{fluid}} = \left(\frac{1}{N} \sum_{t=0}^{N} |\dot{u}[t]| \right)^{-1}, \quad (7.1)$$

where T is the trajectory execution time of the time series $u(t)$, dot specifies the time derivative, and N is the number of time stamps of discrete trajectory $u[t]$. High values of this metric are returned for quick and accurate trajectories, while low values will be generated for slow, nonaccurate, and faulty trajectories and paths with abrupt temporal changes (i.e., task failures and human mistakes). It is generally accepted that the economy of motion contributes to the skills assessment of various activities by reflecting the total energy demand. It is also critical to note that human mistakes such as unintentional motions often have high velocity and large energy injection into the patient-side robot in surgical robotic applications, both of which can lead to dangerous and traumatizing outcomes. It was shown that the kinetic energy of a given trajectory approximates the critical factor of economy of motion. In different configurations of robotic platform in RAS, the total inertia of the patient-side robot remains quite the same, so the economy of motion metric will be calculated as follows:

$$f_{econo} = \left(\frac{1}{2}\int_{t=0}^{T}\dot{u}^2(t)dt\right)^{-1}, \quad \text{or} \quad f_{econo} = \left(\frac{1}{2}\sum_{t=0}^{N}\dot{u}^2[t]\right)^{-1}. \tag{7.2}$$

Smoothness is the most challenging metric. The reason for this is that nonsmooth behaviors can occur at any moment in the trajectory and are relatively insignificant compared to the main variations of the data (e.g., general trends and seasonal patterns). The method that Soleymani et al. used in this chapter is contrastive principal component analysis or cPCA in short. They created two fabricated data sets: one with smooth trajectories (background data set) and another with nonsmooth trajectories (target data set). This study aims to uncover the most notable differences between these sets related to the smoothness of trajectories. At first, the covariance matrices of the target and background data sets are calculated as follows:

$$\boldsymbol{C}_t = \boldsymbol{X}_n\boldsymbol{X}_n^\top, \quad \boldsymbol{C}_b = \boldsymbol{X}_s\boldsymbol{X}_s^\top. \tag{7.3}$$

To highlight the nonsmooth behaviors within the target set relative to the background set, the contrastive covariance matrix $\boldsymbol{C}_c$ and its singular value decomposition were calculated as follows:

$$\boldsymbol{C}_c = \boldsymbol{C}_t - \alpha\boldsymbol{C}_b = \boldsymbol{W}_c\boldsymbol{\Lambda}\boldsymbol{W}_c^\top, \tag{7.4}$$

where hyperparameter α denotes the contrastive strength parameter which represents the importance of target variances versus the irrelevant background variance. Now, the normal PCA can be applied on the projected data. This data–driven approach fully separates smooth and nonsmooth trajectories and can be used together with two other knowledge-based methods for skills evaluation purposes. As highlighted by Soleymani et al. (2022a), such approach reveals label-free information within RAS trajectories and provides more reliable and tangible correction hints to the users. Unfortunately, such hints in inductive models even with high accuracies cannot be fully trusted due to lack of explainability of extracted features. One elegant instance of label-free information in this chapter is classifying one intermediate surgeon close to expert surgeons in JIGSAWS data set. The high global rating score assigned to this surgeon proves the fact that this surgeon performs expertly compared to other participants.

7.3 Surgical skills transfer

Modern robotic surgery systems allow the application of haptic guidance forces to a trainee's hands in order to correct their motion with the aim of improving performance. In haptics and telerobotics, there is a rich literature dealing with the relationship between surgical mentor and the trainee from the perspective of *expert-in-the-loop* and haptics-enabled training (Shahbazi et al., 2016; Sharifi et al., 2017, 2020; Tao et al., 2020; Rossa et al., 2021; Najafi et al., 2020; Zakerimanesh et al., 2019; Shahbazi et al., 2016; Atashzar et al., 2018). As an example, Shamaei et al. (2015) incorporated a trilateral shared control architecture between two users (one mentor and one trainee) for RAS skills training. The platform is composed of one robot located at the patient-side that is simultaneously manipulated by two different user-side robots, one guided by the surgical mentor and one is used by the trainee. The authority level of the surgical mentor over the actions of the RAS trainee is determined by the dominance factor hyperparameter. This training program requires continuous supervision by an expert surgeon. Using a smart DL-based approach for RAS training can provide more opportunities and peace of mind to enable surgical trainees to practice surgery in a safe environment while receiving haptic feedback from a mentor.

In previous sections, it was discussed how incorporating rich skills-related knowledge into RAS training platforms enhances the training quality and reduces the need for the intervention of an expert surgical mentor throughout the entire process. AI can be used to overcome the challenges of human-robot interaction (HRI) and transfer mentorship experiences to trainees by engaging them in a collaborative action with the robot. This intelligent mentorship is even more crucial in complicated tasks, which require long training procedure (more than one session) to become skillful (Sigrist et al., 2013). A framework designed by Ershad et al. (2021), with inspiration from AI-powered surgical training technologies, detects trainees' flawful stylistic behaviors and provides haptic feedback for translational errors in a near-real-time manner. However, the study offers no suggestions for improving an individual user's performance.

Tan et al. (2019) proposed a laparoscopic robotic platform that utilizes both human demonstrations and RL to teach surgical trainees to better manipulate the robotic tool. Tan et al. locally stored expert trajectories in a field-programmable gate array (FPGA) to replay and regenerate an agent for generative adversarial imitation learning. During the training stage, the novice trainee holds the surgical handle and feels the velocity and force patterns and memorizes the skillful translational and rotational trends for better performing the task in the future executions. The limitation of the proposed method is its lack of generalization to new trajectories that the user may want to execute in future applications.

The method presented by Zahedi et al. (2020) for a virtual kinesthetic teaching environment uses machine learning to aim the transfer of skills between mentor and trainee. In the training phase, mentor's demonstrations produce a map specifying the stiffness variations of various bone layers. An estimator model of motion similarity measures how similar a trainee's drilling motion pattern is to a mentor's pattern at different layers of the bone. Trainees are provided with a set of assisting and resisting forces to correct their deficient stylistic behavior while performing operational motions. The generated corrective force is proportional to the similarity of the novice trajectory and the recorded expert demonstration. It is beyond dispute that the resultant model for the given task cannot be generalized to other RAS tasks. Similar to Zahedi et al. (2020), another platform was developed by Fekri et al. (2021) to train novice users for orthopedic surgical drilling task. In Fekri et al. (2021), an ordinary RNN with a LSTM architecture was incorporated to create the model of an expert surgeon which generates a reference trajectory based on the captured demonstrations for guiding a novice trainee toward a better stylistic behavior.

RL, learning from demonstration, and imitation learning are commonly used approaches for transferring expert mentor skills to a robot. As an example, Chi et al. (2020) and Tan et al. (2019) used a model-free generative adversarial imitation learning approach in conjunction with a deep RL model to learn and imitate the skillful behavior within a minimally invasive surgery task with unfamiliar dynamics. Some studies incorporate learning from demonstration (LfD) to adaptively mimic complicated surgical trajectories in various circumstances and then, plan a new online path for tracking in an environment with uncertain and unpredictable factors (Osa et al., 2014). The goal of RL is to capture the contributing skills-related features of a trajectory by using task-specific reward and/or regret functions. As a result, the implementation of these methods becomes difficult or even impossible for similar tasks. In addition, some reward functions such as the completion of a task are based on the completion of the whole process, making it impossible to provide the user with online and fine-grained feedback while the operation is being performed. In addition, it is difficult to design comprehensive, relevant, and meaningful reward/regret functions in complex tasks. Finally, there is always a risk of a distribution mismatch happening due to multimodal behaviors of the user's demonstrations since there are tons of possibilities involved in surgical tasks.

Taking into account the limitations outlined in the aforementioned papers and thanks to recent advances in DL, Soleymani et al. (2021) utilized artificial intelligence to detect the skillful behaviors of surgical experts and inject them into surgical trainees' activities. As a result of this, the human–robot collaboration will be controlled in a more skillful and dexterous manner (i.e., the novice user is performing the surgical operation on his/her own). The authors proposed a deep model named SkillNet to extract skills-related attributes from raw da Vinci kinematics data within each 20 s interval which allows SkillNet to operate in real-time skills evaluation and transfer tasks. The skills transfer algorithm constantly references the mentor's skillful features to generate the desired trajectory which the trainee user's trajectory should follow for the sake of better performance. This architecture design is partially inspired by image style transfer work in the field of computer vision (Gatys et al., 2016). The final objective of the surgical skills transfer algorithm is constructing an optimized trajectory $\vec{C}$ initialized by the novice demonstration $\vec{N}$ such that $\vec{C}$ represents mentor's skillful behaviors with the lowest divergence compared to the initialization $\vec{N}$. In this way, the probability distribution of the novice trajectory will

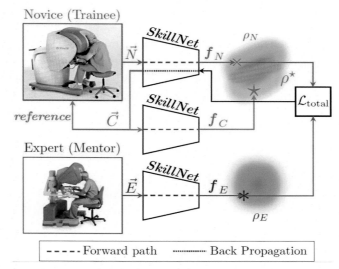

Fig. 7.5 The schematic procedure of skills transfer algorithm presented by Soleymani et al. (2021). First of all, the SkillNet extracts f_N and f_E that is sampled from novice and expert probability distributions ρ_N and ρ_E, respectively. After calculating intension and skill losses, the error is backpropagated through SkillNet to update the novice trajectory $\vec{N}$ to generate optimized trajectory $\vec{C}$ sampled from the probability distribution $\rho^\star$ which is similar to ρ_E than ρ_N. The generated enhanced combined trajectory $\vec{C}$ will be used as a reference for collaborative robots to apply corrective haptic forces to the trainee's hands. *(Used with permission from Soleymani, A., Li, X., Tavakoli, M., 2021. Deep neural skill assessment and transfer: application to robotic surgery training. In: 2021 IEEE/RSJ International Conference on Intelligent Robots and Systems (IROS), pp. 8822–8829.)*

approach to that of the expert trajectory in feature space (see Fig. 7.5). In order to achieve this goal, the following losses need to be minimized: the *skill loss* (i.e., the norm-2 difference between the latent feature distributions of experts and given novice trajectory) and the *intention loss* (i.e., the reconstruction loss between initialization and final optimized trajectory) defined as follows:

$$\mathcal{L}_{\text{skill}}(\vec{E}, \vec{C}) = \| \, \mathcal{G}_{f_E} - \mathcal{G}_{f_C} \|_2, \tag{7.5}$$

$$\mathcal{L}_{\text{intention}}(\vec{C}, \vec{N}) = \| \, \vec{C} - \vec{N} \|_2, \tag{7.6}$$

where for instance $\mathcal{G}_F$ is the Gram matrix of the feature vector F. In this paradigm, skills transformation from an expert trajectory into that of novice trainee simply means generating an optimized trajectory based on minimizing the weighted linear combination of intention and skill losses, that is, minimizing total loss

$$\mathcal{L}_{\text{total}} = \alpha \mathcal{L}_{\text{intention}}(\vec{C}, \vec{N}) + \beta \mathcal{L}_{\text{skill}}(\vec{E}, \vec{C}), \tag{7.7}$$

via gradient descent method where α and β are hyperparameters indicating relative importance of the intention loss versus skill loss in the optimization process.

The presented approach does not impose any restrictions on trainee activities or require field knowledge about the human user, robotic setup, or task. Due to the mentioned properties, the approach can be applied to a variety of robotic platforms and applications. Skill-Net transfers skillful attributes to the novice demonstration $\vec{N}$ in real time and returns $\vec{C}$ as well as the trainee's current performance, ε (see Fig. 7.6). $C[t]$ and $C[t-1]$ represent the last two points of $\vec{C}$ that are used to predict the next point

$$\hat{C}[t+1] \approx 2C[t] - C[t-1] = C[t] + \frac{\Delta T(C[t] - C[t-1])}{\Delta T}. \tag{7.8}$$

Lastly, as shown in Fig. 7.6, the collaborative robot generates a mild correction force F_{co} to user's hand to control him/her toward $\hat{C}[t+1]$ point using variable impedance control method

$$\vec{F}_{co} = \varepsilon K(\hat{C}[t+1] - N[t]), \tag{7.9}$$

where $K = \text{diag}(k_x, k_y, k_z)$ is the virtual compliance coefficients matrix in the Cartesian coordinate system. The skill transfer algorithm presented by Soleymani et al. (2021) makes significant improvements over novice trajectories, makes them more predictable, reduces hand tremors, and cancels signal noise which are all clinically proven factors for RAS skills assessment.

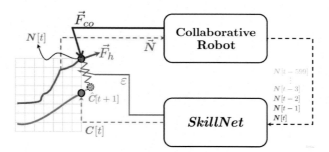

Fig. 7.6 The architecture of RAS skill transfer. The participant's hand (or RAS tooltip) is virtually connected to one end of the spring and SkillNet exerts a corrective force on the other end to guide it to the enhanced trajectory (*green solid line*). With this method, the initialized trajectory (*red solid line*) is guided toward more dexterous stylistic behavior by providing mild and compliant guidance forces. The strength of the correction force F_{co} will be determined via the control gain K and the performance of the trainee in the past. *(Used with permission from Soleymani, A., Li, X., Tavakoli, M., 2021. Deep neural skill assessment and transfer: application to robotic surgery training. In: 2021 IEEE/RSJ International Conference on Intelligent Robots and Systems (IROS), pp. 8822–8829.)*

7.4 Future directions

The core limitation of the mentioned approaches in both skills assessment and transfer is that the trajectory of each hand is viewed as a set of independent data without taking into account potential collaboration between hands (i.e., interaction between different data channels such as vertical displacement of the right hand and that of the left hand in a bimanual lifting task). The true expert in executing complicated bimanual tasks (tasks that require delicate collaboration between both hands), such as surgical operations, is not the one who performs operations with each hand expertly, but the one who plans for future steps and executes a complicated series of correlated movements by using both hands together. The clinically proven relationship between hand coordination (cooperative relationship) and correspondence (lead-follower relationship) in bimanual task performance has not yet been quantitatively assessed in the evaluation of surgical skills. The topic seems to be an excellent direction for future research.

Another direction that still remains open is dexterous autonomous robotic surgery. The results of all mentioned surgical skills assessment and transfer methods can be incorporated into the reward-shaping procedure of training a smart agent developed by RL methods or learning from demonstration paradigm for performing skillful surgical robotics tasks. Beyond ethics and regulatory problems, several issues including but not limited to the high dimensionality of demonstrated trajectories, inconsistency, and minor human error even in expert demonstrations are potential challenges that need to be addressed. It is beyond dispute that a proper trajectory encoding method benefits the tractability of the exploration-exploitation procedure of training a skillful agent.

7.5 Conclusions

In this chapter, a variety of approaches, their advantages, and limitations for extracting the skills-related features of a participant operating on RAS systems were presented. It also explained how one can use these features to classify the level of expertise of the participant and transfer them to a less skillful trajectory to help and teach less-experienced users to perform better in surgical operations. It also explained how to create an optimized trajectory with minimal reconstruction loss compared to the initialized novice trajectory while having more skillful features. The optimized trajectory can be used as a control reference command to generate a virtual corrective force on the RAS platform and guide the participant's hand toward more dexterous stylistic behavior. The enhancement metrics over the trainee's trajectory were introduced to measure the functionality and performance of the skills assessment algorithm. These metrics are including but are not limited to the motion predictability, reduction in hand tremor, and noise cancelation.

References

Ahmidi, N., Tao, L., Sefati, S., Gao, Y., Lea, C., Haro, B.B., Zappella, L., Khudanpur, S., Vidal, R., Hager, G.D., 2017. A dataset and benchmarks for segmentation and recognition of gestures in robotic surgery. IEEE Trans. Biomed. Eng. 64 (9), 2025–2041.

Atashzar, S.F., Shahbazi, M., Tavakoli, M., Patel, R.V., 2018. A computational-model-based study of supervised haptics-enabled therapist-in-the-loop training for upper-limb poststroke robotic rehabilitation. IEEE/ASME Trans. Mechatron. 23 (2), 563–574.

Birkmeyer, J.D., Finks, J.F., O'Reilly, A., Oerline, M., Carlin, A.M., Nunn, A.R., Dimick, J., Banerjee, M., Birkmeyer, N.J.O., 2013. Surgical skill and complication rates after bariatric surgery. N. Engl. J. Med. 369 (15), 1434–1442.

Chi, W., Dagnino, G., Kwok, T.M.Y., Nguyen, A., Kundrat, D., Abdelaziz, M.E.M.K., Riga, C., Bicknell, C., Yang, G.-Z., 2020. Collaborative robot-assisted endovascular catheterization with generative adversarial imitation learning. In: 2020 IEEE International Conference on Robotics and Automation (ICRA), pp. 2414–2420.

Cortes, C., Vapnik, V., 1995. Support-vector networks. Mach. Learn. 20 (3), 273–297.

DiPietro, R., Lea, C., Malpani, A., Ahmidi, N., Vedula, S.S., Lee, G.I., Lee, M.R., Hager, G.D., 2016. Recognizing surgical activities with recurrent neural networks. In: International Conference on Medical Image Computing and Computer-Assisted Intervention, pp. 551–558.

DiPietro, R., Ahmidi, N., Malpani, A., Waldram, M., Lee, G.I., Lee, M.R., Vedula, S.S., Hager, G.D., 2019. Segmenting and classifying activities in robot-assisted surgery with recurrent neural networks. Int. J. Comput. Assist. Radiol. Surg. 14 (11), 2005–2020.

Doughty, H., Damen, D., Mayol-Cuevas, W., 2018. Who's better? Who's best? Pairwise deep ranking for skill determination. In: Proceedings of the IEEE Conference on Computer Vision and Pattern Recognition, pp. 6057–6066.

Doughty, H., Mayol-Cuevas, W., Damen, D., 2019. The pros and cons: rank-aware temporal attention for skill determination in long videos. In: Proceedings of the IEEE Conference on Computer Vision and Pattern Recognition, pp. 7862–7871.

Ershad, M., Rege, R., Fey, A.M., 2021. Adaptive surgical robotic training using real-time stylistic behavior feedback through haptic cues. IEEE Trans. Med. Robot. Bionics 3 (4), 959–969.

Fawaz, H.I., Forestier, G., Weber, J., Idoumghar, L., Muller, P.-A., 2019. Accurate and interpretable evaluation of surgical skills from kinematic data using fully convolutional neural networks. Int. J. Comput. Assist. Radiol. Surg. 14 (9), 1611–1617.

Fekri, P., Dargahi, J., Zadeh, M., 2021. Deep learning-based haptic guidance for surgical skills transfer. Front. Robot. AI 7, 586707.

Funke, I., Mees, S.T., Weitz, J., Speidel, S., 2019. Video-based surgical skill assessment using 3D convolutional neural networks. Int. J. Comput. Assist. Radiol. Surg. 14 (7), 1217–1225.

Gao, Y., Vedula, S.S., Reiley, C.E., Ahmidi, N., Varadarajan, B., Lin, H.C., Tao, L., Zappella, L., Béjar, B., Yuh, D.D., et al., 2014. Jhu-Isi gesture and skill assessment working set (JIGSAWS): a surgical activity dataset for human motion modeling. In: MICCAI Workshop: M2CAI, vol. 3, p. 3.

Gatys, L.A., Ecker, A.S., Bethge, M., 2016. Image style transfer using convolutional neural networks. In: Proceedings of the IEEE Conference on Computer Vision and Pattern Recognition, pp. 2414–2423.

Goh, A.C., Goldfarb, D.W., Sander, J.C., Miles, B.J., Dunkin, B.J., 2012. Global evaluative assessment of robotic skills: validation of a clinical assessment tool to measure robotic surgical skills. J. Urol. 187 (1), 247–252.

He, K., Zhang, X., Ren, S., Sun, J., 2016. Deep residual learning for image recognition. In: Proceedings of the IEEE Conference on Computer Vision and Pattern Recognition, pp. 770–778.

Islam, S.R., Eberle, W., Ghafoor, S.K., Ahmed, M., 2021. Explainable artificial intelligence approaches: a survey. arXiv. ArXiv:2101.09429.

Itzkovich, D., Sharon, Y., Jarc, A., Refaely, Y., Nisky, I., 2019. Using augmentation to improve the robustness to rotation of deep learning segmentation in robotic-assisted surgical data. In: 2019 International Conference on Robotics and Automation (ICRA), pp. 5068–5075.

Jian, Z., Yue, W., Wu, Q., Li, W., Wang, Z., Lam, V., 2020. Multitask learning for video-based surgical skill assessment. In: 2020 Digital Image Computing: Techniques and Applications (DICTA), pp. 1–8.

Judkins, T.N., Oleynikov, D., Stergiou, N., 2009. Objective evaluation of expert and novice performance during robotic surgical training tasks. Surg. Endosc. 23 (3), 590–597.

Kim, T.S., O'Brien, M., Zafar, S., Hager, G.D., Sikder, S., Vedula, S.S., 2019. Objective assessment of intraoperative technical skill in capsulorhexis using videos of cataract surgery. Int. J. Comput. Assist. Radiol. Surg. 14 (6), 1097–1105.

Lea, C., Vidal, R., Reiter, A., Hager, G.D., 2016. Temporal convolutional networks: a unified approach to action segmentation. In: European Conference on Computer Vision, pp. 47–54.

Li, Z., Huang, Y., Cai, M., Sato, Y., 2019. Manipulation-skill assessment from videos with spatial attention network. In: Proceedings of the IEEE International Conference on Computer Vision Workshops.

Liang, K., Xing, Y., Li, J., Wang, S., Li, A., Li, J., 2018. Motion control skill assessment based on kinematic analysis of robotic end-effector movements. Int. J. Med. Rob. Comput. Assist. Surg. 14 (1), e1845.

Liu, D., Jiang, T., 2018. Deep reinforcement learning for surgical gesture segmentation and classification. In: International Conference on Medical Image Computing and Computer-Assisted Intervention, pp. 247–255.

Liu, D., Jiang, T., Wang, Y., Miao, R., Shan, F., Li, Z., 2019. Surgical skill assessment on in-vivo clinical data via the clearness of operating field. In: International Conference on Medical Image Computing and Computer-Assisted Intervention, pp. 476–484.

MacKenzie, L., Ibbotson, J.A., Cao, C.G.L., Lomax, A.J., 2001. Hierarchical decomposition of laparoscopic surgery: a human factors approach to investigating the operating room environment. Minim. Invasive Ther. Allied Technol. 10 (3), 121–127.

Martin, J.A., Regehr, G., Reznick, R., Macrae, H., Murnaghan, J., Hutchison, C., Brown, M., 1997. Objective structured assessment of technical skill (OSATS) for surgical residents. Br. J. Surg. 84 (2), 273–278.

Menegozzo, G., Dall'Alba, D., Zandonà, C., Fiorini, P., 2019. Surgical gesture recognition with time delay neural network based on kinematic data. In: 2019 International Symposium on Medical Robotics (ISMR), pp. 1–7.

Molnar, C., 2020. Interpretable Machine Learning. Lulu.com.

Muralidhar, N., Islam, M.R., Marwah, M., Karpatne, A., Ramakrishnan, N., 2018. Incorporating prior domain knowledge into deep neural networks. In: 2018 IEEE International Conference on Big Data (Big Data), pp. 36–45.

Najafi, M., Rossa, C., Adams, K., Tavakoli, M., 2020. Using potential field function with a velocity field controller to learn and reproduce the therapist's assistance in robot-assisted rehabilitation. IEEE/ASME Trans. Mechatron. 25 (3), 1622–1633.

Nguyen, X.A., Ljuhar, D., Pacilli, M., Nataraja, R.M., Chauhan, S., 2019. Surgical skill levels: classification and analysis using deep neural network model and motion signals. Comput. Methods Programs Biomed. 177, 1–8.

Osa, T., Sugita, N., Mitsuishi, M., 2014. Online trajectory planning in dynamic environments for surgical task automation. In: Robotics: Science and Systems, pp. 1–9.

Reiley, C.E., Hager, G.D., 2009. Decomposition of robotic surgical tasks: an analysis of subtasks and their correlation to skill. In: M2CAI Workshop, MICCAI, London.

Rosen, J., Hannaford, B., Richards, C.G., Sinanan, M.N., 2001. Markov modeling of minimally invasive surgery based on tool/tissue interaction and force/torque signatures for evaluating surgical skills. IEEE Trans. Biomed. Eng. 48 (5), 579–591.

Rosen, J., Solazzo, M., Hannaford, B., Sinanan, M., 2002. Task decomposition of laparoscopic surgery for objective evaluation of surgical residents' learning curve using hidden Markov model. Comput. Aided Surg. 7 (1), 49–61.

Rossa, C., Najafi, M., Tavakoli, M., Adams, K., 2021. Robotic rehabilitation and assistance for individuals with movement disorders based on a kinematic model of the upper limb. IEEE Trans. Med. Robot. Bionics 3 (1), 190–203.

Shahbazi, M., Atashzar, S.F., Tavakoli, M., Patel, R.V., 2016. Robotics-assisted mirror rehabilitation therapy: a therapist-in-the-loop assist-as-needed architecture. IEEE/ASME Trans. Mechatron. 21 (4), 1954–1965.

Shamaei, K., Kim, L.H., Okamura, A.M., 2015. Design and evaluation of a trilateral shared-control architecture for teleoperated training robots. In: 2015 37th Annual International Conference of the IEEE Engineering in Medicine and Biology Society (EMBC), pp. 4887–4893.

Sharifi, M., Salarieh, H., Behzadipour, S., Tavakoli, M., 2017. Tele-echography of moving organs using an impedance-controlled telerobotic system. Mechatronics 45, 60–70.

Sharifi, I., Talebi, H.A., Patel, R.R., Tavakoli, M., 2020. Multi-lateral teleoperation based on multi-agent framework: application to simultaneous training and therapy in telerehabilitation. Front. Robot. AI 7, 538347.

Sigrist, R., Rauter, G., Riener, R., Wolf, P., 2013. Augmented visual, auditory, haptic, and multimodal feedback in motor learning: a review. Psychon. Bull. Rev. 20 (1), 21–53.

Soleymani, A., Li, X., Tavakoli, M., 2021. Deep neural skill assessment and transfer: application to robotic surgery training. In: 2021 IEEE/RSJ International Conference on Intelligent Robots and Systems (IROS), pp. 8822–8829.

Soleymani, A., Li, X., Tavakoli, M., 2022a. A domain-adapted machine learning approach for visual evaluation and interpretation of robot-assisted surgery skills. IEEE Robot. Autom. Lett. 7 (3), 8202–8208.

Soleymani, A., et al., 2022b. Surgical procedure understanding, evaluation, and interpretation: a dictionary factorization approach. IEEE Trans. Med. Robot. Bionics 4 (2), 423–435.

Tan, X., Chng, C.-B., Su, Y., Lim, K.-B., Chui, C.-K., 2019. Robot-assisted training in laparoscopy using deep reinforcement learning. IEEE Robot. Autom. Lett. 4 (2), 485–492.

Tao, L., Elhamifar, E., Khudanpur, S., Hager, G.D., Vidal, R., 2012. Sparse hidden Markov models for surgical gesture classification and skill evaluation. In: International Conference on Information Processing in Computer-Assisted Interventions, pp. 167–177.

Tao, R., Ocampo, R., Fong, J., Soleymani, A., Tavakoli, M., 2020. Modeling and emulating a physiotherapist's role in robot-assisted rehabilitation. Adv. Intell. Syst. 2 (7), 1900181.

van Amsterdam, B., Clarkson, M.J., Stoyanov, D., 2020. Multi-task recurrent neural network for surgical gesture recognition and progress prediction. In: 2020 IEEE International Conference on Robotics and Automation (ICRA), pp. 1380–1386.

van Amsterdam, B., Clarkson, M., Stoyanov, D., 2021. Gesture recognition in robotic surgery: a review. IEEE Trans. Biomed. Eng. 68 (6).

Van der Maaten, L., Hinton, G., 2008. Visualizing data using t-SNE. J. Mach. Learn. Res. 9 (11).

Vedula, S.S., Malpani, A.O., Tao, L., Chen, G., Gao, Y., Poddar, P., Ahmidi, N., Paxton, C., Vidal, R., Khudanpur, S., et al., 2016. Analysis of the structure of surgical activity for a suturing and knot-tying task. PLoS One 11 (3), e0149174.

von Rueden, L., Mayer, S., Beckh, K., Georgiev, B., Giesselbach, S., Heese, R., Kirsch, B., Pfrommer, J., Pick, A., Ramamurthy, R., et al., 2019. Informed machine learning—a taxonomy and survey of integrating knowledge into learning systems. arXiv. ArXiv:1903.12394.

Wang, Z., Fey, A.M., 2018. Deep learning with convolutional neural network for objective skill evaluation in robot-assisted surgery. Int. J. Comput. Assist. Radiol. Surg. 13 (12), 1959–1970.

Zahedi, E., Khosravian, F., Wang, W., Armand, M., Dargahi, J., Zadeh, M., 2020. Towards skill transfer via learning-based guidance in human-robot interaction: an application to orthopaedic surgical drilling skill. J. Intell. Robot. Syst. 98 (3), 667–678.

Zakerimanesh, A., Hashemzadeh, F., Torabi, A., Tavakoli, M., 2019. A cooperative paradigm for task-space control of multilateral nonlinear teleoperation with bounded inputs and time-varying delays. Mechatronics 62, 102255.

CHAPTER 8

A virtual-based haptic endoscopic sinus surgery (ESS) training system: From development to validation

Soroush Sadeghnejad, Mojtaba Esfandiari, and Farshad Khadivar
Bio-Inspired System Design Laboratory, Department of Biomedical Engineering, Amirkabir University of Technology (Tehran Polytechnic), Tehran, Iran

8.1 Introduction

With the integration of robotic systems in surgery, the adaptability and success rate of surgery has noticeably improved, allowing surgeons to automate repetitive tasks, reduce manpower in the OR, and reduce the risk posed to the patient by directly alleviating surgeon fatigue (Taylor et al., 1995; Casals, 1998; Michel et al., 2021). Another critical factor that is addressed through the introduction of robotics in surgery is the high level of skill that is demanded from the surgeon; performing highly delicate surgeries require years of training, in addition to an exceptional understanding of the human anatomy. ESS, characteristically a minimally invasive endoscopic sinus surgery (ESS), is one of such surgeries (Fried et al., 2005; Zhao et al., 2021; Lourijsen et al., 2022). Given the tight spatial and visual constraints, the increased complexity of the procedure demands the ability to navigate around intraoperative issues such as visual perception, anatomy recognition, and nonhomogeneous anatomical makeup, not to mention the real-time identification of the presence of critical regions like the brain tissue, optic nerve, carotid artery, and other intracranial structures (Fried et al., 2004; Amirkhani et al., 2020; Mirbagheri et al., 2020). Thus, the importance of extensive practice and training is undoubtedly high for increasing the success rate of such a surgery.

As such, the need for accurate and adaptable training systems is imperative, in the light of limited opportunities to observe or practice live and/or on cadavers, not to mention the abundance of ethical, monetary, and safety considerations associated with the traditional training methodologies. The suggested training simulation platforms afford repeatable, closely monitored, and adaptable training opportunities, with the added benefit of standardizing a benchmark of acceptable surgical skills that can be applied across any participating group of trainees. The level of training, however, is proportional to how closely the simulated training system matches a real surgical procedure: all the way from the preoperative protocol and the intraoperative tissue

Medical and Healthcare Robotics
https://doi.org/10.1016/B978-0-443-18460-4.00002-0

dynamics to the feel of tool-anatomy interactions (Piromchai, 2014; Samur et al., 2007; Chanthasopeephan et al., 2003).

The use of ESS simulators is a more efficient training method, compared to the conventional methods, and enables both trainers and trainees to safely and frequently practice surgical operations in a standard simulated environment, which results in reduced training time, costs, and risks, to name but a few. In the early stages, most of the VR-based surgical training simulators were mainly based on learning through imitation of simulated surgical operations but lacked surgical skill evaluation and a realistic sense of haptic feedback during tool–tissue interaction (Perez-Gutierrez et al., 2010). Survey results demonstrate that training surgeons and residents with surgical VR-based simulators minimizes patient risks during real-time surgeries (Zhao et al., 2011). However, new technologies in robotic actuation and computational sensing methods have contributed to the advancement of a wide variety of surgical haptic training simulators with different types of feedback modes. In these systems, the human operator interacts with a simulated environment using a haptic interface, usually equipped with some sort of feedback such as force, vibration, sound, etc. The simulated environment can be physical, virtual, or physical-virtual (hybrid) (Lalitharatne et al., 2020).

8.2 History and state of the art

The first virtual reality (VR) sinus surgery simulator was developed in late 2000. The ES3, a noncommercialized prototype, developed by Lockheed Martin, Akron, Ohio (Wiet et al., 2011), offered a series of increasingly challenging simulation exercises in which trainees could master their skills by focusing on the learning courses of a complete sinus surgery procedure. Force feedback and computer graphics were integral to creating a virtual surgical environment, and training on this platform translated into performance improvement in the OR. The system, which comprises a simulation platform, a haptic controller, a voice recognition-controlled platform (to control the simulator), and a dummy setup for human interaction, further records, analyzes, and reports performance metrics in real time (Fried et al., 2007).

The Dextroscope endoscopic sinus simulator (Caversaccio et al., 2003) is one of the other noncommercialized prototypes, developed for supporting training purposes. With two handheld tools (one for controlling precision and the other for volume manipulation) and stereoscopic goggles, the user can interact with both a panel and segmented virtual models of the endonasal region, represented as 3D volumetric data. However, this system did not translate into improved performance in the OR, and the lack of force feedback was also an issue.

To fill the void of acceptability and validity in ESS training, the VOXEL-MAN sinus surgery simulator was developed in 2010 (Tolsdorff et al., 2010). This system is equipped with a 3D model of a sinus and nasal cavity environment based on high-resolution computed tomography, which can be manipulated with virtual surgical tools, controlled with a low-cost haptic device, yet provide haptic rendering and tissue removal visualization for training purposes.

In addition to actual setups, there have been attempts to improve the paradigm of tissue and/or anatomical modeling for these training simulators. In 2009, Parikh et al. (2009) proposed a new method for an automatic construction of a patient's 3D model anatomy for a virtual surgical environment from preoperative CT images.

Subsequently, in 2010, Perez-Gutierrez et al. proposed a 4-DOF endoscopic endonasal simulator prototype in which they simulated a rigid endoscope movement, a simplified nasal tissue collision, and a contact force model for haptic feedback. Users can simulate movements such as insertion of the endoscope into nasal cavities and receive haptic feedback modeled by a damped mass-spring model (Perez-Gutierrez et al., 2010).

The National Research Council of Canada developed a VR simulator, Neuro-Touch (Delorme et al., 2012), for cranial microneurosurgery training with haptic and graphic feedbacks. The mechanical behavior of tissues is modeled similar to that of viscoelastic solids; elasticity is modeled just like that of a hyperplastic solid using a generalized Rivlin constitutive model, whereas viscosity is modeled by a quasilinear viscoelastic constitutive model. This commercialized simulator is used in seven teaching hospitals across Canada to investigate system performance and validate system behavior (Varshney et al., 2014a, b).

Another proposed product, NeuroTouch-Endo VR, is a VR simulator for training neurosurgeons and is built upon the NeuroTouch system. This system also provides haptic feedback and can be used for teaching endonasal endoscopic transsphenoidal surgery, given its ability to record, analyze, and report performance metrics (Rosseau et al., 2013).

A VR simulator of sinus surgery with haptic feedback, the McGill simulator for endoscopic sinus surgery (MSESS), was developed by the collaboration between McGill University, the Montreal Neurological Institute's Simulation Lab, and the National Research Council of Canada. The objectives of this study were to show the acceptability and perceived realism of the developed system among technical users and to evaluate the training progress *via* some performance metrics.

In 2018, Barber et al. developed a modular VR teaching tool for surgical training purposes. They utilized three-dimensional models to represent the critical anatomical structures. A 3D-printed skull model and a virtual endoscope were designed to develop a low-cost educational device (Barber et al., 2018).

Kim et al. proposed a VR haptic simulator using a patient-specific 3D-printed external nostril and an exchangeable caudal septum model that facilitates real surgical simulation for ESS training (Kim et al., 2020).

Computer tomography (CT) images are used to generate graphical models for the virtual environment. The training outcomes of recruited subjects were evaluated by defining some vivid parameters (Pößneck et al., 2022).

Table 8.1 reveals a summary of the developed VR endoscopic sinus and skull base surgery systems' specifications.

As such, many other training simulators have been developed and launched over the years, each trying to address the issue of providing realistic training simulation for sinus surgery. Some research groups aim to achieve realistic tissue and anatomical modeling, in a bid to offer genuine haptic feedback during the training session, whereas some groups focus on enhancing the control schemes implemented in their systems, trying to overcome the paradigm of stability and haptic transparency. This chapter provides a comprehensive look into the various approaches to developing an ESS simulator that offers

Table 8.1 Virtual reality simulators developed for endoscopic sinus surgery.

Simulators	Characteristics	Real-based tissue mechanical model	Force feedback control	Haptic feedback
ES3	Stereoscopic system, haptic tools, voice recognition system	No	No	Yes
Dextroscope	Stereoscopic display, haptic tools, real-based 3D model environment	No	No	No
VOXEL-MAN sinus surgery	Stereoscopic display, haptic tools	No	No	Yes
Stanford	Real-based 3D visuo-haptic models, haptic devices	No	No	Yes
Perez-Gutierrez simulator	User-endoscope-tissue model, haptic rendering	No	No	Yes
NeuroTouch	Stereovision display, haptic device manipulators, high-end computers	Yes	No	Yes
NeuroTouch-Endo VR	Stereovision display, haptic device manipulators, computers	No	No	Yes
McGill simulator	Stereovision display, haptic tool manipulators, computers	No	No	Yes
Barber et al. simulator	Stereoscopic system, real-based 3D model environment	No	No	No
Kim et al. simulator	Haptic devices, patient-specific 3D model environment	Yes	No	Yes
Pößneck et al. simulator	Haptic devices, 3D model environment	Yes	No	Yes

trainees a viable intraoperative experience through different models for haptic feedback of tooltip and anatomy interaction, in addition to offering an opportunity for skill enhancement of the participants *via* a variety of tests.

We will build upon preexisting and established projects revolving around the fundamentals of ESS as standardized training modules for technical skills and provide a roadmap for other academics to produce a realistic ESS training simulator (Sadeghnejad et al., 2016; (Khadivar et al., 2017; Ebrahimi et al., 2016). We will also discuss the broad results, offering the reader an insight into how to develop and validate a well-rounded ESS training simulator (Sadeghnejad et al., 2019a, b).

8.3 Development of an endoscopic sinus training system

The development of an appropriate ESS training simulator is a multifaceted process, starting from identifying the key skills that the trainees will be trained and assessed on (Fig. 8.1). Following this, there needs to be a careful selection of the hardware necessary to deliver the target feature, including but not limited to a virtual reality platform and supporting equipment, a haptic feedback platform, computers with target specifications, and so forth. Another critical element is the methodology that is to be used to model the

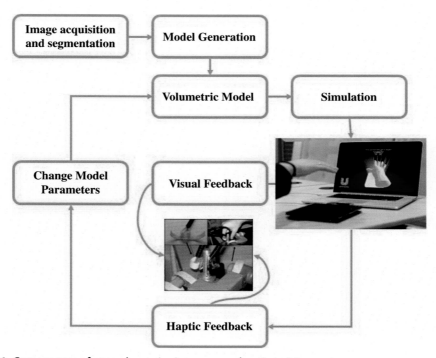

Fig. 8.1 Components of an endoscopic sinus surgery haptic training system.

anatomy/tissue of the endonasal region. The importance of this factor is paramount, given that the selected mathematical model drives the design of virtual interaction force feedback. Furthermore, there also needs to be a specific control structure to manage the force feedback loop. Moreover, there needs to be a comprehensive approach to testing the proposed system, which is essential for evaluating its effectiveness.

8.3.1 Core technical skills identification

The development of an ESS training simulator at the Mechanical Engineering Department, Djavad Mowafaghian Research Center for Intelligent Neuro-Rehabilitation Technologies (DMRCINT), Sharif University of Technology, Tehran, Iran, was the effort of a team focusing on the training of ESS skills and prioritizing performance objectives that involved hands–on techniques. After consulting with experts on the broad categories of the identified skills in the ESS, we highlighted the force feedback hands–on techniques, which would enable the trainees to identify the key haptic feedbacks revolving around the interaction of an operative/surgical tool with the key vital structures as orbits and the carotid artery present through the intricate anatomy of the paranasal sinuses (Fig. 8.2).

Furthermore, it is naturally critical that trainees are given access to the training simulator for an appropriate amount of time to allow for an acceptable level of skill acquisition and become comfortable with the user and haptic interfaces as they pertain to the simulated

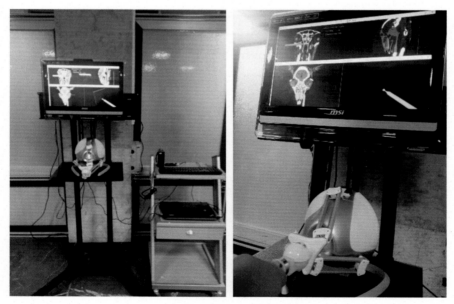

Fig. 8.2 A haptic interface, a virtual complete model, and the endoscopic view being utilized in an ESS training system.

orbital floor removal in an actual ESS. This was achieved by having the participants partake in three tasks: preexperimental learning, training, and evaluation tasks.

8.3.2 Hardware

The VR simulator consists of a computer, a virtual graphics rendering system (for VR application), a haptic user interface, data acquisition instruments, a force sensor, and a video for monitoring the developed environment (Fig. 8.3). The training system can allow the user to manipulate devices and instruments commonly used during endonasal transsphenoidal surgery, e.g., curettes. The interaction force of tools with the tissues, in the virtual environment, is replicated on the user hand through the haptic interface.

8.3.3 Anatomical modeling

The tissue mechanics model computes the force-displacement relation of the coronal orbital floor tissue deformation and fracture (Jafari et al., 2022). Considering the complexity of modern surgical techniques in the fields of otolaryngology and ophthalmology, the development of a model for virtual environment dynamics based on mechanical behavior of the tissue seems to be crucial.

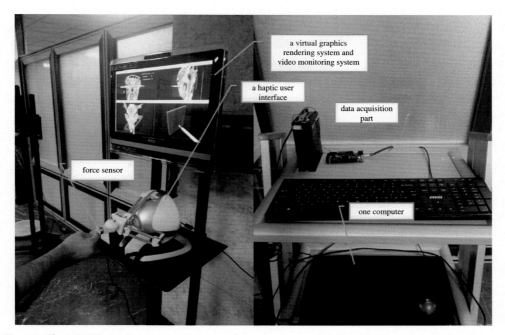

Fig. 8.3 The VR ESS training system consists of a computer, a virtual rendering system, a haptic user device, data acquisition instruments, a force sensor, and a video monitoring system (Sadeghnejad et al., 2019a, b).

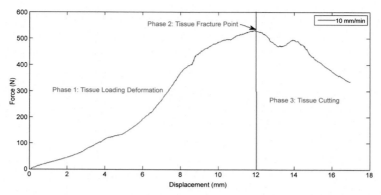

Fig. 8.4 The proposed nonlinear model for predicting the tissue's behavior during fracture (Sadeghnejad et al., 2019a, b).

The nonlinear behavior of the sinus region caused by the simultaneous existence of soft and hard tissues makes tissue dynamic modeling challenging. Therefore, before simulating the interaction of human users with a haptic interface, we tried to select a proper dynamic model that realistically simulates the interaction phenomenon. One such model to be used in a simulation system for ESS purposes, which can also estimate tissue fracture, was suggested by Sadeghnejad et al. (2016, 2019). A nonlinear rate-dependent model was proposed to predict the mechanical behavior of the tissues both before and after the fracture, which is totally dependent on the tool's displacement and velocity of movement. The mechanical behavior of the tissues is modeled as three parts for the measurement of: (1) stiffness, (2) fracture, and (3) cutting forces, based on the tool insertion rate during the indentation of the simulated surgery tool into the specific tissues.

Another model, also proposed by Sadeghnejad et al., proposes hyperelastic modeling of the sinonasal tissue, specifically for haptic neurosurgery simulation. This mechanical modeling approach utilized optimization techniques with inverse finite element models to produce a model of hyperelastic behavior of the sinonasal tissue. Indentation experiments on sheep skulls showed that their force-displacement curves proved to best fit a Yeoh hyperelastic model, given its simplicity and low number of parameters (Sadeghnejad et al., 2020).

For the purposes of the proposed ESS system, phenomenological tissue fracture modeling was employed (Fig. 8.4).

8.3.4 Simulation

CHAI3D was used as the platform of choice for compiling the virtual environment of the proposed ESS simulation system. It is a cross-platform C++ simulation framework that supports a multitude of commercially available haptic devices, giving researchers the freedom to merge both visual and haptic feedbacks, combined with virtual sinonasal tissues with realistic aesthetic and physical properties, thus allowing for a more realistic surgical simulation. CHAI3D supports several libraries to connect some haptic tools, such as Novint Falcon, Omega, Phantom, and Delta, thus making it a versatile and adaptable tool.

Two main tasks were chosen for improving the performance of the trainees, which mimic the stepwise approaches employed in sinus surgery wall removals (which also represents the increasing level of the addressed difficulties). Surgically, the simulated environment can be monitored during training sections, and the relevant force feedback can be imposed by the proposed mechanical models of the tissues, in comparison to the real operating room.

8.3.5 Haptics

Haptic feedback allows users to employ one of their most critical senses during their training with a simulated training system: the sense of touch (Kolbari et al., 2016; Esfandiari et al., 2015, 2017). Be it tactile or kinesthetic, accurate haptic feedback (i.e., feedback corresponding to the realistic interaction between surfaces and objects) gives the trainee a better sense of the work space, thus enabling them to potentially improve the quality and/or speed of procedure completion (Westebringv-van der Putten et al., 2008; Kolbari et al., 2015a, b; Kolbari et al., 2018; Patel et al., 2022).

One potential resource can be the Novint Falcon haptic device (Martin and Hillier, 2009). Falcon is a low-cost parallel impedance-type robot with considerable load capacities and a proper work space. The Falcon's programmable interface enables convenient control of the robot's motion on the three-motion Cartesian axis (Martin and Hillier, 2009). Another potential haptic device is the Phantom Omni, a serial link-based 6-DOF haptic device with a substantial work space and producing a force up to 3.3 N. With its stylus–like end effector, users can simulate the use of a surgical tool and can expect to experience the appropriate haptic feedback, assuming an accurate feedback model (Khadivar et al., 2020). This is employed for developing ESS simulators and is considered a device on the lower end of the quality spectrum, given its relatively low price point. To design a VR–based haptic system based on the model predictive control (MPC) approaches, we need to dynamically identify the haptic interface, by use of a calibrated force sensor, to produce the proper robot impedance (Khadivar et al., 2020). This haptic device was used in the development of our platform. The interface between the utilized sensor and the computer is an Arduino chip, and we used a PC platform for running the control processes (Fig. 8.5).

We used the CHAI3D open software for implementing all the sensor updates and the control system. It should be noted that the mechanical behavior of the virtual environment was identified as a linear parametric variable (LPV)-constrained problem. We further employed an online robust MPC for the proposed system, inspired by a modified MPC algorithm for the LPV (Fig. 8.6).

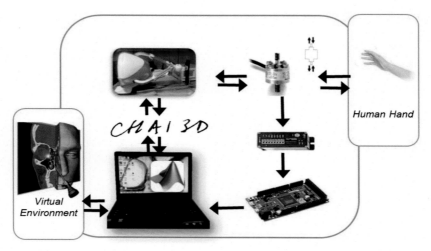

Fig. 8.5 The whole components a VR ESS haptic training system.

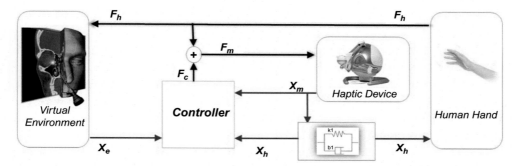

Fig. 8.6 Introduction of the general components utilized in the design of a closed-loop control system.

8.4 Experimental procedures and evaluation

8.4.1 Experimental procedure

The system was tested on 2 teams of 10 members each, with 1 team receiving all the training through all the training tasks and the second team getting limited exposure to training tasks. The participants were given the opportunity to partake in a preexperimental learning session regarding the haptic interface and virtual environment. Following this, they progressed to participating in three specific tasks: pretraining task, training task, and evaluation task. The pretraining task was aimed at familiarizing the participants with the system introduced in the previous sections. The training task involved using a curette (a surgical tool) to indent a simulated orbital floor tissue (which has variable stiffness that can be felt by the use of haptic feedback).

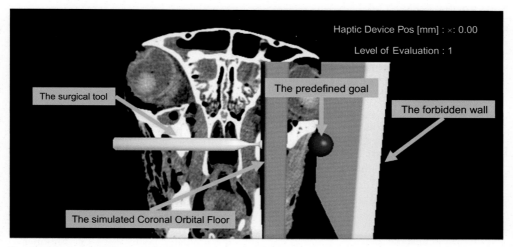

Fig. 8.7 The schematic introductory view of an evaluation task.

Finally, the evaluation task aimed at improving the flexibility of the participants, by allowing them to navigate across a simulated orbital floor while avoiding virtual boundaries (Sadeghnejad et al., 2019a, b). The first group was allowed to partake in both the pretraining and training tasks before moving onto the evaluation task, whereas the second group was directly placed into the evaluation task. During the evaluation task, we recorded the generated interaction forces (both from the force sensor and device's recording), the position of the surgery tooltip, and the task completion time, which is defined as the period during which the tip of the surgery tool is in contact with the coronal orbital floor at the time of hitting the goal or the forbidden wall (Fig. 8.7). The evaluation task has five different kinds of tissue stiffness, representing a spectrum of different performance levels. The participants were randomly assigned a level.

8.4.2 Performance metrics and evaluation methods

As discussed in the previous sections, collecting, analyzing, and reporting the user data is as crucial a step as any other in developing and proposing a new training simulator. Great care must be taken in measuring performance, by introducing metrics that can measure the performance of a trainee in an efficient, reproducible, actionable, and commensurable manner. Based on the task analysis process outlined earlier, our target dimensions of quantitative data generated covered the following areas: **quality** (sense of haptic feedback of forces because of tool/anatomy interaction), **efficiency** (task performance with the least number of unnecessary maneuvers as well as time taken), and **safety** (interaction between the goal and forbidden area).

Table 8.2 Description of the performance metrics.

Metric sphere	Definition	Metric	Units
Quality	Sense of haptic feedback of forces, generated by tool-tissue interaction	• The generated force and realism for orbital floor removal • How it will work for the surgical education curricula	• Percentage • Percentage
Efficiency	Task performance with the least number of unnecessary maneuvers	• Distanced traveled to reach the predefined goal • The necessary time to reach the goal	• Millimeters • Seconds
Safety	Measurement of the simulated collateral damage	• How a regular tool hits the simulated goal • How a regular tool hits the simulated forbidden goal	• Number • Number

The participants were also asked to fill out a questionnaire after the completion of the study, *via* a 10-point rating scale and open-ended questions, aimed at gathering qualitative data regarding their perceptions, opinions, and potential suggestions (Table 8.2).

8.5 Results

The performance of our developed system is examined by analyzing the feedback of the participants in group I and group II regarding the two criteria, namely, the sense of fracture and the sense of tissue stiffness variation. Based on the former criterion, the system was rated (out of 10) as 7.57 ± 1.43 and 6.32 ± 1.02 by group I and group II, respectively, whereas the score associated with the system on the latter criterion was 8.13 ± 1.87 and 6.05 ± 0.95, respectively. Moreover, 77.1% of the participants of group I and 61.1% of those of group II expressed that they could adapt and improve their training level while working with the developed simulation system, which is substantial evidence that the system was appreciated by the users, in addition to the obvious difference in the performance metrics of the group with training vs. the group without training. It should be noted that while 9/10 participants from the trained group reported their appreciation for using the developed simulation system as a supporting tool for hand-eye coordination means, 7/10 members of the untrained group also reported the same belief, which only seeks to improve the credibility of the virtual-based haptic ESS training system (Table 8.3).

Table 8.3 Postevaluation questionnaire assessment.

Evaluation items	Group I		Group II	
	Mean value	Standard deviation	Mean value	Standard deviation
How users will sense the fracture during the operation	7.57	1.43	6.32	1.02
How users will sense the tissue hardening effect of the tool's interaction with the tissues	8.13	1.87	6.05	0.95
How effective will be the developed platform for educational purposes	7.71	1.41	6.11	0.99

The results show that there is a common considerable difference between the two recruited groups in the educational programs. Although the participants of group II did not attend the training task sessions, their total time of evaluation task completion was reported to be much longer than that of those who participated in group I (Fig. 8.8). It was also evident that those with a greater level of training had superior control over the trajectory of the path toward the specific goal, not to mention their ability to maintain steadier hand movements during the task, relative to the untrained group participants. It was also found that the untrained group, on average, exerted more force at the tooltip for the duration of the evaluation task (Figs. 8.9 and 8.10).

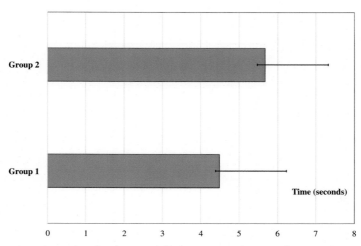

Fig. 8.8 The average time taken by the two different recruited groups for completing the evaluation task.

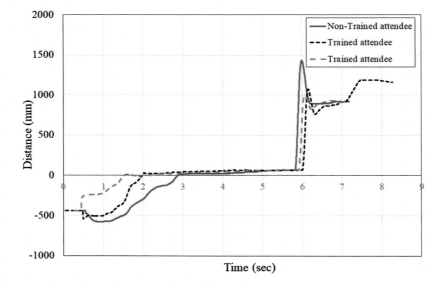

Fig. 8.9 The average distance traveled by the simulated tool.

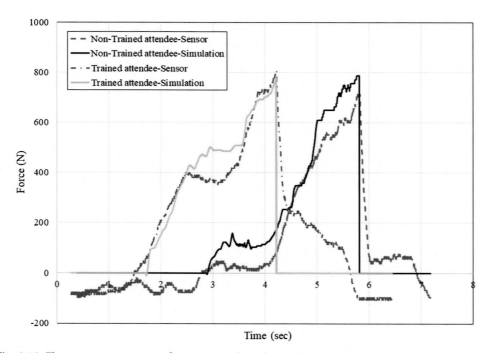

Fig. 8.10 The average interaction force imposed on the tooltips during the simulation tasks.

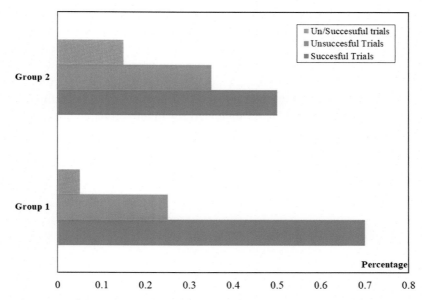

Fig. 8.11 The successful trials percentage of simulated evaluation tasks in reaching to the specified goal through.

The assessment of "safety" was conducted by comparing the number of times the predefined goal, after fracturing through the wall simulating the coronal orbital floor, was touched or by the number of times the forbidden wall was hit. It is seen that most of the participants were generally able to show efficient hand–eye coordination prior to the fracture while maintaining such control over the tool after the fracture proved to be challenging as some participants hit both the predefined goal and the forbidden wall at the same time. In all, 70% of the participants in group I and 50% of those in group II were able to accomplish the evaluation task as expected, 25% in group I and 35% in group II hit the forbidden wall without touching the predefined goal, and 5% in group I and 15% in group II touched both the predefined goal and the forbidden wall at the same time. The results of the safety assessment test demonstrate a significant effect of the training tasks on the performance of the participants working with the system (Fig. 8.11).

8.6 Conclusions and final remarks

In this research, we developed an educational VR system with haptic force feedback from the interaction between the virtual tool and a graphical model of the sinus tissue.

To evaluate the system's performance, we defined three different scenarios that require performing specific tasks by the users. In the first phase, the pretraining phase, the users were asked to work with the system to get used to the virtual environment and the haptic interface. In the second phase, the training phase, the users were asked to exert higher force until fracture of the sinonasal tissue takes place to perceive the fracture force level. Finally, we assessed the users' training levels and performance over the defined scenarios. The virtual environment is simulated in CHAI3D open software. Using the Novint Falcon robot as the haptic interface, we provided haptic feedback to the user hand from the tool–tissue interaction force.

From the hardware setup and the anatomical model to the test study's specifics, we propose a full-scale approach to developing an ESS training system that not only offers haptic feedback based on realistic tissue modeling (phenomenological tissue fracture modeling) but also delivers a variable testing protocol, which involves varying the level of task complexities at different levels. Thus, with a reliable physical and visual environment, the user, or trainee, is exposed to a realistic experience. Moreover, studying different performance metrics enables the researchers to make a more comprehensive, and conclusive, assessment of the users' skills.

We proposed the evaluation of the participants to be based on three distinct principles: quality, efficiency, and safety. This, in conjunction with a postevaluation questionnaire test, has set precedence to enhance the system qualitatively and quantitatively. Based on the reported results, it is possible to develop an educational environment with the ability to reflect the haptic force. Based on the average completion time alone in the evaluation test, it was evident that practice through the pretraining and training tasks had noticeable positive effects on the performance of the participants. This point is further strengthened by the analysis of the applied force, the steadiness of the operating hand during the tasks, and the overall better control over the applied force. It was observed that the group with no access to the two training modules consistently applied a greater magnitude of force at tissue fracture compared to the group that had been trained through the pretraining and training tasks.

It is thus obvious that a surgeon can only benefit from preoperative training, especially one that is conducted in a standardized environment, which can be replicated for other trainees and trainers. In fact, this system reduces the need for a trainer, which can help make hospital operations more economical and efficient while also reducing the overall risk of surgery by introducing a standardized method for training and evaluating any number of trainees.

Overall, ESS training systems with accurate tissue modeling and appropriate haptic feedback have vast potential for shifting the current paradigm of cadaver and observation-based training methodologies to a more sustainable, repeatable, economically viable, and adaptable training system.

Conflict of interest statement

The authors confirm that this book chapter and its corresponding research work involve no conflict of interest.

Acknowledgments

We appreciate the Djavad Mowafaghian Research Center for Intelligent Neuro-Rehabilitation Technologies, the Bio-Inspired System Design Laboratory of Biomedical Engineering Department of Amirkabir University of Technology, and the Research Center of Biomedical Technologies and Robotics at the Research Institute of Medical Technology of Tehran University of Medical Sciences for their support of this research. Moreover, we thank Mr. Ehsan Abdollahi for helping us in conducting the experiments.

References

Amirkhani, G., Farahmand, F., Yazdian, S.M., Mirbagheri, A., 2020. An extended algorithm for autonomous grasping of soft tissues during robotic surgery. Int. J. Med. Robot. Comput. Assist. Surg. 16 (5), 1–15.

Barber, S.R., Jain, S., Son, Y.J., Chang, E.H., 2018. Virtual functional endoscopic sinus surgery simulation with 3D-printed models for mixed-reality nasal endoscopy. Otolaryngol. Head Neck Surg. 159 (5), 933–937.

Casals, A., 1998. Robots in surgery. In: Autonomous Robotic Systems. Springer, London, pp. 222–234.

Caversaccio, M., Eichenberger, A., Häusler, R., 2003. Virtual simulator as a training tool for endonasal surgery. Am. J. Rhinol. 17 (5), 283–290.

Chanthasopeephan, T., Desai, J.P., Lau, A.C., 2003. Measuring forces in liver cutting: New equipment and experimental results. Ann. Biomed. Eng. 31 (11), 1372–1382.

Delorme, S., Laroche, D., DiRaddo, R., Del Maestro, R.F., 2012. NeuroTouch: a physics-based virtual simulator for cranial microneurosurgery training. Oper. Neurosurg. 71 (suppl_1), ons32–ons42 (Vancouver).

Ebrahimi, A., Sadeghnejad, S., Vossoughi, G., Moradi, H., Farahmand, F., 2016. Nonlinear adaptive impedance control of virtual tool-tissue interaction for use in endoscopic sinus surgery simulation system. In: 2016 4th International Conference on Robotics and Mechatronics (ICROM), IEEE, pp. 66–71.

Esfandiari, M., Sadeghnejad, S., Farahmand, F., Vosoughi, G., 2015. Adaptive characterisation of a human hand model during intercations with a telemanipulation system. In: 2015 3rd RSI International Conference on Robotics and Mechatronics (ICROM), IEEE, pp. 688–693.

Esfandiari, M., Sadeghnejad, S., Farahmand, F., Vosoughi, G., 2017. Robust nonlinear neural network-based control of a haptic interaction with an admittance type virtual environment. In: 2017 5th RSI International Conference on Robotics and Mechatronics (ICRoM), IEEE, pp. 322–327.

Fried, M.P., Satava, R., Weghorst, S., Gallagher, A.G., Sasaki, C., Ross, D., Sinanan, M., Uribe, J.I., Zeltsan, M., Arora, H., Cuellar, H., 2004. Identifying and reducing errors with surgical simulation. BMJ Qual. Saf. 13 (Suppl 1), i19–i26.

Fried, M.P., Satava, R., Weghorst, S., Gallagher, A., Sasaki, C., Ross, D., 2005. The use of surgical simulators. Adv. Patient Saf.: Res. Implement. 4, 165.

Fried, M.P., Sadoughi, B., Weghorst, S.J., Zeltsan, M., Cuellar, H., Uribe, J.I., Sasaki, C.T., Ross, D.A., Jacobs, J.B., Lebowitz, R.A., Satava, R.M., 2007. Construct validity of the endoscopic sinus surgery simulator. II. Assessment of discriminant validity and expert benchmarking. Arch. Otolaryngol. Head Neck Surg. 133 (4), 350–357.

Jafari, B., Shams, V., Esfandiari, M., Sadeghnejad, S., 2022. Nonlinear contact modeling and haptic characterization of the ovine cervical intervertebral disc. In: 2022 9th IEEE RAS/EMBS International Conference for Biomedical Robotics and Biomechatronics (BioRob), IEEE, pp. 1–6.

Khadivar, F., Sadeghnejad, S., Moradi, H., Vossoughi, G., Farahmand, F., 2017. Dynamic characterization of a parallel haptic device for application as an actuator in a surgery simulator. In: 2017 5th RSI International Conference on Robotics and Mechatronics (ICRoM), IEEE, pp. 186–191.

Khadivar, F., Sadeghnejad, S., Moradi, H., Vossoughi, G., 2020. Dynamic characterization and control of a parallel haptic interaction with an admittance type virtual environment. Meccanica 55 (3), 435–452.

Kim, H.M., Park, J.S., Kim, S.W., 2020. Virtual reality haptic simulator for endoscopic sinus and skull base surgeries. J. Craniofac. Surg. 31 (6), 1811–1814.

Kolbari, H., Sadeghnejad, S., Bahrami, M., Kamali, E.A., 2015a. Nonlinear adaptive control for teleoperation systems transitioning between soft and hard tissues. In: 2015 3rd RSI International Conference on Robotics and Mechatronics (ICROM), IEEE, pp. 055–060.

Kolbari, H., Sadeghnejad, S., Bahrami, M., Kamali, A., 2015b. Bilateral adaptive control of a teleoperation system based on the hunt-crossley dynamic model. In: 2015 3rd RSI International Conference on Robotics and Mechatronics (ICROM), IEEE, pp. 651–656.

Kolbari, H., Sadeghnejad, S., Parizi, A.T., Rashidi, S., Baltes, J.H., 2016. Extended fuzzy logic controller for uncertain teleoperation system. In: 2016 4th International Conference on Robotics and Mechatronics (ICROM), IEEE, pp. 78–83.

Kolbari, H., Sadeghnejad, S., Bahrami, M., Ali, K.E., 2018. Adaptive control of a robot-assisted tele-surgery in interaction with hybrid tissues. J. Dyn. Syst. Meas. Control. 140 (12).

Lalitharatne, T.D., Tan, Y., Leong, F., He, L., Van Zalk, N., De Lusignan, S., Iida, F., Nanayakkara, T., 2020. Facial expression rendering in medical training simulators: current status and future directions. IEEE Access 8, 215874–215891.

Lourijsen, E.S., Reitsma, S., Vleming, M., Hannink, G., Adriaensen, G.F., Cornet, M.E., Hoven, D.R., Videler, W.J., Bretschneider, J.H., Reinartz, S.M., Rovers, M.M., 2022. Endoscopic sinus surgery with medical therapy versus medical therapy for chronic rhinosinusitis with nasal polyps: a multicentre, randomised, controlled trial. Lancet Respir. Med. 10 (4), 337–346.

Martin, S., Hillier, N., 2009. Characterisation of the Novint Falcon haptic device for application as a robot manipulator. In: Australasian Conference on Robotics and Automation (ACRA). Citeseer, Sydney, Australia, pp. 291–292.

Michel, G., Salunkhe, D.H., Bordure, P., Chablat, D., 2021. Literature review on endoscopic robotic systems in ear and sinus surgery. J. Med. Dev. 15 (4).

Mirbagheri, A., Amirkhani, G., Yazdian, S., Farahmand, F., Sarkar, S., Sina Robotics and Medical Innovators Co Ltd, 2020. Controlling a Laparoscopic Instrument. U.S. Patent Application 16/832,439.

Parikh, S.S., Chan, S., Agrawal, S.K., Hwang, P.H., Salisbury, C.M., Rafii, B.Y., Varma, G., Salisbury, K.J., Blevins, N.H., 2009. Integration of patient-specific paranasal sinus computed tomographic data into a virtual surgical environment. Am. J. Rhinol. Allergy 23 (4), 442–447.

Patel, R.V., Atashzar, S.F., Tavakoli, M., 2022. Haptic feedback and force-based teleoperation in surgical robotics. Proc. IEEE 110 (7), 1012–1027.

Perez-Gutierrez, B., Martinez, D.M., Rojas, O.E., 2010. Endoscopic endonasal haptic surgery simulator prototype: a rigid endoscope model. In: 2010 IEEE Virtual Reality Conference (VR), IEEE, pp. 297–298.

Piromchai, P., 2014. Virtual reality surgical training in ear, nose and throat surgery. Int. J. Clin. Med. 2014.

Pößneck, A., Ludwig, A.A., Burgert, O., Nowatius, E., Maass, H., Cakmak, H.K., Dietz, A., 2022. Development and evaluation of a simulator for endoscopic sinus surgery. Laryngo-Rhino-Otol.

Rosseau, G., Bailes, J., del Maestro, R., Cabral, A., Choudhury, N., Comas, O., Debergue, P., De Luca, G., Hovdebo, J., Jiang, D., Laroche, D., 2013. The development of a virtual simulator for training neurosurgeons to perform and perfect endoscopic endonasal transsphenoidal surgery. Neurosurgery 73 (suppl_1), S85–S93.

Sadeghnejad, S., Esfandiari, M., Farahmand, F., Vossoughi, G., 2016. Phenomenological contact model characterization and haptic simulation of an endoscopic sinus and skull base surgery virtual system. In: 2016 4th international conference on robotics and mechatronics (ICROM), IEEE, pp. 84–89.

Sadeghnejad, S., Farahmand, F., Vossoughi, G., Moradi, H., Hosseini, S.M.S., 2019a. Phenomenological tissue fracture modeling for an endoscopic sinus and skull base surgery training system based on experimental data. Med. Eng. Phys. 68, 85–93.

Sadeghnejad, S., Khadivar, F., Abdollahi, E., Moradi, H., Farahmand, F., Sadr Hosseini, S.M., Vossoughi, G., 2019b. A validation study of a virtual-based haptic system for endoscopic sinus surgery training. Int. J. Med. Robot. Comput. Assist. Surg. 15 (6), e2039.

Sadeghnejad, S., Elyasi, N., Farahmand, F., Vossughi, G., Sadr Hosseini, S.M., 2020. Hyperelastic modelling of sino-nasal tissue for haptic neurosurgery simulation. Scientia Iranica 27 (3), 1266–1276.

Samur, E., Sedef, M., Basdogan, C., Avtan, L., Duzgun, O., 2007. A robotic indenter for minimally invasive measurement and characterization of soft tissue response. Med. Image Anal. 11 (4), 361–373.

Taylor, R.H., Funda, J., Eldridge, B., Gomory, S., Gruben, K., LaRose, D., Talamini, M., Kavoussi, L., Anderson, J., 1995. A telerobotic assistant for laparoscopic surgery. IEEE Eng. Med. Biol. Mag. 14 (3), 279–288.

Tolsdorff, B., Pommert, A., Höhne, K.H., Petersik, A., Pflesser, B., Tiede, U., Leuwer, R., 2010. Virtual reality: a new paranasal sinus surgery simulator. Laryngoscope 120 (2), 420–426.

Varshney, R., Frenkiel, S., Nguyen, L.H., Young, M., Del Maestro, R., Zeitouni, A., Saad, E., Funnell, W.R.J., Tewfik, M.A., 2014a. The McGill simulator for endoscopic sinus surgery (MSESS): a validation study. J. Otolaryngol.-Head Neck Surg. 43 (1), 1–10.

Varshney, R., Frenkiel, S., Nguyen, L.H., Young, M., Del Maestro, R., Zeitouni, A., Tewfik, M.A., National Research Council Canada, 2014b. Development of the McGill simulator for endoscopic sinus surgery: a new high-fidelity virtual reality simulator for endoscopic sinus surgery. Am. J. Rhinol. Allergy 28 (4), 330–334.

Westebringv-van der Putten, E.P., Goossens, R.H., Jakimowicz, J.J., Dankelman, J., 2008. Haptics in minimally invasive surgery—a review. Minim. Invasive Ther. Allied Technol. 17 (1), 3–16.

Wiet, G.J., Stredney, D., Wan, D., 2011. Training and simulation in otolaryngology. Otolaryngol. Clin. N. Am. 44 (6), 1333–1350.

Zhao, Y.C., Kennedy, G., Yukawa, K., Pyman, B., O'Leary, S., 2011. Can virtual reality simulator be used as a training aid to improve cadaver temporal bone dissection? Results of a randomized blinded control trial. Laryngoscope 121 (4), 831–837.

Zhao, R., Chen, K., Tang, Y., 2021. Olfactory changes after endoscopic sinus surgery for chronic rhinosinusitis: a meta-analysis. Clin. Otolaryngol. 46 (1), 41–51.

CHAPTER 9

Medical and healthcare robots in India

Kshetrimayum Lochan[a], Ashutosh Suklyabaidya[b], and Binoy Krishna Roy[c]
[a]IIT Palakkad IHub Foundation, Indian Institute of Technology Palakkad, Palakkad, Kerala, India
[b]Department of General Surgery, Silchar Medical College and Hospital, Silchar, Assam, India
[c]Department of Electrical Engineering, National Institute of Technology Silchar, Silchar, Assam, India

9.1 Introduction

As medical science and engineering have been developed, robotics in the surgical area has made significant progress in terms of safety, accuracy, and efficiency. In the medical field, robots transform surgery procedures, disinfecting, and supply delivery, and enable providers to focus more on caring for their patients. As India has been preparing since some time for the Industrial Revolution 4.0, the rate of adoption for the emerging technologies such as robotics, artificial intelligence (AI), the internet of things (IoT), and various others have increased exponentially. According to a study in 2018 from a leading job portal, the number of people seeking jobs in the area of robotics has increased by 186% (Akshaya, 2019). The analyses state that the initial boost for this demand depends on several factors, including a government push to "Make In India," where the government has invested the significant sum of USD13 billion into robotics. In the Indian medical sector, the critical area where robotics has been deployed within healthcare is surgery. Robotic assistants are prevalent throughout the country in both public and private hospitals. The first case of using a robotic surgical assistant dates back to 1998. A robot called AESOP (Automated Endoscopic System for Optimal Positioning) 3000 was used by a team of doctors from Escorts Heart Institute and Research Centre to fix a hole in a patient's heart (Maurice, 2017).

It was Asia's first robotic surgery. The very idea of robots turning surgeons was/is very frightening. This robotic surgeon outperforms human surgeons by cutting down the risk of tremors in the surgeons' hands and creating a tireless doctor to avail of the benefits round the clock (Maurice, 2017). Reducing the risk to a patient's life by using a thinking robot is a massive leap for robotics and the medical surgeon community. The patient's surgery at the current medical edge, along with AESOP's nonhuman intervention, allows the surgeon to carry out minimally invasive surgery (MIS) that is much more accurate than before. After AESOP made a significant mark by doing the first ever MIS surgery, AESOP and other robots were actively used for surgery for almost 3 years in Europe and the United States. AESOP is also the world's first robot cleared by the US Food and Drug Administration. It is specially designed to carry out complete surgical procedures and can think intelligently.

221

After this, the number of assisted surgery cases has improved appreciably in India. However, a revolution in the Indian medical industry had arisen after the introduction of IBM's AI and cognitive-powered bot called Watson in 2015 (Dogra, 2012). Since its inception, IBM has partnered with private institutions like Manipal hospitals (Karnataka, India) to seek solutions. Its cognitive systems built Watson, with natural language processing (NLP) capabilities, deep insights, and the ability to analyze electronic medical records, which has turned out to be a good resource for surgeons.

Moreover, with as confirmed by another study, India's surgical robotics market is expected to grow fivefold by 2025 due to high capability in the medical field. Enhanced demand for automation solutions will also drive the estimated growth. According to this study, the robotics market in India is expected to grow at a compound annual growth rate (CAGR) of 20% between 2017 and 2025, reaching USD350 million (Puntambekar et al., 2012).

Nowadays, robots are also used in operating rooms and in clinical settings for the support and enhancement of healthcare (Akshaya, 2019). Examples include the deployment of robots in a much wider range to reduce exposure to pathogens during the COVID-19 pandemic. The use of robotics for the automation of manual, high-volume repetitive tasks frees up time in research laboratories so that technicians and scientists can focus their attention on discoveries (Mahesh et al., 2015). There are some risk reduction areas and workflows which are provided by medical robotics that offer good value. These include cleaning by the robots and preparing patients' rooms, helping to manage person-to-person contact in the infectious disease wards, including robots with AI-enabled software to reduce the time for the identification, matching, and distribution of medicines to patients.

With evolving technologies, these robots will function autonomously and automatically. Benefits of robotics in the healthcare section which enable high-level patient care are as follows (Intel, 2020):

1. *High-quality patient care*: These medical robots support the procedures of minimally invasive surgeries, social interactions, and intelligent therapeutic engagements for elderly patients and also frequent monitoring of patients. Moreover, these robots reduce the workloads of nurses and caregivers and can promote long-term well-being.
2. *Streamlined clinical work*: Mobile robots such as autonomous mobile robots (AMRs) simplify the routine tasks of human workers and ensure more consistent processes. These AMRs give details about staff shortages and keep track of timely orders by helping and making more supplies, equipment, and medication as and where needed. These AMRs also help to keep the hospital rooms ready for incoming patients to check in quickly and allow healthcare workers to do the value-driven work.
3. *Maintaining a safe work environment*: For keeping the healthcare workers at a safer zone, mobile robots transport and supply linen to hospitals where there is high risk of infection. These robots limit pathogen exposure while helping to reduce hospital-acquired

infections (HAIs), protecting hundreds of healthcare workers. In addition, social robots and one type of AMR help with heavy lifting, such as moving beds or patients, reducing the physical strain on the healthcare workers.

The rest of this chapter is organized as follows. Section 9.2 describes the types of healthcare robots in India. Section 9.3 focuses on the commercial and research areas of surgical robots. The swindle of robotics in healthcare are discussed in Section 9.4. The future of robotic surgery in India and conclusions are described in Sections 9.5 and 9.6, respectively.

9.2 Types of healthcare robots

Different healthcare robots are discussed in this section.

9.2.1 Surgical-assistance robots

As robotic technologies with motion control adaptations have advanced, robots with surgical assistance have become extremely precise. These surgical robots help surgeons to reach new levels of accuracy while performing complex surgical procedures which involve AI and computer vision-enabled technologies. Some of these surgical robots can even complete the tasks autonomously by allowing surgeons to oversee the operations even at a far comfort (Intel, 2020). Surgeries performed using robotics assistance fall into two main categories:

A. *Minimally invasive surgeries*: These include robotic prostatectomy, hysterectomy, bariatric surgery, and other procedures, primarily focused on soft tissues. A small insertion is made for the incision, and these robots lock themselves in place, creating a stable platform from which surgeries are performed via remote control. Working manually via a button-sized incision is extremely difficult, even for an experienced surgeon (Intel, 2020). These robots make these procedures easy and accurate, reducing infections and other complications.

B. *Orthopedic surgeries*: Some devices are preprogrammed to perform common orthopedic surgeries which are done for knee and hip replacements. By combining smart robotic arms, data analytics, and 3D imaging, these robots can give more predictable results by providing a specified and defined boundary to assist the surgeon. Some AI tools allow the robot to be trained in specific orthopedic surgeries with precise directions on how to perform the procedures. The ability to share the surgery by video feed from the operating room to other locations, far or near, has enabled surgeons to benefit from consultations with other specialists in their respective fields, resulting in patients having the best possible surgeons involved in their procedures (Dogra, 2012). The high-definition 3D computer vision which is installed in the robots can provide surgeons with detailed information and enhance their performance during operations. Eventually, these robotic surgery robots will take over small subprocedures such as suturing under the watchful gaze of the surgeons.

9.2.2 Modular robots

Modular robots can be configured to perform different multiple functions, which includes the exoskeleton and prosthetic robot arms and legs. Therapeutic robots can also help people who have strokes through rehabilitation and/or patients with traumatic brain injuries or with impairments such as multiple sclerosis. Wheelchairs are mounted with robotic arms which can assist patients with spinal injuries in performing their daily tasks. These robots are instilled with AI and depth cameras, and can monitor the posture of a patient when they undergo prescribed exercises by measuring the degree of motion in different positions and tracking the patient with more precision than a human eye (Desai, 2018).

9.2.3 Autonomous mobile robots

Hospitals rely on AMRs, due to their ability to assist in dealing with the critical needs of a patient which includes disinfection, telepresence, and delivery of medical supplies by creating a safe environment for the hospital staff and patients. They are also equipped with light detection and ranging (LiDAR) systems. AMRs can visually compute the mapping capabilities, allowing the AMRs to self-navigate in patients' rooms and allow medical professionals to interact from afar. AMRs can be controlled by remote specialists and accompany doctors for hospital rounds or be used for on-screen consultations regarding patient diagnostics and care (Sequeira, 2020).

9.2.4 Service robots

Service robots relieve the day-to-day burden of healthcare workers by handling routine logistical tasks. These robots function autonomously and can send a report after completing a task. Such robots can also set up patients' rooms by tacking the supplies and the files. They can also transport bed linen to and from the laundry. These robots use the ultraviolet light, hydrogen peroxide vapor, or air filtration to reduce infection and to sanitize all accessible areas uniformly. The controlled robotics can be use in everyday lives, the disabled and the elderly for the assistive and rehabiliation for improving the quality of their lives (Boubaker, 2020).

9.2.5 Social robots

Social robots interact with humans directly. In the long term, such robots are used for patient care and to provide social interaction by monitoring patients. These robots may also help patients with their treatment regimens or provide cognitive engagement, helping to keep patients positive and alert. They can also offer directions to visitors in the hospital environment. Such robots will help reduce the workloads of caregivers and improve patients' emotional well-being.

9.3 Commercial and research areas of surgical robots

With the advent of surgical robots, a new era has started since the millennium in the domain of MIS. Robotic surgery has come a long way from its initial beginning with a RoboDoc, adapted for medical use from its industrial applications, to the current state-of-the-art da Vinci Surgical systems (Puntambekar et al., 2012). The advantages of robotic surgery over laparoscopic surgery are the unprecedented explosion of the use of robotics in this domain. The advantages include enhanced magnification, superior ergonomics, motion scaling, 3D vision, enhanced dexterity, tremor filtering, and control of operating instruments. The robotics platform offers $10\times-15\times$ magnification, which provides meticulous anatomy of the surgical field of interest, and avoids inadvertent injury. The anatomic structures are much better appreciated in the surgical field of interest. Compared to conventional laparoscopy, 3D vision makes working more ergonomic. With this comparison, the main surgeon need not stand and operate but can sit comfortably and perform time-consuming surgeries. Likewise, the nursing assistant and the patient's assistant can remain comfortably seated for the procedure with the main work consisting of exchanging the robotic instruments and intermittently providing suction and sutures.

The seven degrees of freedom (DOFs) of the EndoWrist movements have provided an ideal platform for suturing and reconstructive procedures, including pyeloplasty, prostatectomy, partial nephrectomy, and other surgeries. The robotic systems' motion scaling feature has also reduced physiological tremors, making the fine dissection tasks and suturing ideal to be performed by this technology and more precise at the same time. It has also improved working in deeper body cavities such as the pelvis than traditional open surgery. Furthermore, it has the advantages for the patient of smaller incisions, less pain, decreased blood loss, quicker healing time, and reduced hospital stay (Carbone et al., 2018). In addition, the technique and technology's rapid dissemination and aggressive marketing have captured the imaginations of patients and doctors. For laparoscopic surgery, the surgeon has to be trained at a superior level with great accuracy and precision. Robotic surgery allows a surgeon to do the same with little or no previous laparoscopic training, with the advantage of minimal access required in surgery.

Robotic surgery in India has made a paradigm shift in laying the foundations of surgery. Urologists were the first in the surgical fraternity to realize the immense potential of robotic surgery. Radical prostatectomy using robotics has become the validated treatment in treating prostate cancer. According to unpublished data, in the United States alone, more than 55,000 radical prostatectomies have been performed and more than 70,000 have been carried out worldwide at the time of writing, by using da Vinci robotic assistance. Robot-assisted surgery in various specialties such as cardiothoracic surgery, surgical oncology, gastrointestinal and bariatric surgery, general surgery, gynecology, and otorhinolaryngology has gained momentum worldwide. Robotic surgery is still in its infancy stage in India. There are more than 500 trained robotic surgeons in India with 66 centers and 71 robotic installation as of July 2019. In the last 12 years, more than 12,800 surgeries have been performed with robotic assistance. The All India Institute of Medical Sciences (AIIMS), Delhi, India has been at the forefront of the robotics revolution for India.

Presently, in urology and gynecology, many robot-assisted surgeries are performed. However, robotic surgeries in cardiovascular surgery, oncology, general surgery, and transplants are on the rise (Mahesh et al., 2015). The first program of the Indian urological surgical system was started at AIIMS, New Delhi, in 2006. Since then, there has been an ever-increasing trend in surgical robotics. In India, more than 300 robot-assisted laparoscopic radical prostatectomies have been carried out successfully at the time of writing. The outcome of the perioperative phase has been successfully analyzed by Puntambekar et al. (2012). Robotic assistance for the urological procedures includes extirpative oncological surgeries such as radical cystectomy, radical nephrectomy, pelvic exenteration, ilioinguinal lymph node dissection, and adrenalectomy. Reconstructive surgeries such as vesicovaginal and ureterovaginal fistulae repairs, pyeloplasty, and other stone surgeries including ureterolithotomy and pyelolithotomy are also performed with robotic assistance. Meanwhile, ear, nose, and throat (ENT) surgeons are performing robot-assisted surgery in the oro/hypopharynx and nasopharynx for benign and malignant lesions in order to achieve better functional results than open surgery. Many types of gastrointestinal procedures are also being performed with the help of robotics. Some of these are esophageal fundoplication, pancreaticoduodenal procedures, colorectal surgeries, and bariatric surgeries. In cardiothoracic procedures, robot-assistive surgeries are carried out for mitral valve repairs, coronary artery bypass grafting, lung resection, esophagectomy, and thymectomy.

In the past 3–4 years, the installation of robotic systems across India has been on the rise. Many robotic systems have been installed in all major cities installed by corporate sector hospitals. About 19 surgical robotic systems are currently operating in the country, including New Delhi, Mumbai, Gurgaon, Chennai, Nadiad, Bengaluru, and Hyderabad. These numbers seem low, bearing in mind India's ever-growing population. With the increase in patient demand for MIS, there remains a pertinent question: "is it the need of the hour in the Indian scenario?" Since there has always been an allegation that the technology is market and commercially driven.

In India, robotic surgeries are cheaper than the same ones performed in the Western world. This fact can promote medical tourism in India and lead the country to a new dimension.

9.3.1 Surgical robots in India

Some of the best hospitals that provide the best robotic surgeries in India include the All India Institute of Medical Sciences, Apollo Hospitals, Narayana Hospitals, etc. in different parts of India. Robot-assisted surgery integrates experienced and skilled surgeons with advanced computer technology. The special instruments for surgery are much more flexible and maneuverable than the human hand. Movements by the surgeons can be replicated by robots while minimizing hand tremors. Now, a surgeon can carry out with ease the most complex procedures with enhanced precision, control, and dexterity. A summary of robot-assisted surgery is given in Table 9.1. Robotic surgery systems include the following.

1. *Corindus CorPath GRX Advancing Interventions with Vascular Robotics* (Siemens, 2020): Corindus CorPath GRX pictured by Siemens (2020) is the latest robot, designed for robot-assisted procedures and high-level accuracy. The application of Corindus Cor-Path GRX has provided procedures with accurate and precise device manipulation in safer environments for patients. This robotic system is used for percutaneous coronary intervention (PCI) and peripheral vascular intervention (PVI) procedures to apply precisely the device and stent position for better outcomes. It enables robotic control of the guidewire, guide catheter, and balloon or stent catheter with 1 mm advancement for the control console operated by the surgeon. The advantages of this robotic system include close proximity and ergonomic visualization, robotic precision for the stent placement with 1 mm movement, reduced radiation exposure, optimized stent selection with submillimeter anatomy measurement, and faster patient recovery.

2. *da Vinci Robotic System* (Intuitive, 2022): The da Vinci Surgical Robotic system (Intuitive, 2022) is mainly used in a very traditional surgery with large open incisions or laparoscopies; it uses small incisions and is typically limited to a very straightforward

Table 9.1 Summary of robotic-assisted, minimally invasive surgeries.

State-of-the-art equipment	Departments offering robotic surgery	Advantages	Patient benefits
1. da Vinci Si Surgical System 2. Mazor robotics renaissance	1. Orthopedics 2. Cardiology 3. Gastroenterology 4. Gynecology 5. Urology 6. Pediatrics	1. Greater flexibility and maneuverability 2. Microprecision	1. Minimal blood loss 2. Less tissue damage 3. Less pain 4. Reduced risk of infection 5. Fast recovery 6. Less visible scars

procedure. It is an effective and minimally invasive alternative to open surgery for complex surgical procedures. This robotic surgical system is used in urology (for prostate, bladder, and kidney cancer, ureteropelvic junction obstruction, congenital disabilities, vesicoureteric reflux disease, etc.), gynecology (multiple fibroids, uterine and cervical cancer, uterine and vaginal prolapse, endometriosis, vesicovaginal fistula, ovarian cyst, etc.), cardiology (atrial septal defects, mitral and aortic disease, coronary artery disease, etc.), and gastroenterology and hepatology (liver disease, colon and rectal cancer, obesity and metabolic disorders, gastric cancer, esophageal disorders, etc.). Advantages include faster recovery, reduced hospital stays, reduced risk of wound infection, minimal blood loss during surgery, less visible scars, long-term weight loss, resolution of comorbidities, better cancer control, less damage to healthy tissue, faster return to continence, and faster recovery of sexual function.

3. *ExcelsiusGPS Spine Surgery Robot* (GlobusMedicals, 2022): ExcelsiusGPS is a next-generation revolutionary spine surgery robot. The combination includes a rigid robotic manipulator with full navigation abilities into an adaptable platform for the precise platform in the surgery of the spine. Using this navigation system, the surgeon places the spinal implants with highly precise and accurate movements. Using the ExcelsiusGPS surgery robot, key spinal procedures can be carried out, such as transformational lumbar interbody fusion (TLIF), postscoliosis correction, spondylodiscitis treatment, high-grade spondylolisthesis treatment, spinal tumor excision and stabilization, adult spinal deformity correction with pelvic fixation, C1, C2 fusion procedures, and cervical decompression with stabilization. Advantages include screw placement accuracy, reduced radiation exposure, smaller scars, less tissue damage, less blood loss, and faster recovery.

4. *Mako Robotic-Arm-Assisted Technology* (Ochsner, 2022): Mako Robotic-Arm-assisted surgery has transformed how knee replacement surgeries are performed. This surgery is designed for orthopedic patients and helps them by providing a customized surgical plan for each patient based on their unique anatomy and specific diagnosis. The surgeon guides the manipulator for removing the bone and the cartilages, then places the implants with greater accuracy. Mako Robotic-Arm-assisted surgery provides stereotactic and haptic guidance during orthopedic surgical procedures with specific applications of software and hardware. This procedure is done by using the patient's computed tomography (CT) scan data, which acts as an intelligent tool holder and then helps the operating surgeon with presurgical planning, placement of implants, and intraoperative navigation of the patient's anatomy. Advantages of this robotic technology include generating a patient-specific 3D preoperative plan based on the CT scan report, protecting the surrounding soft tissue from damage, allowing the surgeon to execute surgical actions more accurately, and providing data in real-time throughout the procedure. This is done by providing the dynamic joint balance for successful long-term outcomes, minimally invasive procedures, reduced postoperative pain, less blood loss, and faster recovery.

5. *Navio Surgical System* (SmithNephew, 2020): Similar to the Mako Robotic-Arm-assisted technology, the Navio Surgical System assists orthopedic surgeons in positioning the implants accurately. All the procedures are based on the precise plan of the patient's unique anatomy, and the system works in conjunction with the skilled surgeon's hands. This robot-assisted technology relays information about the knee to the hand-held robotic piece used by the surgeon. This surgical procedure is used to perform both partial and total knee replacement surgeries. Such advanced technology creates a 3D knee image without a preoperative CT scan. With these details, the surgeon achieves a virtual and successful customized surgery plan. Advantages of this technology are customized planning for surgery, no CT scan required, smaller surgical incision, less pain due to less cutting of tissue, near-to-natural knee motion, accurate placement of the implant, reduced hospital stay, and quicker rehabilitation and recovery.

6. *Renaissance Robotic Surgical System* (Neurosurgical Associates PC, 2022): The Renaissance Robotic Surgical System is the only technology designed especially for spine surgery. The Apollo Hospitals in India were the first in the Asia-Pacific to offer this surgical guidance system, which provides minimally invasive robotic-guided spine surgery. This technology transforms spine surgery from freehand procedures to highly accurate state-of-the-art robotic procedures. It also ensures less radiation during procedures, including scoliosis, and other complex spinal deformities. A total of 175 successful surgeries have been performed, and several patients have already benefited from this procedure. Clinical applications of these procedures include MIS, percutaneous posterior thoracolumbar approaches, scoliosis, and other complex spinal deformities, pedicle screws for short and long fusions, transfacet screws and translaminar facet screws, osteotomies, biopsies, and single vessel and multivessel small thoracotomy. Advantages include higher accuracy, lower radiation, and fast learning curve of the data.

7. *Hugo Robot-Assisted Surgery (RAS) System* (Jonathon, 2022): The Hugo Robotic-Assisted Surgery (RAS) System is designed to perform a wider range of robot-assisted laparoscopic procedures. This system includes an integrated wrist instrument, 3D visualization, and cloud-based surgical video capture with the management solution. It allows surgeons to meet the unique needs of each patient by transforming surgical approaches. Clinical application of the RAS system offers a modular and multiquadrant platform to perform a range of soft-tissue procedures.

9.3.2 Robotic surgical procedures in India

Some of the robotic surgical procedures which are being carried out in India are as follows:

1. *Robotic Cancer Surgery* (Jonathon, 2020): Cancer treatment via surgery is traditionally done with an open surgery approach and involves large wounds and delayed recovery.

It is also done with laparoscopic surgery for a safe and sound option for certain cancers, including colon, endometrial, cervical, and esophageal cancer. However, it has limitations. These limitations are imposed by 2D images, limited range of motion of the instruments, and dependence on a trained assistant to hold the camera. Robotic surgery using the da Vinci system provides high definition and magnified 3D vision with the camera controlled by the surgeon. In cancer surgery, the robot enables radical operations to be performed with the preservation of nerves and other critical structures, due to better visualization. This is important, particularly in rectal, gynecologic, and prostate cancer surgeries. As all the nerves and vessels are magnified, efforts are being made to spare the nerves and to help retain their use. Benefits from the patients' view include precise removal of cancerous tissues, significantly reduced pain, less blood loss, less scarring, shorter hospital stays, faster return to normal daily activities, and equivalent cancer cure rates to those of open surgery. Specific types of cancer in which robotic surgery can provide excellent outcomes include colon and rectal cancer, endometrial and cervical cancer, esophageal and stomach cancer, early cases of pancreatic cancer, kidney, bladder and prostate cancer, and transoral robotic surgery such as throat cancer, tongue cancer, and tonsil cancer.

2. *Robotic Surgery in Gynecology* (Rooma et al., 2015): Since the inception of robotic surgery, the perspective on women's surgical procedures has changed entirely. Robotic surgery not only offers ergonomic advantages for a surgeon but also increases the precision of the surgery based on 3D vision and the EndoWrist instrument. This reduces surgical complications and also the overall operation time required. Some gynecological surgeries that are performed using robotic technology relate to endometriosis, bleeding or other problems, uterine fibroids which cause pain, uterine prolapse (i.e., sliding of the uterus from its normal position into the vaginal canal), chronic pelvic pain, abnormal vaginal bleeding, and adenomyosis or thickening of the uterus. Such surgery involves myomectomy, sacrocolpopexy, radical hysterectomy, and surgery for complex endometriosis and tubal anastomosis.

3. *Robotic Surgery in Urology* (Robotic Surgery, 2017): Urology is a forerunner in the domain of robotic surgery. The number of radical prostatectomy robotic surgeries has increased remarkably since the advent of this technology. Advantages include its precision and accuracy, such as in partial nephrectomy, which has resulted in kidney function preservation by enabling nephron-sparing surgery. Robot-assisted adrenalectomy, radical nephrectomy, donor nephrectomy, and pyeloplasty are also performed more commonly.

4. *Robotics in Colorectal Surgery* (Jitender et al., 2020): In colorectal surgery, there is a new category of MIS that is an alternative to both open surgery and laparoscopy. Here, da Vinci's robotic surgical system enables surgeons to offer a minimally invasive option for a complex surgical procedure. It gives the surgeon a 3D high-definition inside view, enhanced vision, precision, and control, and wristed instruments that bend

and rotate much more than human hands. Some conditions requiring robotic surgery are diverticulitis, inflammatory bowel diseases such as ulcerative colitis and Crohn's disease, and colon and rectal cancer.

5. *Robotics in Spine Surgery* (Hina, 2019): The Renaissance Robotic technology has been specifically designed for spine surgery, transforming spine surgery from a freehand procedure to a state-of-the-art, highly accurate, and less radiative robotic procedure. Various spine surgery procedures include MIS, correction of scoliosis, and other complex spinal deformities, complex reconstructive surgery to treat childhood deformities, minimally invasive surgeries for lower back pain, osteotomies, and biopsies (Yashoda, 2022). The Renaissance Robotic technology overcomes the challenges of the traditional method, making it the standard care for MIS.

6. *Robotics in Kidney Transplantation* (Nambala, 2021): Robot-Assisted Kidney Transplant (RAKT) is a minimally invasive technique that uses robotic support to perform kidney transplants. Very high levels of skill and expertise are required in kidney transplants, with extensive training and experience in transplant surgery and robotics. Technical advantages include replicating a surgeon's hand movement via the robot while minimizing hand tremors. Surgeons use their skills and experience and operate with precision and dexterity with proper control procedures. In addition, this technology relates to a valuable and appropriate operative field when the surface is deep and narrow, requiring fine dissection and microsuturing. RAKT surgery reduces the chances of complications, especially in the cases of immunocompromised, and those with end-stage renal disease. It is also safer for obese patients as it involves less blood loss, shorter hospital stays, good recovery time, and smaller surgical scars.

7. *Robotic Surgery in Cardiology* (Wei et al., 2014): The Department of Cardiology, Advanced Cardiac Centre, PGI, as an innovational procedure, carried out the first ever robotic-assisted bioresorbable stent (BRS) implantation in the world. This BRS, made of dissolving polymer, similar to dissolving sutures, allows the artery to pulse and flex naturally. Robotic-assisted surgery was also performed for PCIs. A newer BRS with thinner struts of 100 μm was developed in India. These stents are dissolved in the body over 2–3 years, leaving the natural artery intact. Older generations of the bioresorbable stent had a strut thickness of 150 μm. Center has also achieved negligible mortality (6.8%) for acute coronary syndrome, comorbidities, and cardiogenic shock in all age groups.

9.3.3 Robotic surgery and healthcare robotics research in India

Apart from the applications of robotics in clinical and other peripheral settings of hospitals, some institutes in India are also carrying out research in robotic surgery and healthcare robotics. Karthik and Asokan (2017) presented a monobloc design for a two-degree-of-freedom compliant tool tip which will be used to power surgical tools. This toolbox has been externally powered that includes a drive box with a stainless steel tool tip at the distal end. The drive box includes a thumb joystick specially for the command inputs, a microcontroller, and a three-servo actuator. It involves experimentation to determine the maximum pinching force that the compliant grasper will exert and includes a prototype of the complete surgical tool, which was demonstrated. A master-slave-based free environment online trajectory control of a redundant surgical robot was presented by Mahesh et al. (2015). This work gives a solution in terms of precise positioning of the surgical tool tip using an intelligent controller for the accurate placement of a radiation source during brachytherapy, deployment of internal organs, radiosurgery of the gastrointestinal tumor, etc.

A master-slave-based robotic system with haptic feedback was presented by Mahesh et al. (2015). This work offers a natural orifice transluminal endoscopic surgery (NOTES) using an in vivo robot. An adaptive control technique is used with optimal control parameters. An impedance control is implemented, which is given as haptic feedback to the surgeon for the interaction of the forces between the slave robot and the soft tissue. Furthermore, the system can be applied in the actual teleoperated surgery and training of individuals in NOTES. Mahesh et al. (2015) discussed MIS of bevel-tip needles for medical interventional procedures. The main objective was to maintain and ensure the safety of the needle steering of the target organ. Different types of sliding mode control were considered for the control of needle steering along with a kinematic model of the asymmetric flexible bevel-tip needle. Suraj et al. (2013) discussed an optimization strategy for an active remote center mechanism (RCM) to minimize the extracorporeal workspace. The lengths of the active RCM were optimized, and the dexterity of the tool was analyzed for various trocar positions.

Inertia compensation and static balancing for the master manipulator was discussed by Suraj et al. (2013) for teleoperated surgical robot applications. Here, the primary focus was on the design for the reduction of mass and strategically positioning the joint axes. An optimal design was achieved through proper design and modifications, and any residual imbalance was then corrected through inertia compensation. Mahesh et al. (2015) discussed the design of a flexure-based compliant grasper for the master arm of a surgical robot. The design, along with the analysis of the master arm grasper based on a partially compliant mechanism, was considered.

The Indian Institute of Technology Delhi and the All India Institute of Medical Sciences Delhi collaborated and developed a telerobotic ultrasound system (IIT Delhi, 2021) (Fig. 9.1). In traditional ultrasound, the physicians stand close to the patient for a significant period of time. However, the telerobotic ultrasound system enables remote ultrasound access through a robotic arm. Karthik et al. (2021) discussed a practical approach of designing and developing teleoperated surgical robots in a constrained environment. This approach focuses on reducing the balancing masses by utilizing a compliant mechanism

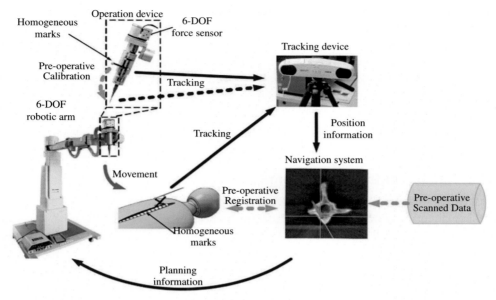

Fig. 9.1 Architecture of robot-assisted spine surgery (Wei et al., 2014).

for several system mechanisms to reduce complexity and mitigate biofouling. Anup et al. (2016) discussed the total robotic radical rectal resection with the da Vinci Xi system for the single docking and single-phase technique. With this new system and technology, the hybrid technique of laparoscopic and the robotic surgery of rectal malignancies will be replaced by four-arm robotic rectal surgery. More unique port positions in a straight line were used by Karthik and Asokan (2020).

9.4 Disadvantage/downside of robotics in health

Major hindrances to the proper utilization of robotic technology to a great extent are cost and training of skilled personnel.

9.4.1 The cost factor

The equipment cost is prohibitively high, and the market is under monopoly due to the substantial initial investment. The capital costs and yearly maintenance are enormous. Almost all the instruments are disposable, adding to the final cost. The entire cost of the da Vinci system and its annual maintenance is prohibitively high. The maintenance and depreciation costs of the robotic system can be minimized if the number of robotic surgery cases is increased. The substantial or extra expenses of the robot and its consumables have made the robotic-assisted surgeries much more expensive. Since the availability of this robotic technology is limited to only some hospitals, surgical training opportunities have also been reduced.

Another disadvantage of the current Indian scenario in terms of robotic surgery is the lack of robotic surgery fellowships. As the technology has not yet entered primary healthcare systems, there is less access to technology and educational opportunities. Another drawback is the lack of evidence-based evaluation of robotic surgery outcomes for the centers of India. A practical solution to reduce the cost of robotic surgery is developing the robotic systems indigenously, which seems to be arduous at present, although some institutes are trying to build prototypes.

9.4.2 Training of the robotic surgical team

The critical attributes in the development of a robotic program are the commitment and intellectual curiosity of all stakeholders, from the head to the nurses. Effective surgeons mostly rely on the skill sets of the surgical team. It is more apparent when surgeons work from the console and rely on the assistants present near the bedside to carry out critical tasks. It is important to educate not only the operating room staff but also the anesthetists and other staff. Training everyone requires proper methods and modules, as discussed by Hiten et al. (2009).

Considering India's physical and population size, considerable efforts are required to meet the demand for trained workforces in robotic surgery. Several training programs in different formats have been organized for this purpose. The Association of Robotic and Innovative Surgeons of India conducted their first training cum Fellowship for 4 months (in online and physical modes) for all specialists, including gynecology, urology, oncology, pediatric surgery, and other surgical specialties. The Indian Association of Gastro Intestinal Endosurgeons conducted their second FALS (fellowship in advanced laproscopic surgery) robotics for surgeons using robotics in 2021. The International Institute of Laparoscopic and Robotic Surgery (India), the Institute of Medical and Minimal Access Surgery Training, Mumbai, the World Laparoscopy Hospital Gurugram, Delhi, and others have been conducting various training related to robotic surgeries in India. Several webinars have been conducted to enhance awareness of various aspects of robotic surgery. An example is a technical webinar titled "Capsule Robot: Importance in Medical Science" by the Institution of Engineers (India).

9.5 The future of robotic surgery in India

As many specialties have started to accept robotic-assisted surgery in India, the robotic surgery incorporated in the field of multidisciplinary areas hence creates opportunities to reduce the maintenance and costs, which in turn becomes more cost-effective. The availability and lower price of robotic surgery in India compared to Western hospitals have contributed to medical tourism in India. In addition, one of the good factor is being

India becoming a sought-after destination from all over the globe for the patients and it is still continuing to grow more for the robotic surgeries. At present in India, the medical tourism is worth USD2 billion. The availability of this robotic surgery has attracted more foreign patients, which will increase further along with examinations that do not involve robotic surgery. New robotic systems are on the way for clinical applications, benefiting from a decrease in initial investment and maintenance. Thus, India's future appears to be bright for medical and healthcare robotics.

9.6 Conclusions

Robotics in medical and healthcare settings offers many advantages but is still in its initial phase. The best is yet to come in terms of cost-effectiveness and availability in India. Robotic systems' motion scaling features and high precision and accuracy have led to widespread applications in different surgical settings. Along with the previously mentioned advantages, robotics has led to an ever-increasing number of procedures being performed in various parts of India. It is a motivating sign that our country has accepted robotic innovations and evolvement, thereby increasing the real-life applications of robotic platforms for healthcare in many Indian cities, leading to growing patient demand. However, the challenge is providing access to innovative surgical techniques to the wider population of India at a reasonable cost.

References

Akshaya, A., 2019. How surgical robot assistants are becoming a reality in Indian hospitals and healthcare sector. Opinions 1, 241–265.

Anup, S.T., Sudhir, J., Avanish, S., 2016. Total robotic radical rectal resection with da Vinci Xi system: single docking, single phase technique. Int. J. Med. Rob. Comput. Assist. Surg. 12, 642–647.

Boubaker, O., 2020. Control Theory in Biomedical Engineering: Applications in Physiology and Medical Robotics, first ed. Elsevier.

Carbone, G., Marco, C., Doina, P., 2018. New Trends in Medical and Service Robotics: Advances in Theory and Practice, sixty fifth ed. Springer.

Desai, J.P., 2018. Encyclopedia of Medical Robotics, Vol. 4: Rehabilitation Robotics, fourth ed. World Scientific.

Dogra, P.N., 2012. Current status of robotic surgery in India. J. Int. Med. Sci. Acad. 25, 145–146.

GlobusMedicals, 2022. The world's first revolutionary robotic navigation platform. Available from: https://www.globusmedical.com/musculoskeletal%20solutions/excelsiustechnology/excelsiusgps/.

Hina, Z., 2019. First in India: Delhi hospital successfully performs complex spinal surgeries through robotic system. Available from: https://speciality.medicaldialogues.in/first-in-india-delhi-hospital-successfully-performs-complex-spinal-surgeries-through-robotics-system.

Hiten, R.H.P., Ana, L., Jean, V.J., 2009. Robotic and laparoscopic surgery: cost and training. Surg. Oncol. 61, 242–246.

Delhi, I.I.T., 2021. IIT Delhi, AIIMS New Delhi and adverb co-develop telerobotic ultrasound system during COVID times. Available from: https://home.iitd.ac.in/show.php?id=10&in_sections=Research.

Intel, 2020. Robotics in healthcare: the future of robots in medicine. Available from: https://www.intel.in/content/www/in/en/healthcare-it/robotics-in-healthcare.html.

Intuitive, 2022. About da Vinci systems. Available from: https://www.davincisurgery.com/da-vinci-systems/about-da-vinci-systems.

Jitender, R., Praveen, K., Anadi, P., Ashwin, S., Avanish, S., 2020. Evolution of robotic surgery in a colorectal cancer unit in India. Indian J. Surg. Oncol. 11, 633–641.

Jonathon, R.B., 2020. Robot onco surgery. Available from: https://apollocancercentres.com/treatments/robotic-onco-surgery/.

Jonathon, R.B., 2022. Robotic technology in spinal surgery. Available from: http://spinesurgeon.sydney/robotic-technology-in-spinal-surgery/.

Karthik, C., Asokan, T., 2017. Design of a two degree-of-freedom compliant tool tip for a handheld powered surgical tool. J. Med. Dev. 11, 014502.

Karthik, C., Asokan, T., 2020. Design of a tether-driven minimally invasive robotic surgical tool with decoupled degree-of-freedom wrist. Int. J. Med. Rob. Comput. Assist. Surg. 16, 642–647.

Karthik, C., Suraj, P., Srikar, A., Sourav, C., Ramalingam, M., Asokan, T., 2021. A practical approach to the design and development of tele-operated surgical robots for resource constrained environments—a case study. J. Med. Dev. 15, 011105.

Mahesh, D., Jaspreet, C., Arvind, P.G., 2015. Robotic surgery is ready for prime time in India: for the motion. J. Minim. Access Surg. 11, 2–4.

Maurice, O., 2017. A shifting global economic landscape: update to the world economic outlook. Available from: https://www.imf.org/en/Blogs/Articles/2017/01/16/a-shifting-global-economic-landscape-update-to-the-world-economic-outlook.

Nambala, S., 2021. Robotic heart surgery in India. Available from: https://www.micsheart.com/robotic-heart-surgery-in-india/.

Neurosurgical Associates PC, 2022. Neurosurgical Associates PC and Mazor Robotics Renaissance Guidance System. Available from: https://neurosurgicalassociatespc.com/mazor-robotics-renaissance-guidance-system/.

Ochsner, L.G., 2022. Mako robotic-arm assisted technology (MAKOplasty). Available from: https://ochsnerlg.org/services/orthopedics/makoplasty.

Puntambekar, S., Agarwal, G., Joshi, S.N., Rayate, N.V., Puntambekar, S.S., Sathe, R.M., 2012. Robotic oncological surgery: our initial experience of 164 cases. Indian J. Surg. Oncol. 3, 96–100.

Surgery, Robotic, 2017. Centre for Robotic Surgery—a dedicated urology centre in India. Available from: http://www.mpuh.org/centreforroboticsurgery/category/robotic-surgery/.

Rooma, S., Madhumati, S., Rupa, B., Samita, K., 2015. Robotic surgery in gynecology. J. Minim. Access Surg. 11, 50–59.

Sequeira, J.S., 2020. Robotics in Healthcare: Field Examples and Challenges, 1170th ed. Springer Nature.

Siemens, 2020. Corindus corpath GRX advancing interventions with vascular robotics. Available from: https://www.siemens-healthineers.com/en-in/angio/endovascular-robotics/precision-vascular-robotics?content=hospital-benefits.

SmithNephew, 2020. Navio Surgical System. Available from: https://www.smith-nephew.com/key-products/robotics/navio-detail/.

Suraj, P., Karthik, C., Sourav, C., Asokan, T., 2013. Optimisation of an active remote centre of motion mechanism for minimal extracorporeal workspace for robotic surgery. In: International Conference Proceeding Series (ICPS), pp. 1–6.

Wei, T., Xiaoguang, H., Bo, L., Yajun, L., Ying, H., Xiao, H., Yunfeng, X., Mingxing, F., Haiyang, J., 2014. A robot-assisted surgical system using a force-image control method for pedicle screw insertion. PLoS One 9, 1–9.

Yashoda, H., 2022. Robotic kidney transplantation. Available from: https://www.yashodahospitals.com/diseases-treatments/robotic-kidney-transplant-benefits-risks-cost-success-rate/.

CHAPTER 10

Trend of implementing service robots in medical institutions during the COVID-19 pandemic: A review

Isak Karabegović[a], Lejla Banjanović-Mehmedović[b], Ermin Husak[c], and Mirza Omerčić[d]

[a]Academy of Sciences and Arts of Bosnia and Herzegovina, Sarajevo, Bosnia and Herzegovina
[b]Faculty of Electrical Engineering, University of Tuzla, Tuzla, Bosnia and Herzegovina
[c]Technical Faculty of Bihać, University of Bihać, Bihać, Bosnia and Herzegovina
[d]iLogs Gmbh, Klagenfurt, Austria

10.1 Introduction

It is well-known that in the period from 1960 to 1980, industrial robots were introduced in manufacturing processes across all industrial branches with the aim of automating and rationalizing certain tasks. The development and progress in sensor technology, materials, microtechnology, microprocessors, information communication technologies, signal processing, and navigation technologies led to the progress in robotic technology, thus enabling the development of a new generation of robots named service robots. They have been designed to have the capability to perceive the environment through integrated sensors or computer vision and to autonomously carry out particular tasks. Nowadays, service robots have a significant role in medicine. Modern medicine will greatly benefit from the development of robotics. Medical robotics is still a relatively novel field of research in which a great number of institutes and companies endeavor to develop a variety of styles and techniques in order to improve different medical procedures. One of the main purposes of mobile robots is to reduce person-to-person contact, disinfect, sterilize, and provide support in healthcare institutions. This was extremely important during COVID-19.

10.2 Trend of implementing service robots in medical institutions

The COVID-19 pandemic first emerged in late December of 2019 in the megacity of Wuhan in China's Hubei province. As of February 26, 2020, multiple new infections were reported outside China for the first time. The World Health Organization (WHO) declared the COVID-19 outbreak as a global pandemic since the virus spread

Medical and Healthcare Robotics
https://doi.org/10.1016/B978-0-443-18460-4.00001-9

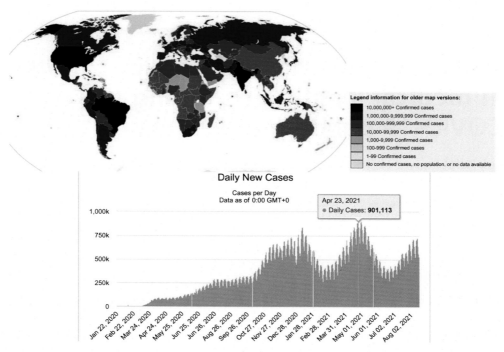

Fig. 10.1 Worldwide spread of the pandemic as of August 17, 2021 (Worldometers, 2021).

from China to other countries, causing multiple incidents everywhere. On August 17, 2021, 208,941,288 people were infected with the COVID-19 virus worldwide, 4,387,956 people died, and 187,303,647 patients recovered. Spread of the COVID-19 infection in 2021 is shown in Fig. 10.1 (Worldometers, 2021).

The COVID-19 virus that spread to all countries almost halted normal work and development. Service robots played an important role during the pandemic, so we provide an analysis of their application here.

The annual supply of service robots to medical institutions for the period 2009–19 and estimates until 2023 are shown in Fig. 10.2 (Doleček and Karabegović, 2012; World Robotics 2020-Service Robots, 2021). Obviously, the number of service robots in medicine is increasing year by year. Unfortunately, service robots in medicine made up only 6% of the implemented professional service robots in 2020, which is shown in Fig. 10.3. Predictions are that these values will change in the future. The aging population will exert significant pressure on medical institutions, which will have to seek solutions offered by medical robots. This can also be seen by the increasing investments made for the development of medical robots.

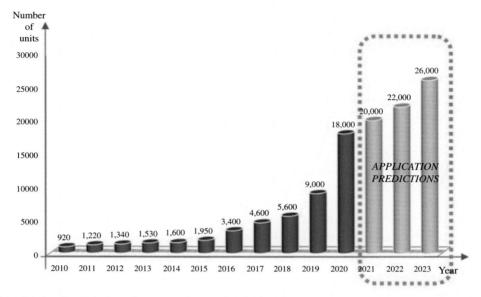

Fig. 10.2 Implementation of service robots in medical institutions on an annual basis for the period 2010–20 and estimates of implementation until 2023.

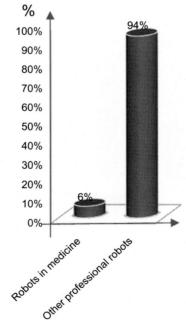

Fig. 10.3 Comparative percentage of implementation of service robots in medical institutions in relation to other professional service robots in 2020 (World Robotics 2020-Service Robots, 2021).

In 2010, the application of medical service robots was about 920 units, as can been seen in Fig. 10.2. This number changed to 18,000 units of service robots in just 10 years. Due to the COVID-19 pandemic in 2019 and 2020, there was an increase in the use of robots in medical institutions so that 5600 units of robots in 2018 increased to 9000 and 18,000 units of robots in 2019 and 2020, respectively. The application therefore increased almost 3.2 times in almost 3 years. This trend of growth is expected to continue in the coming years and reach about 26,000 robot units by 2023. It can be concluded that the implementation of service robots for medical services has been growing exponentially over the last 10 years.

Fig. 10.3 shows the difference between the implementation of service robots in medical institutions and other professional service robots in 2020. However, 94% of them were used to perform other tasks, which are listed in the introduction of this chapter, whereas only 6% of service robots were used in medical institutions to perform a variety of medical services, as shown in Fig. 10.3.

Fig. 10.4 shows the percentage shares of medical service robots by the area of application in medicine. Service robots are greatly implemented in small intervention surgeries (28%), orthopedics (13%), neurosurgery (8%), endoscopy (12%), point radiation (15%), colonoscopy (7%), disinfection and cleaning (12%), and the others (5%) (Karabegović and Banjanović-Mehmedović, 2021).

Table 10.1 shows the relevance factors for types of service robots used in medicine. The fields with two + have the highest relevance, and the empty fields have no relevance.

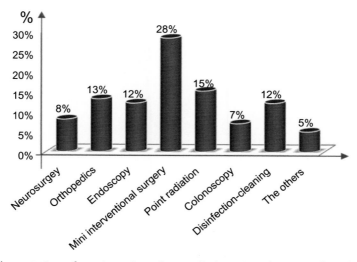

Fig. 10.4 Implementation of service robots for medical services by area of medical application (Karabegović, 2021; World Robotics 2020-Service Robots, 2021).

Table 10.1 Assessment of the factor of relevance for types of service robots in medicine (Karabegović and Banjanović-Mehmedović, 2021).

Medical robotics	High–quality labor productivity	Increasing security, avoiding risk
Diagnostic systems	+ +	+
Robot-assisted surgery or therapy	+ +	+
Rehabilitation systems	+	+
Other medical robots		

10.3 Implementation of service robots in medical institutions during the Covid-19 pandemic

During the COVID-19 pandemic, doctors around the world recommended practicing social distancing as one of the measures to prevent the spread of the virus. Another recommendation was to work remotely as much as possible. However, the most important objective was the protection of those who were on the first line of defense against the infection because they could not stay at home. In order to protect doctors and medical personnel and help them remain at a distance from patients and perform certain tasks and their duties, many service robots were developed.

This conclusion was reached by several research groups around the world that focused on developing a strategy against the COVID-19 pandemic (Boston Dynamics, 2020a, b; Seidita et al., 2020; Marín et al., 2021). Various constructions of service robots have been developed and designed, which are able to perform clinical observation, disinfection and cleaning, rapid diagnosis of diseases, remote treatment, and logistics in medical institutions, as shown in Fig. 10.5. Although service robots for carrying out sanitary measures, delivering medicines and food, and other tasks have already been implemented, they proved to be essential for the quick recovery of patients.

Likewise, robotic professors from the Center for Help and Research with the Help of Service Robots of Texas A&M University processed 120 reports from around the world and came to the conclusion that service robots and their implementation played a highly significant role in almost all aspects of COVID-19 crisis management. The primary goal of all robotic research centers and laboratories in the world was to provide efficient tools, to take advantage of the application of service robotics or technology in the context of a pandemic, which mainly strove to avoid the spread of the COVID-19 pandemic, to support doctors and medical personnel, and to provide clean spaces for doctors and medical staff as well as patients.

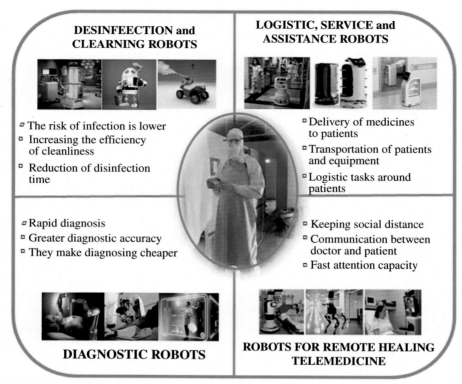

Fig. 10.5 The advantages of service robot applications.

10.3.1 Service robots for disinfection in medical institutions

We are currently witnessing the presence of the COVID-19 virus in all countries. In addition to other regulations for the prevention of the disease, they all emphasize the necessity of disinfection of all objects that people touch, especially in medical institutions. Similarly, in order to prevent the transmission of the virus, disinfection of facilities and communications in medical institutions are also stressed upon, as illustrated in Fig. 10.6. In terms of rooms in medical institutions where patients infected with the COVID-19 virus are located, as shown in Fig. 10.6, there are countless elements that need to be disinfected for the safety of the doctors and staff who treat the patients (Karabegović et al., 2021a, b, c; Karabegović and Banjanović-Mehmedović, 2021; Demaitre, 2020).

Ultraviolet light has been used to clean and disinfect rooms in medical institutions for years. The three main types of ultraviolet (UV) rays are UV-A (315–400 nm), UV-B (280–315 nm), and UV-C (100–280 nm). UV-B rays cause burns to people. UV-C rays kill bacteria and viruses, also known as pathogens, since they have the shortest wavelength and the highest energy. UV-C light has a wavelength between 100 and 280 nm.

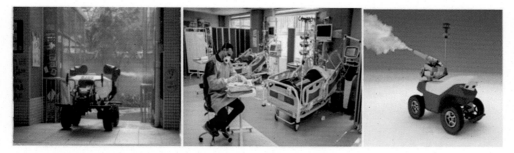

Fig. 10.6 Disinfecting streets and equipment in rooms for examining patients infected by the COVID-19 virus.

However, UV-C rays are more intense and are usually blocked by Earth's ozone layer. A UV-C lamp, used to clean and disinfect, can emit $20 J/m^2 s$ at a distance of 1 m, of 254-nm light, which kills 99.99% of bacteria.

UV-C rays have a narrow spectrum and kill bacteria and viruses without penetrating the outermost cellular layer of the human skin. When it comes to UV-C light, many studies are in progress. A study from 2017 showed that 222-nm UV-C light killed methicillin-resistant *Staphylococcus aureus* (MRSA) bacteria as effectively as UV-C light of 254 nm, which is toxic to humans. This study was repeated in 2018 on the H1N1 virus, and narrow-spectrum UV-C light was again found to be effective in eliminating the virus (Bolton and Cotton, 2008, Otter et al., 2013; Anderson et al., 2013). This has particularly important implications for public health, as the possibility of nontoxic overhead UV lighting in public spaces could drastically reduce disease transmission. Autonomous mobile robots (AMRs) have been deployed to help emergency services and healthcare workers manage the COVID-19 crisis. A number of constructive solutions for disinfection robots have been developed. New solutions of service robots for COVID-19 have been developed, equipped with sources of ultraviolet light or sprayers for disinfection, which perform autonomous disinfection of the object. These robots are capable of operating autonomously in a hospital facility, disinfecting hospital and operating rooms without any human interaction. Since UV light is harmful to humans along with bacteria and viruses, installing it on an AMR platform enables disinfection to be carried out without exposing the hospital staff to risk. UV light must be emitted long enough to kill viruses and bacteria. To disinfect all surfaces, UV light must be moved around the room during the disinfection cycle. Fig. 10.7 shows some service robots used for UV light disinfection designed by companies across the globe (Bolton and Cotton, 2008; Karabegović et al., 2021a, b, c; Anderson et al., 2013; McGinn et al., 2021).

Global companies such as (A) UVD Robots, (B) RoboCop, (C) Milvus Robotics, (D) Wellwit Robotics, (E) Keenon Robotics, (F) Aitheon Robotics, (G) Enabled

Fig. 10.7 Different constructions of service robots for disinfection with UV light.

Robotics, (H) UBTECH Robotics, (I) August Robotics, (J) Badger Technologies, and (K) ZetaBank have developed service robots for disinfection with UV light, which are shown in Fig. 10.7. All designs of autonomous service robots are equipped with software and safety functions based on sensors and a Wi-Fi connection, with an easy-to-use control panel, to cover 360° of disinfection. Service robots, which are safe, reliable, and user-friendly, are designed to be operated by support staff. It takes 10–15 min for a robot to disinfect approximately 99.99% of viruses and bacteria, in a room of up to 30 m^2, with UV rays. Therefore, service robots are perfect for disinfection and cleaning in medical institutions. In addition to UV light, the ZetaBank service robot uses ethanol to disinfect hard-to-reach parts of the room, sterilize, and filter the air. Fig. 10.8 shows disinfection with UV light of a patient room in a medical facility by different service robot constructions (Karabegović and Đukanović, 2017; Karabegović et al., 2017; Surgical and Diagnostic Center, 2022; Shen et al., 2021).

Service robots for disinfection have numerous advantages such as faster cleaning of the entire room, disinfection of all areas, improved productivity, easier integration in the EVS cleaning protocol, highly effective treatments, highly pathogenic killing, etc.

We must single out the service robot called TMiRob, developed by TMI Robotics Technology Co. Ltd, which is shown in Fig. 10.9 (Oitzman, 2020; Karabegović et al., 2021a, b, c; Karabegović and Banjanović-Mehmedović, 2021). The TMiRob intelligent robot for disinfection is the only service robot on the market that has three disinfection technologies: UVC light disinfection, an atomizer, and a plasma air atomizer. This service robot is fully autonomous and can move around the entire facility on its own. It is mainly designed to reset surgical rooms, intensive care units, isolation wards, and laboratories, particularly rooms where patients suffering from the COVID-19 virus are treated.

Fig. 10.8 Disinfection of rooms with UV light in medical institutions.

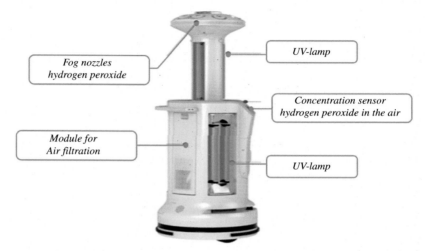

Fig. 10.9 Functional modules of the intelligent robot TMiRob from TMI Robotics Technology Co. Ltd (Oitzman, 2020, Karabegović et al., 2021a, b, c, Karabegović and Banjanović-Mehmedović, 2021).

Service robots showed great efficiency in disinfecting entire rooms during the COVID-19 pandemic, taking less than 10–15 min to achieve maximum disinfection of more rooms necessary for use in a short period of time. They are suitable for medical institutions because they are quick, safe, and easy to use. One service robot can disinfect 64 rooms in a day.

10.3.2 Application of service robots for remote treatment

In order to combat the COVID-19 pandemic, it was necessary for medical institutions to adopt new methods and approaches to treat patients. As the authors state in their research, medical institutions succeeded in their efforts because they very quickly started using remote treatment with existing electronic technology ranging from video visits, virtual applications, and communication through online portals (Shen et al., 2021). At the very beginning, they were not used enough, due to the approach and preference for face-to-face interaction between the patient and the service provider. Experience has shown that

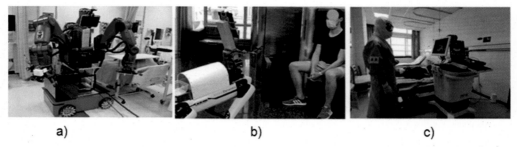

a)   b)   c)

Fig. 10.10 Providing healthcare services to patients at a distance (Hauser and Shaw, 2020; Boston Dynamics, 2020a, b; Wang et al., 2020).

the COVID-19 pandemic has a high rate of infection, so remote examinations, triage, and monitoring of the patient are beginning to be accepted. Robots are being used, which consist of a mobile platform that may or may not include navigation and other autonomous functions that enable them to move freely to perform tasks. The TRINA robot that is currently under development is used as a remote nurse (Fig. 10.10A) and can be used for supervisory control.

Likewise, the Boston Dynamics Spot robot is deployed in medical facilities for patient triage (Fig. 10.10B). In Wuhan, China, a robot is used (Fig. 10.10C) to examine patients at a distance of more than 2 m. Adding remote functionality to the existing equipment would allow hospitals not only to manage costs but also to quickly adopt new capabilities. These upgrades to the existing equipment allow staff to avoid excessive contact and thereby minimize exposure to infectious patients. Many constructions of autonomous systems, or service robots, are used, including robots for remote examination and treatment of patients. They deliver continuous monitoring of each robot, enabling immediate quality control and maximum uptime.

The robot Remote Presence-7 (RP-7) is an innovation by the private company InTouch Health located at Santa Barbara, California. Remote presence (RP) is the ability to project oneself onto another location (without leaving one's current location) and the ability to see, hear, and talk as if one were actually there. The activity of distance communication between doctors and patients is extremely important in times of pandemics because there is no direct contact (Doleček and Karabegović, 2012; Anderson et al., 2013). The system operates on a "many-to-many" system architecture (Fig. 10.11).

Although the physical presence of a doctor with the patient cannot and should not be replaced, service robots have been developed for use in telemedicine. There are many reasons for such an approach, some of which include an extremely old population in certain countries, a limited number of doctors, especially in rural areas where there are no doctors and where the application of such robots is ideal, and prevention of the spread of the COVID-19 virus pandemic. The service robot RP-7 is shown in Fig. 10.12 (Doleček and Karabegović, 2012, Anderson et al., 2013).

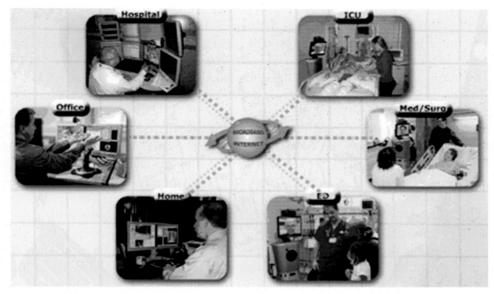

Fig. 10.11 A "many-to-many" architectural system.

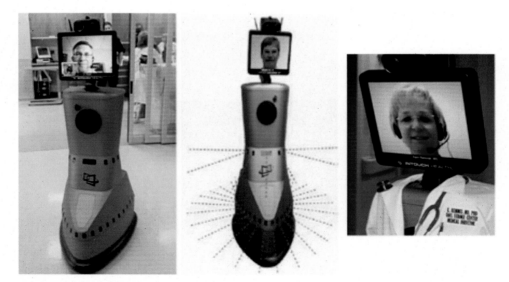

Fig. 10.12 The service robot RP-7 for treating patients remotely.

The RP-7 service robot allows doctors to see, talk, and consult with their patient, observe visual signs, monitor and check the patient's heart rate and blood pressure, review the findings, direct the technician to pass the probe at specific points on the patient, and conduct an ultrasound examination of the patient in real time. This service robot enables the doctor to perform a cursory neurological examination of the patient, who is located kilometers away from him. The doctor is able to conduct an urgent examination of the patient, giving instructions for further treatment.

RP (remote presence) connectivity service is the basis of the infrastructure, providing reliable connectivity between the robot and the control station (Doleček and Karabegović, 2012; Lovo et al., 2017; Karabegović and Ðukanović, 2017; Karabegović et al., 2017). It delivers continuous monitoring of each robot, enabling immediate quality control and maximum uptime.

Fig. 10.13 shows examination of a patient and communication between the patient and medical professional. The doctor uses the control units to maneuver the RP-7 robot and examines the results presented on the screen. The RP-7 robot is positioned near the patient who can see the doctor who is examining, communicating, and giving instructions to him/her, the nurse, or the doctor who is with the patient. In addition to the RP-7 service robot, many others have been developed, as shown in Fig. 10.14.

Doctors can now talk to patients and other doctors located in rural areas, sharing their knowledge and consulting on diagnoses in real time so that there is no need to travel. In other words, the annual checkup could be performed with a remote-controlled tablet instead of physical contact. The Boston Dynamics company has developed the "Spot robot" that helps doctors in the treatment of patients suffering from COVID-19, as illustrated in Fig. 10.15. Doctors use it to evaluate people who believe that they have the

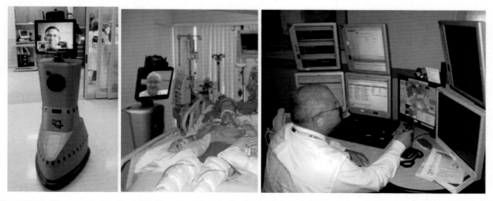

Fig. 10.13 The service robot RP-7—examination of a patient and communication between the patient and the specialist *via* the robot.

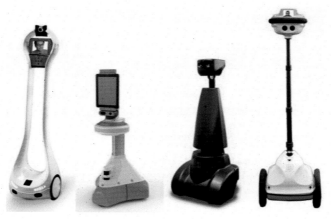

Fig. 10.14 Different constructions of service robots for remote treatment.

Fig. 10.15 Construction of the Spot robot by the Boston Dynamics' company to help treat patients suffering from COVID-19 (Boston Dynamics, 2020a, b; Hughes, 2020; Shen et al., 2021).

symptoms of COVID-19. They use the robot to measure the patient's vital signs at a distance of 2 m.

The robot is equipped with cameras and a tablet and uses computer vision technology. Doctors are able to measure the patient's skin temperature, breathing rate, blood oxygen saturation, and pulse, without being in the same room as the patient. An algorithm has been developed that, using an infrared camera, enables measurement of elevated skin temperature and breathing rate while the camera measures the temperature of the skin on the face. The algorithm then connects this temperature with the temperature of the core of the body and determines the body temperature.

The Spot robot is equipped with three monochrome cameras, which filter different wavelengths of light—670, 810, and 880 nm—and allow researchers to measure small color changes that occur when the hemoglobin in blood cells binds to oxygen and flows through the blood vessels. The measurements are used to calculate the pulse rate and blood oxygen saturation. The researchers at Boston Dynamics have already used the robot to conduct tests and measurements on healthy volunteers, and they are now making plans to use the robot to test people with COVID-19 symptoms in a hospital

Fig. 10.16 The service robot "Tommy" installed at the Circolo Hospital in Varese to help doctors in treating patients with the COVID-19 pandemic (Lynette, 2017; By Reuters, 2020; Scalzo, 2020).

emergency department. In the long term, robots are expected to be deployed in patient rooms, allowing them to continuously monitor patients. These robots, controlled by a handheld remote device and equipped with a tablet, would allow the doctor to interact with the patient and evaluate symptoms with no direct contact.

Fig. 10.16 illustrates the Circolo Hospital in the city of Varese, Italy, which has installed six robots to help medical personnel and doctors fight the COVID-19 pandemic.

The child-sized robot is equipped with all the equipment that enables the doctor to communicate visually and audibly with the patient from another room in order to prevent the spread of the virus. In addition to visual and audio communication, the robot is equipped with equipment that can measure blood pressure and oxygen saturation. In other words, the doctor receives two parameters that provide key information on the patient's health condition, including patients who are on a ventilator. The latest service robot technology reduces the number of direct contacts between doctors, medical staff, and infected patients, which reduces the risk of infection for doctors and medical staff. When it is considered that during the COVID-19 pandemic, more than 4000 health workers were infected in Italy, one can understand the advantage of implementing robots to combat the pandemic. Moreover, the medical facility uses a smaller number of masks and protective clothing, which reduces the costs.

10.3.3 Application of service robots for determining a patient's diagnosis

These service robots have found application in diagnostics, therapy, and patient care. They allow clinicians to determine the anatomical location for the placement of the catheter, regardless of the place in the body, and to diagnose the patient, as shown in Fig. 10.17 (Doleček and Karabegović, 2012).

Robotic diagnostic devices can be located at some distance from the human body either directly on or inside the patient's body. Today, PillCam microrobots are increasingly used in gastroscopy (Doleček and Karabegović, 2012). Tiny pill-shaped microcameras travel through the digestive tract, record, and transfer images to a computer, allowing the doctor to examine the patient (Fig. 10.18). One such system is known as

Fig. 10.17 Service robots at a diagnostic center (Doleček and Karabegović, 2012).

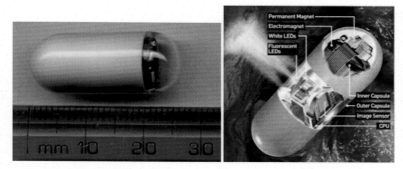

Fig. 10.18 A Sayaka endoscopic capsule.

the Sayaka endoscopic capsule developed by RF System Lab (Japan). Sayaka does not need a motor to move. It requires 50 mW, which is provided by the patient using a vest with coils, while the energy is transferred through magnetic induction. This energy is used to illuminate and photograph the inside of the patient.

In conventional methods, the patient must be under adequate pressure (balloon method). These methods can be uncomfortable for patients and are performed under anesthesia. Compared to conventional methods, the capsules stay longer in the body and thus collect more useful information about the patient (Doleček and Karabegović, 2012) (Fig. 10.19).

This method eliminates the possibility of infection, whereas this is not the case with conventional methods. During the passage of the capsule through the digestive tract, it rotates 360° in its housing and records the entire digestive tract (Fig. 10.19). The photographs taken are transferred to a computer, and a realistic image of the tract (3D model) is created. The corresponding images are developed in high resolution so that the desired places can be enlarged up to 75 times on the screen. This technology ensures the accuracy of diagnosing individual damage sites with an accuracy of 0.1 mm.

The development of new technologies that represent the foundations of Industry 4.0 has led to the implementation of artificial intelligence, especially in the field of medicine.

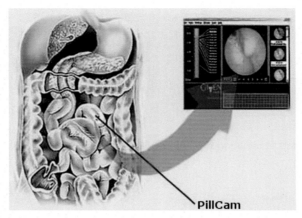

Fig. 10.19 Path of PillCam capsules through the digestive tract of the patient.

Artificial intelligence is already the central technology of the new digital era, which has the potential to fundamentally change and transform our society. Our very development depends on our ability to adapt to new technologies and artificial intelligence, without which there is no progress in the future. Artificial intelligence in medical institutions can be used in two ways. One way is software and virtual systems, i.e., software for image analysis, search, and system of face and speech recognition (Fig. 10.20) (Hopkins, 2016; Karabegović et al., 2021a, b, c). The second way is artificial intelligence embedded in service robots used in medical institutions.

This is where service robotics will mostly contribute to useful implementation in medicine. Using machine learning, researchers can train AI—artificial intelligence—to perform tasks better than a human by providing basically thousands of examples (Badnjevic et al., 2021). The very use of these tools in diagnosing diseases is far-reaching, such as the FDNA system that uses a facial recognition software to screen a patient for 8000 diseases, as well as rare genetic disorders with an extremely high degree of accuracy. Researchers at the Curie Institute have developed an artificial intelligence algorithm that has the ability to identify cancer of unknown primary origin. This deep learning algorithm sequences all the genes expressed in a tumor in order to match its

Fig. 10.20 Use of AI—artificial intelligence—to diagnose diseases.

profile to a specific organ or tissue. The researchers at the Curie Institute provided the algorithm with genetic data on 48 tumors of unknown origin, and, in 79% of cases, the algorithm was able to identify the tissue of origin. Among the 11 newly diagnosed patients, 8 were able to receive treatment selected according to this score (Sauer and Hoff, 2020). All patients benefited greatly from the treatment.

When breast cancer is diagnosed early, it is much easier to treat the disease, so regular examinations and accurate diagnosis are crucial. It has been shown that an artificial intelligence program is better than doctors at identifying breast cancer. Mammogram results are reviewed by two independent doctors, but, from time to time, the diagnosis is not made or someone who does not have cancer is misdiagnosed. To find whether this error can be avoided using technology, researchers from Google Health trained an AI (artificial intelligence) model to recognize breast cancer from mammograms of women in the United States and United Kingdom. AI reduced the number of misdiagnosed cancers by 5.7% for women in the United States, whereas the drop for the UK patients was 1.2%. The number of cases that went undetected decreased by 9.4% and 2.7% in women in the United States and the United Kingdom, respectively. The application of artificial intelligence in the coming years will be increasingly widespread in determining patient diagnoses and other medical procedures.

10.3.4 The application of service robots in the distribution of medicines and patient assistance

Service robots in medicine can, among other things, help deliver or distribute drugs, whereas patient assistance robots can help lift and position patients who are difficult to handle. A large part of the activities related to the provision of professional medical assistance is of a logistical nature. It includes meeting the daily needs of patients (delivering mail or personal care products, environmental cleaning tasks) as well as providing patients with medicines, food, drink, removing dirty linen, delivering fresh linen, transporting regular and contaminated waste, etc. within the medical facility (Fig. 10.21) (Koscinski, 2022; Grellmann, 2020; TIAGo, 2022).

Fig. 10.21 Examples of different constructions of service robots for the distribution of food and medicines to patients.

Fig. 10.22 "TUG system" service robots for patient and laboratory logistics in medical institutions (Doleček and Karabegović, 2012; Karabegović et al., 2021a, b, c; TUG, 2022).

The TUG AMR is an example of an automated robotic delivery system (TUG, 2022). It automates and improves the delivery and tracks most hospital equipment and supplies such as drugs, linens, blood samples, medical records, and IV pumps. The TUG system manages the operations of the internal supply chain within the hospital. It needs wireless network access to communicate with the base computer and elevators, especially on the ground floor and in areas where multiple TUG systems meet one another. It uses a detailed map of the hospital and sophisticated navigation software to plan routes, avoid obstacles, using its sensors, and constantly monitor its location. It can transport several types of hospital carts and can be used for patient care, pharmacy, laboratory, basic supply, delivery of medical records, etc. (Fig. 10.22)

Blood sample courier is Matsushita's robotic system that has been designed as a group comprising AMRs that work together as a team to handle blood sample delivery and courier work in hospitals and laboratories. A computer controlling a group of robots assigns various tasks to individual robots that collect blood samples, deliver them to automated analyzers, and collect the samples, which is shown in Fig. 10.23 (Doleček and Karabegović, 2012; Karabegović et al., 2021a, b, c; Robot of the Year, 2007).

Fig. 10.23 A robotic system for distributing blood samples in the laboratory.

Fig. 10.24 Constructions of service robots used to deliver food and medicines to patients during the COVID-19 virus pandemic in China and Singapore.

The automated battery charging system enables continuous operation of the system, thus preventing all robots from running out of power at the same time (Khan et al., 2020).

Similarly, many constructions, designed by different global companies, have been used to supply food and medicines to various patients in medical institutions, as shown in Fig. 10.24 (Wang and Wang, 2021a, b; Khan et al., 2020; Panasonic, 2022; Swisslog Healthcare, 2022).

These are some examples of service robots that are used for patients treated in medical facilities. A large number of service robots in medical facilities around the world are used by different companies engaged in the development, research, and implementation of service robots. An increase in the use of robots is to be expected in the field of human care and social assistance. This application of robots requires a high level of technology to ensure the safety of humans, preventing them from injuring a human being. It is expected that the next generation of robots will be developed and commercialized for use in facilities such as hospitals, social assistance centers, and private homes. This idea aims to develop robots that meet real needs in the fields of social assistance and human care where a high level of technology is required.

10.4 AI-based mobile disinfection robots against COVID-19

Artificial intelligence for service robot applications can be used for different purposes such as a follow-a-person task, a different form of classification and recognition, optimization, etc. The latest research in the field of mobile robotics with application in disinfection deals with the optimization of the path with regard to the limitations of navigation and the area of disinfection. The basic task in the path planning of a mobile robot is the ability to create collision-free points in the path to the final destination point, taking into account some optimization criteria such as minimum time or the shortest distance.

The path planning module is usually divided into the global planner and the local planner. The former uses the previous information on the environment to create the best possible path, and the latter recalculates the initial plan to avoid possible dynamic obstacles (Marin-Plaza et al., 2018).

Our solution combines the particle swarm optimization (PSO) algorithm with the dynamic window approach (DWA) for optimal path planning and obstacle avoidance (Banjanovic-Mehmedovic et al., 2021). The PSO as a population method has many advantages compared to other evolutionary AI methods: it has few parameters, a relatively small population, and fast convergence (Kubota and Sulistijono, 2006). Instead of providing a specific solution, our solution presents a concept based on ROS components (Quigley et al., 2021), which can be easily transferred to a different hardware platform as an inexpensive and convenient solution, suitable for specific applications in this domain.

The suggested concept of our approach is presented in Fig. 10.25.

For the proposed concept of optimal path planning of the mobile robot for disinfection, we used the GMapping SLAM algorithm due to its robustness in indoor environments and limited computational use. The planner uses a map to create a kinematic trajectory for the robot to move from the start to the final location. If obstacles appear

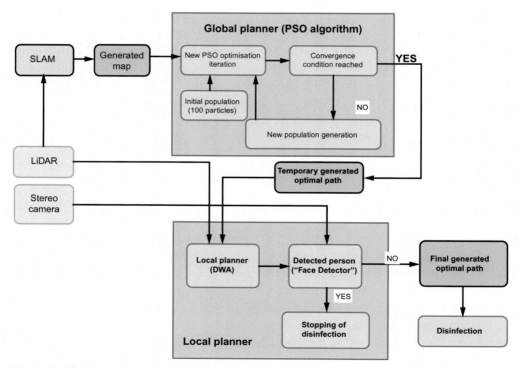

Fig. 10.25 The suggested concept of the disinfection mobile robot relies on ROS components.

Fig. 10.26 The ROS package, "face detector," for detection of people.

on the preplanned path, then the local path planner creates a partial secondary path plan for the robot to avoid the obstacles and arrive at the target position where disinfection is carried out (Fig. 10.25). The DWA to reactive collision avoidance, which exists as an ROS robot local navigation packet, was employed as the local planner in our research. The DWA algorithm was used because its calculation time is short, the local path can be updated in real time, and there is no dead angle obstacle or local optimization in the disinfection scenes (Ruan et al., 2021).

If the robot detects people, for which the ROS package for face recognition, "face detector," from a stereo camera data is used, then it must stop its disinfection process (Fig. 10.26). The face detector employs an OpenCV face detector, based on a cascade of Haar-like features to obtain an initial set of detections (ROS Face Detector, 2021).

The PSO path planner implementation with disinfection of the chosen environment is shown in Fig. 10.27. The PSO path planner implementation with the canceled disinfection of the chosen environment is shown in Fig. 10.28. The mobile robot enters the room, recognizes people, and stops the disinfection process.

10.5 Challenges and open issues for service robotics

Nowadays, artificial intelligence (AI) and robotics are integrated into customer experiences in various settings (Murphy et al., 2017). AI has the ability to improve human

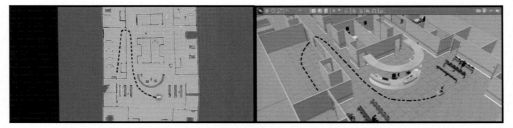

Fig. 10.27 PSO-based optimal path planning with the finished disinfection task.

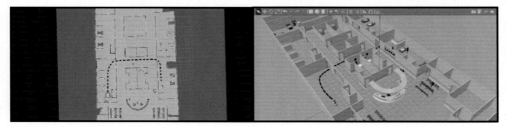

Fig. 10.28 PSO-based optimal path planning with the canceled disinfection task.

cognitive abilities, which is important in many service robot applications in medicine as well as in rehabilitation and surgery applications.

Human–robot collaboration (HRC) is a challenging field that studies the natural means by which a human and a robot can work with each other (Saleh et al., 2015). The role of human–robot interactions is to improve the robotic perception of humans such as understanding human activities, nonverbal communications, and navigation in an environment along with humans (Banjanović-Mehmedović, 2021). The challenges for the framework of human-robot collaboration are robot design, perception, communication, reactive control-based approaches to motion generation, learning, optimization, and explainable robotics (Banjanovic-Mehmedovic et al., 2021).

A new research area called explainable robotics presents a new generation of robots, which can completely interact with humans, requiring adequate intelligence to offer explanations that are understandable by humans (Setchi et al., 2020). The focus is on developing new algorithms, which will provide explanations that allow robots to operate autonomously at different levels and more actively communicate with humans. Explainable robotics supports a wide range of applicable scenarios in medicine. The research challenge in the context of human-robot interactions involves medical management of elderly patients with dementia. In the area of assistive robotics, a robot may act as a mentor to an autistic child or as a guide assisting a blind person.

10.6 Conclusions

This chapter presents an overview of service robot applications in the healthcare sector with a special emphasis on preventing the spread of the COVID-19 virus. We have presented only certain applications of service robots in the prevention of the spread of the COVID-19 virus pandemic, which might change the economy, healthcare, standard of living, and the world we live in. Their implementation in medical institutions is one of the most important roles of service robots today. Only one part of the tasks that could be performed by service robotics during the COVID-19 pandemic is described, such as delivering medicines, food, or mail to hospitals and medical institutions and more complex tasks that appeared during the COVID-19 pandemic, e.g., monitoring patients, control and sampling for control, disinfection of spaces in medical institutions and access roads, assistance in determining the patient's diagnosis, etc. This review proves that the introduction of medical robotics has significantly increased the safety and quality of health management systems compared to manual systems owing to healthcare digitalization. In the coming period, the perspectives for the implementation of service robots in medical institutions are positive and the number of robots will constantly increase, and the reasons are:

- There are many jobs that are dirty, dangerous, monotonous, and hazardous to health in medical institutions; these tasks can be performed by service robots, and thus we can improve the health, safety, and satisfaction of workers in medical institutions.
- The implementation of service robots for the care of the elderly will create an additional need to relieve employees from physical tasks.
- The implementation of the basic technologies of Industry 4.0 in robotic technology will further increase the application of service robots in medicine.
- The implementation of collaborative robots and human–robot applications will complement today's application of service robots in medicine.
- Extremely simple programming and management of service robots will facilitate their implementation in medicine.

By implementing service robots in medical institutions, we have been able to manage the COVID-19 pandemic more effectively. The introduction of as many service robots as possible in medical institutions will significantly increase the safety and quality of the system of health institutions, which is the ultimate goal of the digitalization of healthcare because technical staff can then work remotely from anywhere on the planet and perform tasks productively and safely.

References

Anderson, J., Gergen, M.F., Smathers, E., Sexton, D.J., Chen, L.F., Weber, D.J., Rutala, W.A., 2013. CDC prevention epicenters program. decontamination of targeted pathogens from patient rooms using an automated ultraviolet-C-emitting device. Infect. Control Hosp. Epidemiol. 34 (5), 466–471.

Badnjevic, A., Avdihodžić, H., Gurbeta-Pokvic, L., 2021. Artificial intelligence in medical devices: past, present and future. Psychiatr. Danubina 33 (Suppl. 2), 101–106.

Banjanović-Mehmedović, L., 2021. Artificial intelligence advancement in service robots applications. In: SERVICE ROBOTS: Advances in Research and Application. NOVA Science Publisher, New York, USA, pp. 165–205.

Banjanovic-Mehmedovic, L., Karabegovic, I., Jahic, J., Omercic, M., 2021. Optimal path planning of a disinfection mobile robot against COVID-19 in a ROS-based research platform. Adv. Prod. Eng. Manage. 16 (4), 405–417.

Bolton, J.R., Cotton, C.A., 2008. The Ultraviolet Disinfection Handbook. American Water Works Association, USA.

Boston Dynamics, 2020a. Boston Dynamics COVID-19 Response. 23 April.

Boston Dynamics, (2020b), https://nypost.com/2021/03/19/mit-robot-doctors-see-patients-fight-spread-of-covid-19/;   https://www.bostonglobe.com/2020/04/23/business/robot-will-see-you-now/ (Accessed 15 August 2022).

By Reuters (2020) Robot Nurse Helps Keep Doctors Safe From Coronavirus in Italy; Tommy the Robot Nurse Helps Keep Italy Doctors Safe From Coronavirus (nypost.com) (Accessed 20 November 2022).

Demaitre, E., 2020. Covid-19 Pandemic Prompts More Robot Usage Worldwide. https://www.therobotreport.com/covid-19-pandemic-prompts-more-robot-usage-worldwide/.   (Accessed 12 November 2022).

Doleček, V., Karabegović, I., 2012. Service Robotic. University of Bihać, Technical Faculty, Bihać, Bosna and Herzegovina.

Grellmann, S., 2020. Swiss Post Robots Taking the Lift in the Hospital. https://post-medien.ch/en/swiss-post-robots-taking-the-lift-in-the-hospital/. (Accessed 24 August 2022).

Hauser, K., Shaw, R., 2020. How Medical Robots Will Help Treat Patients in Future Outbreaks. https://spectrum.ieee.org/medical-robots-future-outbreak-response. (Accessed 15 August 2022).

Hopkins, J., 2016. Medicine: Study Suggests Medical Errors Now Third Leading Cause of Death in the U.S. 3 May.

Hughes, O. (2020) Open Source: Boston Dynamics Just Opened Up This Robot Tech to Help Tackle COVID-19. Tech Republic; (Accessed 20 November 2022).

Karabegović, I., 2021. Distribution and Implementation of Service Robotic Systems in Medicine, Book: SERVICE ROBOTS: Advances in Research and Application. NOVA Science Publisher, New York, USA.

Karabegović, I., Banjanović-Mehmedović, L., 2021. SERVICE ROBOTS: Advances and Applications. NOVA Science Publisher, New York, USA.

Karabegović, I., Đukanović, M., 2017. The tendency of development and application of service robots for defense, rescue and security. Int. J. Adv. Eng. Res. Sci. 4 (9), 063–068.

Karabegović, I., Felić, M., Đukanović, M., 2017. Desing and application of service robots in assisting patients and rehabilitations of patients. Int. J. Eng. Technol. 13 (02), 11–17.

Karabegović, I., Karabegović, E., Mahmić, M., Husak, E., 2021a. Service robots and artificial intelligence for faster diagnostics and treatment in medicine. In: Proc. of the NT-2021 Seventh Int. Conf. Advanced Technologies, Systems, and Applications, Sarajevo, 24–26 June.

Karabegović, I., Husak, E., Isić, S., Banjanović-Mehmedović, L., Badnjevic, A., 2021b. Implementation of service robots for space disinfection in medical institutions: a review of control of corona virus infection. In: Proc. of the CMBEBIG 2021 Frist Int. Conf. on Medical and Biological Engineering, Mostar, 21–24 April.

Karabegović, I., Husak, E., Mehmedović-Banjanović, L., Isić, S., 2021c. Research on the application of mobile robots for disinfection of contaminated space with virus COVID-19. In: XXII International Conference, Meeting Point of the Science and Practice in the Fields of Corrosion, Materials and Environmental protection, Serbia, 13–16 September.

Khan, Z.H., Siddique, A., Lee, C.W., 2020. Robotics utilization for healthcare digitization in global COVID-19 management. Int. J. Environ. Res. Publ. Health 17 (3819), 1–17. https://doi.org/10.3390/ijerph17113819. https://www.mdpi.com/journal/ijerph.

Koscinski, K., 2022. Autonomous Robots Used at UPMC and Other Hospitals are Found to Have Been Vulnerable to Hackers. https://www.wesa.fm/health-science-tech/2022-04-13/autonomous-robots-used-at-upmc-and-other-hospitals-are-found-to-have-been-vulnerable-to-hackers.   (Accessed 22 August 2022).

Kubota, N., Sulistijono, I.A., 2006. A comparison of particle swarm optimization and genetic algorithm for human head tracking. In: Joint 3rd International Conference on Soft Computing and Intelligent Systems and 7th International Symposium on Advanced Intelligent Systems, pp. 2204–2209.

Lovo, S.G., Brenna, B., Bustamante, L., Mendez, I., 2017. Using a remote presence robot to improve access to physical therapy for people with chronic back disorders in an underserved community. Physiother. Can. 69 (1), 14–19.

Lynette F. (2017) AI Devices that Walk, Roll and Fly—and Tacos—Draw Developers to NVIDIA HQ|NVIDIA Blog; (Accessed 20 November 2022).

Marín, D.S., Gomez-Vargas, D., Céspedes, N., Múnera, M., Roberti, F., Barria, P., Ramamoorthy, S., Becker, M., Carelli, R., Cifuentes, A.C., 2021. Expectations and perceptions of healthcare professionals for robot deployment in hospital environments during the COVID-19 pandemic. Front. Robot. AI 8, 1–15. https://doi.org/10.3389/frobt.2021.612746. Article 612746.

Marin-Plaza, P., Hussein, A., Martin, D., De la Escalera, A., 2018. Global and local path planning studyin a ROS-based research platform for autonomous vehicles, *Hindawi*. J. Adv. Transp. 2018.

McGinn, C., Scott, R., Donnelly, N., Roberts, K.L., Bogue, M., Kiernan, C., Beckett, M., 2021. Exploring the applicability of robot-assisted UV disinfection in radiology. Front. Robot. AI 7, 1–17. https://doi.org/10.3389/frobt.2020.590306.

Murphy, J., Hofacker, C.F., Gretzel, U., 2017. Robots in Hospitality and Tourism: A Research Agenda. vol. 8 eReview of Tourism Research, ENTER. Research Notes.

Oitzman, M., 2020. Robotic Solution for Covid-19, Autonomous Mobil Robotic Solution for Covid-19. A Special Report by The Mobile Roboic Guide.

Otter, J.A., Yezli, S., Salkeld, J.A., French, G.L., 2013. Evidence that contaminated surfaces contribute to the transmission of hospital pathogens and an overview of strategies to address contaminated surfaces in hospital settings. Am. J. Infect. Control 41 (5 Suppl), S6–S11.

Panasonic, 2022. Panasonic Autonomous Delivery Robots—HOSPI—Aid Hospital Operations at Changi General Hospital (Accessed 15 August 2022).

Quigley, M., Gerkey, B., Conley, K., 2021. ROS: An Open-Source Robot Operating System. http://www.robotics.stanford.edu/~ang/papers/icraoss09-ROS.pdf. (Accessed 11 November 2021).

Robot of the Year, 2007. Robot Blood Sample Courier System. http://pinktentacle.com/2007/12/2007-robot-of-the-year/. (Accessed 25 August 2022).

ROS Face Detector, 2021. wiki.ros.org/face_detector (Accessed 11 November 2021).

Ruan, K., Wu, Z., Chio, I., Zhang, Y., Xu, Q., 2021. Design and development of a new autonomous disinfection robot combating COVID-19 pandemic. In: 2021 6th IEEE International Conference on Advanced Robotics and Mechatronics (ICARM), pp. 803–808.

Saleh, S., Sahu, M., Zafar, Z., Berns, K., 2015. A multimodal nonverbal human–robot communication system. In: Cerrolaza, M., Oller, S. (Eds.), VI International Conference on Computational Bioengineering ICCB 2015.

Sauer, I., Hoff, S., 2020. Artificial Intelligence in Health. The Netherlands Enterprise Agency, AC The Hague, Netherlands.

Scalzo, F.L., 2020. Tommy the Robot Nurse Helps Keep Italy Doctors Safe Form Coronavirus Patients. https://www.reuters.com/article/us-health-coronavirus-italy-robots-idUSKBN21J67Y.   (Accessed 22 November 2022).

Seidita, V., Lanza, F., Pipitone, A., Chella, A., 2020. Robots as intelligent assistants to face COVID-19 pandemic. Brief. Bioinform., 1–9. https://doi.org/10.1093/bib/bbaa361.

Setchi, R., Dehkordi, B., Khan, J.S., 2020. Explainable robotics in human–robot interactions. Elsevier Procedia Comput. Sci. 176, 3057–3066.

Shen, Y., Guo, D., Ding, H., Long, F., Xiu, L., Mateos, Z., et al., 2021. Robots under COVID-19 pandemic: a comprehensive survey, IEEEAccess, multidisciplinary. Rapid Rev. 9, 1590–1615.

Surgical & Diagnostic Center, 2022. The XenexLightStrike Germ-Zapping Robot. http://thalheimer-kuhlung.com/en/does-uv-light-kill-germs-muvgi-8-uv-c-against-covid-19.   https://lsdc.net/lsdc-news/xenex-lightstrike-germ-zapping-robot/. (Accessed 12 August 2022).

Swisslog Healthcare, 2022. Relay Autonomous Service Robot for Hospitals. Swisslog. Available online: https://www.swisslog-healthcare.com/en-us/products-and-services/transport-automation/relayautonomous-service-robot. (Accessed 15 August 2022).

TIAGo, 2022. Delivery Makes an Impact in Hospitals Tackling Covid-19 Thanks to DIH-HERO. https://blog.pal-robotics.com/tiago-delivery-impact-hospitals-covid19/. (Accessed 24 August 2022).

TUG, 2022. Autonomous Mobile Robot for Healthcare and Hospitality. https://aethon.com/products/. (Accessed 26 August 2022).

Wang, X.V., Wang, L., 2021a. A literature survey of the robotic technologies during the COVID-19 pandemic. J. Manuf. Syst. 60, 823–836.

Wang, X.V., Wang, L., 2021b. A literature survey of the robotic technologies during the COVID-19 pandemic. J. Manuf. Syst. 60, 823–836.

Wang, J., Peng, C., Zhao, Y., Ye, R., Hong, J., Huang, H., Chen, L., 2020. Application of a robotic tele-echography system for COVID-19. J. Ultrasound Med. 40 (2), 385–390.

World Robotics 2020-Service Robots, 2021. The International Federation of Robotics. Statistical Department, Frankfurt am Main, German.

Worldometers, 2021. https://www.worldometers.info/coronavirus/worldwide-graphs/. https://www.worldometers.info/coronavirus/#countries. (Accessed 17 August 2021).

Further reading

Ackerman, E., Why Boston Dynamics is Putting Legged Robots in Hospitals_IEEE Spectrum. Available at: https://spectrum.ieee.org/automaton/robotics/medical-robots/boston-dynamics-legged-robots-hospitals. (Accessed 15 August 2022).

Banjanović-Mehmedović, L., Gurdić, A., 2021. Collaborative service robots: challenges, paradigms and applications. In: Book: SERVICE ROBOTS: Advances in Research and Application. NOVA Science Publisher, New York, USA, pp. 165–205.

Bertalan, M., 2017. The Guide to the Future of Medicine: Technology and The Human Touch. Webicina Kft, USA.

Doleček, V., Karabegović, I., 2016. The role of service robots and robotic systems in the treatment of patients in medical institutions. In: Proc. of the NT-2016 the Eighth Int. Conf. Advanced Technologies, Systems, and Applications, Neum, 25–27 May.

Kaicheng, R., Zehao, W., Qingsong, X., 2021. Smart cleaner: a new autonomous indoor disinfection robot for combating the COVID-19 pandemic. J. Robot. 10 (87), 2–16.

Reed, N.G., 2010. The history of ultraviolet germicidal irradiation for air disinfection. Public Health Rep. 125 (1), 15–27.

Simmons, S., Morgan, M., Hopkins, T., Helsabeck, K., Stachowiak, J., Stibich, M., 2013. Impact of a multi-hospital intervention utilising screening, hand hygiene education and pulsed xenon ultraviolet (PX-UV) on the rate of hospital associated methicillin resistant *Staphylococcus aureus* infection. J. Infect. Prev. 14 (5), 172–174.

Sun, K.X., Allard, B., Buchman, S., Williams, S., Byer, R.L., 2006. LED deep UV source for charge management of gravitational reference sensors. Class. Quant. Grav. 23 (8), S141.

Zilinska, T., 2012. History of service robots. In: Ceccarelli, Marco, Service Robots and Robotics: desing and Application. IGI Global, Pennsylvania, USA.

CHAPTER 11

Conclusions

Olfa Boubaker
University of Carthage, National Institute of Applied Sciences and Technology, Tunis, Tunisia

To summarize, the book *Medical and Healthcare Robotics* offers a comprehensive review of medical robotics, starting with the basic definitions and major requirements of these devices to important achievements, the most recent advances, and still open issues. By providing an exclusive overview of current trends and recent advances, this volume highlights the importance and limitations of these applications.

On the whole, the different chapters of this volume have covered all application areas of medical and healthcare robotics as well as the major research issues related to this exciting field. The considered applications include rehabilitation devices and assistive technologies to which three chapters have been devoted, surgical robots, body part simulators, exploration, and therapy devices to which two chapters have been devoted, and, finally, service robotics that has been considered in the last chapter of this book. This book highlights the crucial role that these devices play in profoundly impacting the lives of a large part of the human population by increasing their independence and in strongly assisting doctors and medical technical staff. Generally, it may be said that the end users of these machines are principally hospitals and that the progress of these devices considerably benefits the geriatric population.

Alternatively, the increasing demand for laparoscopic surgeries has made the market of surgical robotics one of the fastest growing ones among others. We should note here that in 2023, the price of the most used surgical robotic system, the da Vinci system, may reach USD 3.0 million. Nevertheless, we have shown that building reliable body part simulators for training students in surgery remains a crucial matter in order to standardize medical procedures in surgery. In this particular detail, the application of artificial intelligence and software for analyses, in combination with soft materials, contractile actuators, and flexible sensors, has effectively contributed to the design of this category of medical robots in order to generate realistic healthcare scenarios.

As demonstrated through several chapters, the key challenge for medical robotics remains building reliable systems with the greatest levels of autonomy and safety. Safety considerations, including the compliance of a robot with its operating environment, absolutely remain the most important exigence of medical robotics compared to robotic systems used in other application fields. Indeed, medical robots may operate inside a human body by either substituting or interacting with a human organ and then ensuring a certain minimum level of compatibility during such applications.

Soft robotics may include rehabilitation robots like gloves and wearable devices, cardiac assist devices, surgical robots, drug delivery systems, and so on. We have shown that safety considerations should involve various softness aspects, including modeling, actuation, sensing, control, simulation, and implementation of these machines.

Taking everything into account, it is important to note that robotics for medical applications is one of the fastest growing sectors in engineering and technology. Indeed, the global medical robotic systems market is projected to grow at a rate exceeding 17% between 2022 and 2030. North America (especially the United States and Canada) and Europe (especially Germany and the United Kingdom) are leading in the field of medical robotics, whereas the Asia–Pacific region has retained the principal revenue share of more than 50.0% since 2021. The analysis shows that India is one of the leading countries in this region. The penultimate chapter of this book is devoted to understanding the evolution of this field in this country.

Finally, let us note that the field of service robotics has witnessed the fastest growth among all medical robotic applications with a share of more than 65% in 2022, even if this area is considered to be relatively new. The last chapter of this book analyzes the situation and explains it using the example of the COVID-19 pandemic where the worry of disinfecting hospitals and medical centers and the integration of artificial intelligence were the key factors behind such an expansion.

In the end, we hope that this inaugural volume of the book series *Medical Robots and Devices: New Developments and Advances* has awoken the need to know more about one of the fastest growing fields in engineering and technology. Therefore, we hope that you will have the curiosity to discover more about this field in the next volumes of the same series.

Index

Note: Page numbers followed by *f* indicate figures and *t* indicate tables.

A

Abduction/adduction (AB/AD), 160
Acoustic feedback, 94
Active-assisted movement, 3–4
Activities of daily living (ADL), 28, 144
Actuation method, 42, 50*t*
Adverse events (AEs), 111, 114
Aibo, 17–18
Alumina, 165
Ambulation-Assisting Tool for Human
 Rehabilitation (ARTHuR), 103–104
American Heart Association, 123
Anklebot, 99–100
Ankle-foot orthosis (AFO), 73–74
ANYexo 2.0, 144–145, 144*f*
Arduino chip, 209
ArmeoPower, 135–137, 135*f*
ARMin rehabilitation robot, 72, 72*f*
Armotion device, 133–135, 133*f*
Artificial intelligence (AI), 63, 193–196, 251–253,
 252*f*, 257–258
 mobile disinfection robots against COVID-19,
 255–257, 256*f*
Assistance-only-as-needed control algorithms,
 75–77
Assisted Rehabilitation and Measurement Guide
 (ARM-GUIDE), 69–70, 129–130, 129*f*
Assistive robotics, 64–68, 67*f*
Augmented performance feedback (APF), 6, 6*f*
AutoAmbulator, 93
Automated Endoscopic System for Optimal
 Positioning (AESOP), 221
Automated navigation system, 68
Autonomous mobile robots (AMRs), 222, 224, 243

B

Balloon method, 251
Bi-Manu-Track, 131–133, 132*f*
Biomaterials, in total hip arthroplasty, 161–165, 163*t*
Bionic Leg, 100
Bioresorbable stent (BRS), 231
Body part simulators, 263

Body weight supported treadmill therapy
 (BWSTT), 88
Body weight support systems (BWSSs), 88, 96, 103
Bousquet system, 174

C

CABLEankle device, 106, 106*f*
Cable-driven lower limb rehabilitation robot,
 104–106
CABXLexo-7, 145–146, 146*f*
Cancer surgery, robotic, 229
Cardio-assistive devices, 35–37, 36*f*
Cardiology, robotic surgery in, 231
Cassino Tracking System (CaTraSys), 107, 108*f*
Ceramic matrix composites (CMCs), 176
Ceramics, 165
Cerebral palsy (CP), 65
CHAI3D, 208–209
Child-sized robot, 250
Cobalt–chromium alloys, 164
Colorectal surgery, robotics in, 230
Compound annual growth rate (CAGR), 222
Computer-assisted navigation, 161
Computer tomography (CT), 203
Concentric tube robots, 2
Continuous passive movement (CPM), 3–4
Continuum robot, 10
Contrastive covariance matrix, 192
Contrastive principal component analysis
 (cPCA), 192
Convolutional neural network (CNN), 7–8, 185
Corindus CorPath GRX Advancing Interventions
 with, vascular robotics, 227
Cost factor, healthcare robotics, 233–234
Covariance matrices, 192
Current controller, 91–92
Cybernic Autonomous Control System, 98–99
Cybernic Voluntary Control System, 98–99

D

Data-driven learning process, 191–192
da Vinci Surgical Robotic system, 227
Deep learning models, 185, 187

Dexterous autonomous robotic surgery, 197
Dextroscope endoscopic sinus simulator, 202
Direct cardiac compression (DCC), 35
Dynamic window approach (DWA), 256

E

Ekso Bionics, 75
EksoGT, 99
EksoNR, 99
Emotional therapy, 65
End effectors, 3–4
End effector-type upper limb rehabilitation robots,
    124f, 125–135
  Armotion device, 133–135, 133f
  Assisted Rehabilitation and Measurement Guide
    (ARM-GUIDE), 129–130, 129f
  Bi-Manu-Track, 131–133, 132f
  InMotion2, 126–128, 126f
  Mirror Image Movement Enabler (MIME),
    130–131, 130f
Endoscopic sinus surgery (ESS) training system, 202
  anatomical modeling, 207–208, 208f
  core technical skills identification, 206–207, 206f
  development of, 205–209, 205f
  experimental procedure, 210–211, 211f
  haptics, 209, 210f
  hardware, 207, 207f
  performance metrics and evaluation methods,
    211–212, 212t
  postevaluation questionnaire assessment, 212,
    213t
  results, 212–215, 213–215f
  simulation, 208–209
ETS-MARSE robot system, 141–143, 142f
Evaluation task, 211, 211f
ExcelsiusGPS Spine Surgery Robot, 228
ExoBoot, 33–34
Exo-Glove, 29f, 30
Exoskeleton-based lower extremity rehab robots
    (EBLERRs), 6
Exoskeleton robots, 3–4
Exoskeleton-type upper limb rehabilitation robots,
    124f, 125, 135–148
  ANYexo 2.0, 144–145, 144f
  ArmeoPower, 135–137, 135f
  CABXLexo-7, 145–146, 146f
  ETS-MARSE and SREx, 141–143, 142f
  EXO-UL8, 137–138, 137f

FLEXO-Arm1, 138–139, 139f
  harmony, 139–141, 140f
  SUEFUL-7, 147–148, 147f
Exoskeleton with electric actuators, 104
EXO-UL8, 137–138, 137f
Exowalk, 98
Explainable robotics, 258
Exploration devices, 263
External/internal rotation (EIR), 160

F

Face detector, 257, 257f
Fast Fourier transforms (FFTs), 190–191
Fiber Bragg grating (FBG), 45–47
Field-programmable gate array (FPGA), 193
Finite element analysis (FEA), 38
Flexible fluidic actuators (FFAs), 40–41
Flexion/extension (F/E), 160
FLEXO-Arm1, 138–139, 139f
Footplate trajectories, 96
Force-aware control, 113
Frontal planes (FP), 158
Fugl-Meyer assessment (FMA), 127, 134
FUM knee exoskeleton, 73, 73f
Functional independence measure (FIM), 128

G

Gait analysis, 107
Gait-Assistance Robot (GAR), 94
Gait cycle, 160
Gait Enhancing and Motivating System
    (GEMS), 100
Gait Exercise Assist Robot (GEAR), 94
GaitMaster4, 97
Gait rehabilitation devices, stroke, 88–90
  challenges, 113–116
  characteristics, 109–110t
  control techniques, 112–113
  robot-assisted gait training (RAGT)
    (see Robot-assisted gait training (RAGT))
  safety aspects, 111–112
  stationary end-effector gait training devices,
    88–90
  stationary exoskeleton gait training devices, 88
  wearable exoskeleton devices, 90
Gait rehabilitation exoskeleton, 73–74, 74f
Gait Trainer, 70
Gait Trainer GT II, 95–96

G-EO System, 96
Gradient descent method, 194–196
Gynecology, robotic surgery in, 230

H

Haptics, 192–193, 209
Harmony, 139–141, 140*f*
Healthcare robots, in India, 223–225
    autonomous mobile robots (AMRs), 224
    cost factor, 233–234
    modular robots, 224
    research, 231–233
    robotic surgery, future perspectives, 234–235
    robotic surgical procedures, 229–231
    robotic surgical team training, 234
    service robots, 224
    social robots, 224–225
    surgical-assistance robots, 223–224
    surgical robots, 164, 227–229
Hefei University of Technology (HFUT), 145
Hephaestus Systems, 68
Hidden Markov models (HMMs), 184–185
High-fidelity robotic patient simulators, 2
Hip joint mobility, 160*f*
Host personal computer (PC), 91–92
Hugo Robot-Assisted Surgery (RAS) System, 229
Human-robot collaboration (HRC), 258
Human-robot interfaces, 65
Hybrid Assistive Limb (HAL), 98–99
Hydroxyapatite, 164

I

Image feedback, 94
Image-guided intervention (IGI), 175
Impedance control, 113, 133–134
Implant fracture, 172–173, 173*f*
Indego, 100–101
InMotion2, 126–128, 126*f*
Intention loss, 194–196
Interactive Motion Technologies (IMT), 126
International Conference on Rehabilitation
        Robotics (ICORR), 69
Interventional robotics, 2
Interventricular septum (IVS), 36–37
Intraprosthetic dislocation (IPD), 174–175

K

Kidney transplantation, robotics in, 231

L

Laparoscopic robots, 8–10
LARM Wire driven EXercising device (LAWEX),
        106, 107*f*
LD-UVC, 13
Light detection and ranging (LiDAR) systems, 224
LightStrike, 13
Liquid metal, 47–48
LokoHelp, 6
Lokomat, 6, 74–75, 91–92
Lower extremity, 73–75, 76–77*t*
LOwer extremity Powered ExoSkeleton (LOPES)
        device, 95
Lower extremity rehabilitation robot (LERR), 5*t*
Lower limb assistive device, 32–35, 33*f*
Lower limb rehabilitation exoskeleton robots
        (LLRERs), 73–74

M

Machine learning, 193–194
Mako Robotic-Arm-Assisted Technology, 228
Manorobots, 11–13, 12*f*
Matsushita's robotic system, 254, 254*f*
McGill simulator for endoscopic sinus surgery
        (MSESS), 203
Medical Research Council Motor Power scale, 128
Medical robotics, 1, 222, 237, 263–264.
        See also Healthcare robots, in India
    applications, 2–18
    laparoscopic robots, 8–10
    micro robots, 2, 11–13
    nanorobots, 11–13, 12*f*
    nonlaparoscopic robots, 10–11
    rehabilitation robots, 2–6
    service robots, 13–14
    social robots, 15–18
    soft robots, 11, 12*f*, 12*t*
    surgical robots, 2, 7–11, 7–8*f*, 9*t*
Metabolic control, 113
Micro robots, 2, 11–13
Minimally invasive surgery (MIS), 37–38, 221, 223,
        227*t*
Mirror Image Movement Enabler (MIME),
        130–131, 130*f*
MIT-MANUS, 69–70, 124–125, 127–128, 127*f*
MIT-Skywalker, 102–103
Mobile manipulators, 2
Model predictive control (MPC), 209

Modular robots, 224
Morning Walk, 96–97
Motore, 133
Motricity Index (MI), 134–135
Multiple input, multiple output (MIMO), 38

N

Natural hip joint, 158–161, 159*f*
Natural orifice transluminal endoscopic surgery
  (NOTES), 232
Navio Surgical System, 229
Nonlaparoscopic robots, 10–11
NREL rehabilitation exoskeleton, 75*f*

O

Online classification method, 34–35
Orthopedic surgeries, 223
Orthoses, 65
Overground exoskeleton type devices, 97–101
  Anklebot, 99–100
  Bionic Leg, 100
  EksoGT, 99
  EksoNR, 99
  Exowalk, 98
  Gait Enhancing and Motivating System
    (GEMS), 100
  Hybrid Assistive Limb (HAL), 98–99
  Indego, 100–101
  ReWalk P6.0, 101
  Walking Assist Device with Stride Management
    Assist, 97–98
Oxford Intelligent Machines (OxIM), 68

P

PARO robot, 17–18
Particle swarm optimization (PSO), 256–257,
  258*f*
Pattern recognition classifier, 112
Pelvic Assist Manipulator (PAM), 103–104
Pepper robot, 17–18
Percutaneous coronary intervention (PCI), 227
Peripheral vascular intervention (PVI), 227
PillCam capsules, 252*f*
Pleated pneumatic interference actuator (PPIA), 34
Polyethylene wear, 166–169, 167–168*f*, 169*t*
Polyurethane (PU), 176
Predictive control, 113
Pretrained ResNet50 model, 190–191

Pretraining task, 210
Prosthetics, 65

R

RAGT. *See* Robot-assisted gait training (RAGT)
RECOVER device, 104
Recurrent neural network (RNN), 185–186
Red, green, blue, depth (RGBD) sensors, 45–46
Rehabilitation robotics, 2–6, 28–30, 124–125,
  148–150, 149–150*t*, 264
  applications, 70–81
  assistive robotics, 66–68, 67*f*
  control techniques, 75–78
  evaluation standards, 78–79
  history of, 66–70
  lower extremity, 73–75, 76–77*t*
  overview, 63–64
  safety assessments, 80–81
  soft rehabilitation gloves, 28–30
  taxonomy, 64–65
  terminology, 64–65
  therapeutic applications, 69–70
  upper extremity, 71–72
Reinforcement learning (RL), 51–53, 194
Remote center mechanism (RCM), 232
Remote Presence-7 (RP-7), 246, 247–248*f*, 248
Renaissance Robotic Surgical System, 229
ReoAmbulator, 6
ReWalk, 75, 111–112
ReWalk P6.0, 101
Rhythm pattern generator, 98
Right heart failure (RHF), 36–37
Right ventricular ejection device (RVED), 36–37
Rivlin constitutive model, 203
Robear, 17–18
Robogait, 93
Robot-assisted gait training (RAGT), 87,
  90–110, 116
  Ambulation-Assisting Tool for Human
    Rehabilitation (ARTHuR), 103–104
  cable-driven lower limb rehabilitation robot,
    104–106
  exoskeleton with electric actuators, 104
  gait analysis, 107
  MIT-Skywalker, 102–103
  overground exoskeleton type devices, 97–101
  Pelvic Assist Manipulator (PAM), 103–104
  RECOVER device, 104

Robowalk, 101–102
single degree of freedom devices, 107–110
stationary end-effector-type devices, 95–97
stationary exoskeleton-type devices, 91–95
Robot-Assisted Kidney Transplant
    (RAKT), 231
Robot-assisted minimally invasive surgery
    (RAMIS), 44–45
Robot-assisted spine surgery, 233*f*
Robot-assisted surgery (RAS) skills assessment,
    183–192
domain-adapted models, 189–192
domain knowledge-based models, 184, 187–189
inductive learning-based models, 184–187
Robotic-assisted surgery, 225–226, 227*t*
Robotic medical capsules, 10–11
Robotic surgical team, training of, 234
Robot transparency, 75–77
Robowalk, 101–102
Root mean square error (RMSE), 32–33
Rule-based classifier, 112

S
Sagittal planes (SP), 158
Sayaka endoscopic capsule, 250–251, 251*f*
Sensing method, 43–50, 50*t*, 52*t*
Service robots, 13–14, 224, 264
Service robots in medical institutions, during
    COVID-19 pandemic, 237–240,
    238–240*f*, 241*t*
advantages of, 242*f*
AI-based mobile disinfection robots, 255–257,
    256*f*
challenges and issues, 257–258
disinfection, 242–245, 243–245*f*
distribution of food and medicines to patients,
    253–255, 253*f*, 255*f*
implementation of, 241–255
overview of, 259
patient's diagnosis, 250–253, 251*f*
remote treatment, 245–250, 246–247*f*, 249–250*f*
Single degree of freedom devices, 107–110
Skill loss, 194–196
SkillNet, 194–196
Slidercrank mechanism, 97
Smart Robotic Exoskeleton (SREx) robot system,
    141–143, 142*f*
Smoothness, 192

Smoothness evaluation, 189
Social robots, 15–18, 224–225
Soft lift assister for the knee (SLAK), 34
Soft rehabilitation gloves, 28–30, 29*f*
Soft robotics, in medical applications, 26*f*, 264
actuation method, 42, 50*t*
advantages, 27
applications, 27
cardio-assistive devices, 35–37, 36*f*
challenges, 49, 51*f*
control method, 42–50, 50*t*
defintion, 25–26
disadvantages, 27
lower limb assistive device, 32–35, 33*f*
modeling, 41–42
rehabilitation, 28–30
robots, 11, 12*f*, 12*t*
sensing method, 43–50, 50*t*, 52*t*
simulation, 51–53, 54*t*
and soft actuators, 49, 51*f*
soft rehabilitation gloves, 28–30, 29*f*
soft wearable robots, 30–35
surgical robots, 37–41
types, 26*f*
upper limb assistive device, 31–32, 31*f*
Soft wearable robots, 30–35
Sonar navigation system, 68
Spatial attention-based approach, 185–186
Spinal cord injury (SCI), 65
Spine surgery, robotics in, 231
Spot robot, 248–250, 249*f*
Stainless steel, 162–164
Stationary end-effector-type devices, 88–90
GaitMaster4, 97
Gait Trainer GT II, 95–96
G-EO System, 96
Morning Walk, 96–97
Stationary exoskeleton-type devices, 88
AutoAmbulator, 93
Gait-Assistance Robot (GAR), 94
Gait Exercise Assist Robot (GEAR), 94
Lokomat, 91–92
LOwer extremity Powered ExoSkeleton
    (LOPES) device, 95
Robogait, 93
Walkbot_S, 93–94
Stroke, 65, 87, 123
SUEFUL-7, 147–148, 147*f*
Surgical-assistance robots, 223–224

Surgical robots, 2, 7–11, 7–8f, 9t, 37–41, 263
Surgical robots, in India, 225–233
   robotic surgery and healthcare robotics research,
      231–233
   robotic surgery, future perspectives, 234–235
   robotic surgical procedures, 229–231
Surgical skills transfer, 192–196

T

Target personal computer, 91–92
Telemanipulators, 66
Telemetry-controlled manipulators, 66–68
Telerobotics, 192–193
THA. *See* Total hip arthroplasty (THA)
Therapeutic robotics, 64–65
Therapy devices, 263
Therapy Wilmington robotic exoskeleton
      (T-WREX), 72
Time delay estimation (TDE), 142–143
Titanium alloys, 164
TMiRob, 244, 245f
Total hip arthroplasty (THA), 157–165, 158f, 162f
   biomaterials in, 161–165, 163t
   ceramics, 165
   cobalt–chromium alloys, 164
   dislocation of, 173–175, 174f
   failures of, 165–175
   implant fracture, 172–173, 173f
   installation defects, 171–172, 172f
   loosening, 169–171, 170–171f
   natural hip joint, 158–161, 159f
   polyethylene wear, 166–169, 167–168f, 169t
   stainless steel, 162–164
   titanium alloys, 164
   total hip replacements (THRs), 161
   ultrahigh-molecular-weight polyethylene
      (UHMWPE), 164–165
Total hip replacements (THRs), 157, 161
Training task, 210
Transverse planes (TP), 158
TRINA robot, 245–246
TUG system service robots, 254, 254f

U

ULD. *See* Upper limb dysfunction (ULD)
Ultrahigh-molecular-weight polyethylene
      (UHMWPE), 164–165, 168t
Ultraviolet (UV) rays, 242–243
Upper extremity, 71–72
Upper limb assistive device, 31–32, 31f
Upper limb dysfunction (ULD), 123–124
Upper limb impairments, 123–124, 151–152
Upper limb rehabilitation, 123, 125, 151
Upper limb rehabilitation exoskeleton robots
      (ULRERs), 71–72
Urology, robotic surgery in, 230

V

VALOR robot, 4–6
Veebot robot, 13–14
Ventricular assist devices (VADs), 35
Virtual fixtures, 183
Virtual reality (VR) sinus surgery simulator, 202,
      204t, 207
Vision control, 113

W

Walkbot_S, 93–94
Walking Assist Device with Stride Management
      Assist, 97–98
Walking cycle, 160, 161f
Walk ratio, 97
Wearable exoskeleton devices, 90
Weight support system (WSS), 108
World Health Organization (WHO), 237–238
World Stroke Organization, 123

Y

Yeoh hyperelastic model, 208
Young's modulus, 25–26
YuMi robot, 13–14

Z

Zirconia, 165
Zirconia-hardened alumina (ZTA), 165

Printed in the United States
by Baker & Taylor Publisher Services